OXFORD STUDIES IN PROBABILITY

SERIES EDITORS

L. C. G. ROGERS

with

P. BAXENDALE   P. GREENWOOD   F. P. KELLY
J.-F. LE GALL   E. PARDOUX   D. WILLIAMS

OXFORD STUDIES IN PROBABILITY SERIES

1. F. B. Knight: *Foundations of the prediction process*
2. A. D. Barbour, L. Holst, and S. Janson: *Poisson approximation*
3. J. F. C. Kingman: *Poisson processes*
4. V. V. Petrov: *Limit theorems of probability theory*

OXFORD STUDIES IN PROBABILITY · 4

Limit Theorems of Probability Theory
Sequences of Independent Random Variables

VALENTIN V. PETROV

St Petersburg University
St Petersburg
Russia

CLARENDON PRESS · OXFORD
1995

Oxford University Press, Walton Street, Oxford OX2 6DP

Oxford New York
Athens Auckland Bangkok Bombay
Calcutta Cape Town Dar es Salaam Delhi
Florence Hong Kong Istanbul Karachi
Kuala Lumpur Madras Madrid Melbourne
Mexico City Nairobi Paris Singapore
Taipei Tokyo Toronto
and associated companies in
Berlin Ibadan

Oxford is a trade mark of Oxford University Press

Published in the United States
by Oxford University Press Inc., New York

© Valentin V. Petrov, 1995

All rights reserved. No part of this publication may be
reproduced, stored in a retrieval system, or transmitted, in any
form or by any means, without the prior permission in writing of Oxford
University Press. Within the UK, exceptions are allowed in respect of any
fair dealing for the purpose of research or private study, or criticism or
review, as permitted under the Copyright, Designs and Patents Act, 1988, or
in the case of reprographic reproduction in accordance with the terms of
licences issued by the Copyright Licensing Agency. Enquiries concerning
reproduction outside those terms and in other countries should be sent to
the Rights Department, Oxford University Press, at the address above.

This book is sold subject to the condition that it shall not,
by way of trade or otherwise, be lent, re-sold, hired out, or otherwise
circulated without the publisher's prior consent in any form of binding
or cover other than that in which it is published and without a similar
condition including this condition being imposed
on the subsequent purchaser.

A catalogue record for this book is available from the British Library

Library of Congress Cataloging in Publication Data
Petrov, V. V. (Valentin Vladimirovich)
Limit theorems of probability theory: sequences of independent
random variables/Valentin V. Petrov.
p. cm.—(Oxford studies in probability; 4)
Includes bibliographical references and indexes.
1. Limit theorems (Probability theory) I. Title. II. Series.
QA273.67.P47   1994   519.2—dc20   94-28802
ISBN 0 19 853499 X

Typeset by Integral Typesetting Ltd, Gt. Yarmouth, Norfolk
Printed in Great Britain on acid-free paper
by Bookcraft (Bath) Ltd, Midsomer Norton, Avon

Preface

Limit theorems of probability theory form an evergreen field of probability theory. Its methods and results continue to have great influence on other fields of probability theory, mathematical statistics, and their applications.

In 1975 an English translation of my book *Sums of independent random variables* was published. Since then, limit theorems of probability theory have developed greatly. The desire to take into account many modern investigations led to my next Russian book, *Limit theorems for sums of independent random variables* (1987). The present book differs much from my previous books. It contains many recent (as well as classical) limit theorems and probability inequalities for sums of independent random variables.

The aim of the book is twofold. First, the book can be a basis of a course in probability for graduate or advanced undergraduate students. Only some fundamentals of probability theory are needed. The corresponding prerequisite is given in Sections 1.1–1.4. The exposition in the basic sections of the book is self-contained, with detailed proofs. Much work was done in searching for shorter proofs and methodological novelties.

Second, the book can serve as a reference book for specialists in probability and statistics. Every chapter has addenda with precise formulations of many (mostly recent) results which are usually inaccessible in book form. These addenda border on the basic text (which is independent of the addenda) and make it possible for the reader to look at the frontier of modern investigations. I hope that everyone will find something in the basic sections and/or the addenda.

Much attention is paid to works of the Russian probabilistic school headed by A. N. Kolmogorov and Yu. V. Linnik. Of course, there are many regretful omissions, owing to the impossibility of reflecting a very large area of investigations in a small book.

Writing and publishing a book is always an adventure, but the present book was involved in an unusual adventure. On my route to the Amsterdam airport, my suitcase which contained the manuscript of the book was stolen. It was the only complete copy. In several days, the broken suitcase was found by good people on another railway platform. The manuscript was inside; however, it was wet. As the Russian author Mikhail Bulgakov wrote in his novel *The Master and Margarita*, manuscripts don't burn.

This book was written on the basis of my lectures at St Petersburg University and Tulane University in New Orleans, Louisiana. I am using the opportunity to thank my teachers and students as well as the authors of many beautiful results which are cited in the book.

I owe much to my friends and colleagues A. A. Borovkov, I. A. Ibragimov, A. I. Martikainen, Yu. V. Prohorov and A. V. Skorohod for their help and stimulating conversations.

A large portion of the book was read by Dr J. L. Mijnheer. I am grateful to him for useful remarks and all his trouble connected with returning my stolen manuscript.

This work was done with the partial support of Tulane University and the Dutch Scientific Foundation, which is gratefully acknowledged. I am greatly indebted to Professors A. D. Barbour, I. Berkes, N. H. Bingham, P. Diaconis, P. Embrechts, P. Erdös, C. M. Goldie, P. Hall, A. Levine, T. Mikosch, P. Révész, and L. C. G. Rogers for help and moral support in this endeavour.

I express my deep gratitude to the staff of Oxford University Press for their excellent work and understanding.

V.V.P.

Contents

Notation and abbreviations   x

1. Some basic concepts and theorems of probability theory   1
    1.1   Random variables and their distributions   1
    1.2   Moments and quantiles   5
    1.3   Characteristic functions   9
    1.4   Convergence of distributions   16
    1.5   Concentration functions   22
    1.6   Infinitely divisible distributions   28
    1.7   Bibliographical notes   36
    1.8   Addenda   37

2. Probability inequalities for sums of independent random variables   50
    2.1   Inequalities for the maximum of sums of independent random variables   50
    2.2   Exponential bounds   54
    2.3   Inequalities for moments of sums of independent random variables   58
    2.4   Inequalities for the concentration functions of sums of independent random variables   63
    2.5   Bibliographical notes   77
    2.6   Addenda   77

3. Weak limit theorems: convergence to infinitely divisible distributions   88
    3.1   The condition of infinite smallness   88
    3.2   Infinitely divisible distributions as limit laws   91
    3.3   Necessary and sufficient conditions for convergence to a given infinitely divisible distribution   99
    3.4   Limit distributions of class L and stable distributions   101
    3.5   Bibliographical notes   107
    3.6   Addenda   108

viii | Contents

4. Weak limit theorems: the central limit theorem and the weak law of large numbers — 112
4.1 The central limit theorem for a sequence of series of independent random variables — 112
4.2 Classical forms of the central limit theorem — 120
4.3 The weak law of large numbers for a sequence of series of independent random variables — 127
4.4 Classical forms of the weak law of large numbers — 131
4.5 Bibliographical notes — 134
4.6 Addenda — 135

5. Rates of convergence in the central limit theorem — 142
5.1 Estimating the difference of distribution functions by the nearness of characteristic functions — 142
5.2 Esseen's inequality — 147
5.3 Generalizations of Esseen's inequality — 150
5.4 Upper and lower estimates having the same order — 157
5.5 Non-uniform estimates — 163
5.6 Asymptotic expansions in the central limit theorem: formal construction of the expansions — 169
5.7 Asymptotic expansions in the central limit theorem: the i.i.d. case — 172
5.8 Limit theorems for large deviations — 176
5.9 Bibliographical notes — 184
5.10 Addenda — 186

6. Strong limit theorems: the strong law of large numbers — 199
6.1 The Borel–Cantelli lemma — 199
6.2 Convergence of series of independent random variables — 204
6.3 The strong law of large numbers — 208
6.4 The strong law of large numbers: the i.i.d. case — 212
6.5 The strong law of large numbers: the necessary and sufficient conditions — 216
6.6 Estimates of the growth of sums of independent random variables in terms of sums of their moments — 219
6.7 Bibliographical notes — 226
6.8 Addenda — 227

7. Strong limit theorems: the law of the iterated logarithm — 239
7.1 Kolmogorov's theorem — 239
7.2 The Hartman–Wintner theorem — 248
7.3 The generalized law of the iterated logarithm — 250
7.4 Bibliographical notes — 255
7.5 Addenda — 255

Bibliography	**265**
Author index	**287**
Subject index	**291**

Notation and abbreviations

a.s.	almost surely
c.f.	characteristic function
d.f.	distribution function
i.i.d.	independent identically distributed (random variables)
i.o.	infinitely often
□	end of a proof
$\mathbb{R}$	the set of all real numbers
$\mathbb{R}_+$	the set of all non-negative numbers
$\mathbb{Z}$	the set of all integers
$\mathbb{N}$	the set of all positive integers
$I(B)$	the indicator of the set B
$\theta, \theta_1, \theta_3, \ldots$	quantities whose modulus does not exceed unity

Unless otherwise specified, $A, A_1, A_2, \ldots$ denote absolute positive constants, and $C, C_1, C_2, \ldots, c, c_1, c_2, \ldots$ denote positive constants.

$\log x$ is the natural logarithm of $x \in \mathbb{R}_+$

$\sup_x g(x)$ means $\sup_{x \in \mathbb{R}} g(x)$

$$\Phi(x) = (2\pi)^{-1/2} \int_{-\infty}^{x} e^{-t^2/2} \, dt$$

Unless otherwise specified, all limits are taken as $n \to \infty$.

$a_n = O(b_n)$	means that $\limsup a_n/b_n < \infty$
$a_n = o(b_n)$	means that $a_n/b_n \to 0$
$a_n \asymp b_n$	means that $0 < \liminf a_n/b_n \leqslant \limsup a_n/b_n < \infty$
$a_n \sim b_n$	means that $a_n/b_n \to 1$
$\xrightarrow{P}$	means convergence in probability

1

Some basic concepts and theorems of probability theory

This chapter contains a summary of some basic concepts and theorems of probability theory that are needed in the following chapters. The proofs of many statements are omitted; they can be found in the books of, among others, Chung [78], Feller [133], Gnedenko [146], Loève [256], Lukacs [257] and Shiryaev [424]. Some more specialized results that border on the basic facts are also presented here.

1.1 Random variables and their distributions

Let Ω be a non-empty set of elements. These elements are called points, or elementary events, and will be denoted by the symbol ω with or without an index. The set Ω is called the space of elementary events.

Let $\mathscr{A}$ be some set of subsets of the space Ω having the following properties: (1) if $A \in \mathscr{A}$, then $\Omega \setminus A \in \mathscr{A}$; (2) if $A_1, A_2, \ldots$ is a finite or infinite sequence of subsets belonging to $\mathscr{A}$, then $\bigcup_n A_n \in \mathscr{A}$. The set $\mathscr{A}$ is called a σ-algebra of events, or a Borel field of events, and its elements are called events.

If $\mathscr{A}$ is a σ-algebra of events, it is easy to see that $\Omega \in \mathscr{A}$. Furthermore, the empty set $\emptyset$ (which is called the impossible event) and the intersection of a finite or countable set of events belonging to $\mathscr{A}$ also belong to $\mathscr{A}$.

A non-negative and countably additive function $P(A)$ defined on the events $A \in \mathscr{A}$ and normalized by the relation $P(A) = 1$ is called a probability measure. The value of $P(A)$ is called the probability of the event A. The triplet $(\Omega, \mathscr{A}, P)$ is called a probability space.

A real function $X = X(\omega)$, defined on Ω, maps the space of elementary events Ω onto the real line $\mathbb{R}$. Let B be some set of points of $\mathbb{R}$. We put $X^{-1}(B) = \{\omega : X(\omega) \in B\}$. The set $X^{-1}(B)$ is a subset of the space of elementary events Ω and is called the inverse image of the set B. If $X^{-1}(B) \in \mathscr{A}$ for an arbitrary Borel set B of points of the real line,* the function $X(\omega)$ is said to be measurable.

A real finite measurable function $X : \Omega \to \mathbb{R}$ is called a random variable. A function $P_X(B) = P(\{\omega : X(\omega) \in B\})$ that is defined for all Borel sets B of

---

* The class of Borel sets on the real line is defined as the smallest σ-algebra containing all intervals.

points on the real line is called a probability function of the random variable X. We shall use the shorter notation $P(X \in B)$ instead of $P(\{\omega: X(\omega) \in B\})$.

Let $(\Omega, \mathscr{A}, P)$ be a probability space on which a random variable X is defined. The random variable X generates a new probability space $(\mathbb{R}, \mathscr{B}, P_X)$, where $\mathscr{B}$ is the σ-algebra of Borel sets on the real line $\mathbb{R}$.

Let us consider the probability $P(X \in B)$ when B is the interval $(-\infty, x)$, that is, the interval consisting of the points y on the real line satisfying the inequality $y < x$. We write $F(x) = P(X < x) = P(\{\omega \in \Omega: X(\omega) < x\})$. The function $F(x)$ is defined for all real x; it is called the distribution function (d.f.) of the random variable X.

A d.f. $F(x)$ has the following properties: (1) $F(x)$ is non-decreasing and is continuous on the left; (2) $\lim_{x \to -\infty} F(x) = 0$, $\lim_{x \to +\infty} F(x) = 1$. The converse is also true: an arbitrary function $F(x)$ having these properties is the distribution function of some random variable defined on some probability space.

We shall use the terms 'probability distribution of the random variable X' and 'distribution of the random variable X' for either the probability function $P_X(B)$ or the d.f. $F(x)$ of X.

The random variable X is said to have a discrete distribution if there exists a finite or countable set of points B on the real line such that $P(X \in B) = 1$. If X is a random variable with a discrete distribution and $P(X = x) > 0$, the point x is said to represent a possible value of the variable X. A random variable X is said to have a lattice distribution if with probability 1 it takes on values of the form $a + kh$ ($k = 0, \pm 1, \pm 2, \ldots$), where a and $h > 0$ are constants. The number h is called a span of the distribution. If there are no numbers a_1 and $h_1 > h$ such that the values taken on with probability 1 by X can be expressed in the form $a_1 + kh_1$ ($k = 0, \pm 1, \pm 2, \ldots$), the span h is called maximal.

The distribution of the random variable X is said to be continuous if $P(X \in B) = 0$ for any finite or countable set B of points of the real line. It is said to be absolutely continuous if $P(X \in B) = 0$ for all Borel sets B of Lebesgue measure zero. It is said to be singular if it is continuous and if there exists a Borel set B of Lebesgue measure 0 such that $P(X \in B) = 1$.

The distribution of a random variable X is discrete if and only if the distribution function $F(x) = P(X < x)$ is purely discontinuous. It is continuous if and only if $F(x)$ is continuous everywhere. The distribution F is absolutely continuous if and only if

$$F(x) = \int_{-\infty}^{x} p(t)\, dt$$

for every x, where $p(x)$ is a non-negative function that is integrable on the real line; $p(x)$ is called the density of the distribution. (The integral is understood to be taken in the sense of Lebesgue.)

By the Lebesgue decomposition theorem, an arbitrary d.f. $F(x)$ can be uniquely represented as the sum of three components:

$$F(x) = c_1 F_1(x) + c_2 F_2(x) + c_3 F_3(x), \tag{1.1}$$

where $c_k \geq 0$ ($k = 1, 2, 3$), $c_1 + c_2 + c_3 = 1$, and $F_1(x)$, $F_2(x)$, $F_3(x)$ are respectively discrete, absolutely continuous, and singular distribution functions.

The point x is called a growth point, or point of growth of the d.f. $F(x)$ if $F(x + \varepsilon) - F(x - \varepsilon) > 0$ for every $\varepsilon > 0$. The set of all growth points of a distribution F is called the spectrum of F.

The following distributions play exceptionally important roles. The random variable X has a degenerate distribution if there exists a number c such that $P(X = c) = 1$. The distribution function $F(x)$ of this random variable equals 0 for $x \leq c$ and 1 for $x > c$.

Let n be a positive integer and let p satisfy the condition $0 < p < 1$. The random variable X has a binomial distribution with parameters (n, p) if

$$P(X = m) = \binom{n}{m} p^m (1 - p)^{n-m}$$

for $m = 0, 1, 2, \ldots, n$. If $n = 1$ we obtain a Bernoulli distribution.

Let λ be a positive number, and a and $b \neq 0$ be real numbers. The random variable X has a Poisson distribution with parameters (a, b, λ) if

$$P(X = a + bm) = \frac{\lambda^m}{m!} e^{-\lambda}$$

for every non-negative integer m. This definition is somewhat wider than the customary definition, in which $a = 0$ and $b = 1$.

Let a be a real number and let σ be a positive number. The random variable X has a normal distribution with parameters (a, σ), or a normal (a, σ) distribution, if it has the density

$$p(x) = \frac{1}{\sigma \sqrt{2\pi}} e^{-(x-a)^2/2\sigma^2} \quad (-\infty < x < \infty).$$

The normal $(0, 1)$ distribution function is called the standard normal distribution function. It will be denoted by $\Phi(x)$. Thus,

$$\Phi(x) = \frac{1}{\sqrt{2\pi}} \int_{-\infty}^{x} e^{-t^2/2} \, dt.$$

If $X_1, \ldots, X_n$ are random variables defined on a common probability space $(\Omega, \mathcal{A}, P)$, the vector $X = (X_1, \ldots, X_n)$ is called a random vector or n-dimensional random variable. The range of values of such a random vector is the n-dimensional Euclidean space $\mathbb{R}^n$. For every Borel set B of $\mathbb{R}^n$ the probability

$$P(X \in B) = P(\{\omega : (X_1(\omega), \ldots, X_n(\omega)) \in B\})$$

is defined; it is called the probability function of the random vector X. In particular, for all real values $x_1, \ldots, x_n$ the function

$$F(x_1, \ldots, x_n) = P\left(\bigcap_{k=1}^{n} \{\omega: X_k(\omega) < x_k\}\right)$$

is defined. The expression on the right is usually written as $P(X_1 < x_1, \ldots, X_n < x_n)$. The function $F(x_1, \ldots, x_n)$ is called the distribution function of the random vector $X = (X_1, \ldots, X_n)$.

Let $(\Omega, \mathcal{A}, P)$ be a probability space, and let $A_k \in \mathcal{A}$ $(k = 1, \ldots, n)$. The events $A_1, \ldots, A_n$ are called mutually independent if

$$P\left(\bigcap_{s=1}^{k} A_{i_s}\right) = \prod_{s=1}^{k} P(A_{i_s})$$

for every integer k $(2 \leq k \leq n)$ and every integer $i_1, \ldots, i_k$ satisfying the condition $1 \leq i_1 < \cdots < i_k \leq n$.

Let $X_1, \ldots, X_n$ be random variables defined on a common probability space $(\Omega, \mathcal{A}, P)$. These random variables are called mutually independent or, for short, independent, if the events $\{\omega: X_k(\omega) \in B_k\}$ $(k = 1, \ldots, n)$ are mutually independent for arbitrary Borel sets $B_1, \ldots, B_n$ on the real line. The random variables $X_1, \ldots, X_n$ are independent if and only if

$$F(x_1, \ldots, x_n) = \prod_{k=1}^{n} F_k(x_k)$$

for every real $x_1, \ldots, x_n$. Here $F(x_1, \ldots, x_n) = P(X_1 < x_1, \ldots, X_n < x_n)$ and $F_k(x) = P(X_k < x)$. The independence of the random variables $X_1, \ldots, X_n$, having discrete distributions with the sets of possible values $\{x_k^{(1)}\}, \ldots, \{x_k^{(n)}\}$ $(k = 1, 2, \ldots)$ respectively, is equivalent to the fulfilment of the equalities

$$P(X_1 = x_{k_1}^{(1)}, \ldots, X_n = x_{k_n}^{(n)}) = \prod_{m=1}^{n} P(X_m = x_{k_m}^{(m)})$$

for arbitrary integers $k_1, \ldots, k_n$.

If the random variables X and Y are independent and have distribution functions F_1 and F_2, the sum $X_1 + X_2$ has the distribution function

$$F(x) = \int_{-\infty}^{\infty} F_1(x - y) \, dF_2(y). \tag{1.2}$$

The integral on the right is called the convolution or the composition of the distributions F_1 and F_2, and it is denoted by $F_1 * F_2$. We may also consider the convolution of functions of bounded variation on the real line, say F_1 and F_2, that are not necessarily distribution functions. In this case we define the convolution $F = F_1 * F_2$ by the same equality (1.2). The n-fold convolution of a function of bounded variation F will be denoted by F^{*n}.

A sequence of random variables $X_1, X_2, \ldots$ defined on a common probability space is said to be a sequence of independent random variables if the random variables $X_1, \ldots, X_n$ are independent for every n. For an arbitrary sequence of distribution functions $F_1, F_2, \ldots$ there exist a probability space and a sequence of independent random variables $X_1, X_2, \ldots$ defined on it and such that for every n the distribution function of the random variable X_n is F_n.

1.2 Moments and quantiles

Let $(\Omega, \mathscr{A}, P)$ be a probability space, and let X be a random variable. Since a probability space is a measurable space with a measure, we may introduce the notion of the integral. If $\int_\Omega |X| \, dP < \infty$, we say that the mathematical expectation, or mean, of X exists; we denote it by EX and define it by the equality

$$EX = \int_\Omega X \, dP.$$

We have $EX = \int_{-\infty}^{\infty} x \, dF(x)$ where the integral on the right is a Stieltjes integral; F is the distribution function of X.

Suppose that g is a Borel function.* If one of the two following conditions is satisfied: (1) the mean $Eg(X)$ exists, or (2) $\int_{-\infty}^{\infty} |g(x)| \, dF(x) < \infty$, then the other is satisfied, and moreover

$$Eg(X) = \int_{-\infty}^{\infty} g(x) \, dF(x).$$

Let k be a positive number. The mathematical expectation of the random variable X^k, if it exists, is called the kth order moment of X about the origin; we shall denote it by α_k. Thus

$$\alpha_k = EX^k = \int_{-\infty}^{\infty} x^k \, dF(x),$$

where F is the d.f. of the random variable X. If the moment α_k exists, then the absolute moment of order k about the origin is finite; this is denoted by β_k and defined by the equality

$$\beta_k = E|X|^k = \int_{-\infty}^{\infty} |x|^k \, dF(x).$$

---

* A real function g defined on the real line is called a Borel function if for all a the set $\{x: g(x) < a\}$ is a Borel set.

If $0 < m \leqslant k$, then $|x|^m \leqslant 1 + |x|^k$ for all real x. If we here replace x by X and integrate over Ω, we arrive at the inequality $E|X|^m \leqslant 1 + E|X|^k$. Therefore, $\beta_m \leqslant 1 + \beta_k$ if $0 < m \leqslant k$. The existence of the moment α_k implies the existence of the moment α_m for any positive $m \leqslant k$.

The central moment and the absolute central moment of order k are defined respectively by the equalities

$$\mu_k = E(X - EX)^k = \int_{-\infty}^{\infty} (x - \alpha_1)^k \, dF(x)$$

and

$$v_k = E|X - EX|^k = \int_{-\infty}^{\infty} |x - \alpha_1|^k \, dF(x).$$

The central moment of order 2 is called the variance. The variance of the random variable X is also denoted by Var X.

If $X_1, \ldots, X_n$ are independent random variables having mathematical expectations, then $E(X_1 \cdots X_n) = EX_1 \cdots EX_n$. If $X_1, \ldots, X_n$ are pairwise independent random variables having variances, then $\text{Var}(X_1 + \cdots + X_n) = \text{Var } X_1 + \cdots + \text{Var } X_n$.

If X is a non-negative random variable having a mathematical expectation, then $P(X \geqslant t) \leqslant EX/t$ for every $t > 0$. It follows from this that for any random variable X possessing the second-order moment and for every $\varepsilon > 0$ we have

$$\left. \begin{array}{l} P(|X| \geqslant \varepsilon) \leqslant \varepsilon^{-2} EX^2, \\ P(|X - EX| \geqslant \varepsilon) \leqslant \varepsilon^{-2} \text{Var } X \end{array} \right\} \quad (1.3)$$

(Chebyshev's inequalities).

Let X and Y be random variables, $r > 1$, $1/r + 1/s = 1$. Then

$$E|XY| \leqslant (E|X|^r)^{1/r}(E|Y|^s)^{1/s} \quad (1.4)$$

(Hölder's inequality).

If $r \geqslant 1$ then

$$(E|X + Y|^r)^{1/r} \leqslant (E|X|^r)^{1/r} + (E|Y|^r)^{1/r} \quad (1.5)$$

(Minkowski's inequality).

A consequence of Hölder's inequality is the Cauchy–Buniakowsky inequality

$$E|XY| \leqslant (EX^2)^{1/2}(EY^2)^{1/2}. \quad (1.6)$$

Let I be a finite or infinite open interval on the real line $\mathbb{R}$. Let g be a real function which is continuous and convex on I. Let X be a random variable taking on the values from the interval I with probability 1. If there exist means EX and $Eg(X)$, then

$$g(EX) \leq Eg(X) \qquad (1.7)$$

(Jensen's inequality).

The following theorem contains one more Chebyshev inequality.

Theorem 1.1 *Let (a, b) be a finite or infinite interval on $\mathbb{R}$ (possibly coinciding with $\mathbb{R}$). Let u and v be both non-decreasing or both non-increasing functions defined on (a, b). Let X be a random variable taking on values from the interval (a, b) with probability 1. Then*

$$Eu(X)\,Ev(X) \leq E(u(X)v(X)) \qquad (1.8)$$

if these means exist.

Proof For every $x, y \in (a, b)$ we have $(u(x) - u(y))(v(x) - v(y)) \geq 0$. For arbitrary random variables X_1 and X_2 with values from the interval (a, b) we obtain as a corollary of this inequality the following result:

$$E[(u(X_1) - u(X_2))(v(X_1) - v(X_2))] \geq 0$$

or

$$Eu(X_1)v(X_1) - Eu(X_2)v(X_1) - Eu(X_1)v(X_2) + Eu(X_2)v(X_2) \geq 0$$

(we suppose that all means here exist). Let X_1 and X_2 satisfy the following condition: the random vector (X_1, X_2) has the same distribution as the vector (X_2, X_1). Then

$$Eu(X_1)v(X_1) - 2\,Eu(X_1)v(X_2) + Eu(X_1)v(X_1) \geq 0$$

and

$$Eu(X_1)v(X_1) \geq Eu(X_1)v(X_2).$$

Now let X_1 and X_2 be independent random variables having the same distribution as X. We obtain the inequality

$$Eu(X)v(X) \geq Eu(X)\,Ev(X). \qquad \square$$

If u does not increase and v does not decrease, then the inequality (1.8) holds with the sign $\geq$ instead of $\leq$. It is easy to prove this by making the necessary changes in the proof of Theorem 1.1.

If there exists the moment α_s of order s of the random variable X, then

$$\beta_r^{1/r} \leq \beta_s^{1/s} \qquad (1.9)$$

for every positive $r \leq s$ (Lyapunov's inequality). It follows that $v_r^{1/r} \leq v_s^{1/s}$ ($r \leq s$), $\beta_l \beta_m \leq \beta_{l+m}$, and $v_l v_m \leq v_{l+m}$ for every l and m. We can strengthen Lyapunov's inequality.

Theorem 1.2 *Let X be a random variable, $\beta_r = E|X|^r$ and $0 < r < s$. Then*

$$\beta_r^{1/r} \leq \gamma^{1/r - 1/s} \beta_s^{1/s} \tag{1.10}$$

where $\gamma = P(X \neq 0)$.

In the case when $\gamma < 1$ or, equivalently, $P(X = 0) > 0$, inequality (1.10) is stronger than (1.9). If X is a random variable with two values 0 and 1 and the corresponding probabilities $1 - p$ and p, where $0 < p < 1$, then we have equality in (1.10).

Proof If $\gamma = 0$ then $P(X = 0) = 1$ and the inequality (1.10) is obvious. Let $\gamma > 0$. Let I_A be the indicator function of a set A of the points ω. We have $I_A(\omega) = 1$ if $\omega \in A$, $I_A(\omega) = 0$ if $\omega \notin A$. We shall also use the shorter notation $I(A)$ instead of $I_A(\omega)$. Obviously, $E|X|^r = E|X|^r I(X \neq 0)$. If $0 < r < s$, then

$$E|XY| \leq (E|X|^{s/r})^{r/s} (E|Y|^{s/(s-r)})^{1-(r/s)}$$

for arbitrary random variables X and Y by Hölder's inequality. If we here replace X by X^r and Y by $I(X \neq 0)$, then we obtain

$$E|X|^r I(X \neq 0) \leq (E|X|^s)^{r/s} \gamma^{1-r/s}.$$

This implies (1.10). □

In turn, as a consequence of (1.10) we get $\beta_l \beta_m \leq \gamma \beta_{l+m}$ for every l and m.

The moment-generating function of the random variable X is defined by the equality $M(t) = Ee^{tX}$. The mathematical expectation on the right-hand side always exists for $t = 0$, but it does not always exist in a non-degenerate interval. If it does exist in the interval $|t| < a$, then in that interval we have

$$M(t) = 1 + \sum_{k=1}^{\infty} \alpha_k \frac{t^k}{k!}.$$

If there exists a constant C such that $P(|X| \leq C) = 1$, the moment-generating function of X exists for all t.

If the moment-generating function of a random variable X exists in some non-degenerate interval with the centre at the origin, then we say that X satisfies the Cramér condition. This condition implies the existence of moments of all orders of the random variable X. Equivalent formulations of Cramér's condition can be found in Lemma 2.2 (Chapter 2).

If there exists a non-negative constant b^2 such that the moment-generating function of the random variable X satisfies the condition $M(t) \leq e^{b^2 t^2}$ for every real t, then the random variable X is called subgaussian.

Moments do not always exist, therefore some other numerical characteristics of random variables are introduced that are free from this disadvantage.

Let X be a random variable, $0 < q < 1$. A quantile of order q of the random variable X is called any number κ_q satisfying the inequalities

$$P(X \leqslant \kappa_q) \geqslant q, \quad P(X \geqslant \kappa_q) \geqslant 1 - q.$$

The following statement is true: either the random variable has just one quantile of order q, or the set of all quantiles of order q of this random variable coincides with some closed interval on the real line. The random variable X has just one quantile of arbitrary order q ($0 < q < 1$) if the distribution function of this random variable is strictly increasing on the real line.

A quantile of order $\frac{1}{2}$ is called a median. Thus a median of the random variable X is any number m for which $P(X \geqslant m) \geqslant \frac{1}{2}$ and $P(X \leqslant m) \geqslant \frac{1}{2}$.

Lemma 1.1 *Let X be a random variable. Let a and b be real numbers, $0 < q < 1$. If $P(X \leqslant a) > q$ then every quantile κ_q of the random variable X satisfies the condition $\kappa_q \leqslant a$. If $P(X \geqslant b) > q$ then every quantile κ_{1-q} satisfies the condition $\kappa_{1-q} \geqslant b$. If $P(X \leqslant a) \geqslant q$ then there exists a quantile κ_q of the random variable X satisfying the condition $\kappa_q \leqslant a$. If $P(X \geqslant b) \geqslant q$ then there exists a quantile κ_{1-q} such that $\kappa_{1-q} \geqslant b$.*

Proof We shall begin with the first proposition. Suppose that $P(X \leqslant a) > q$. If there exists a quantile $\kappa_q > a$, then $P(X \geqslant \kappa_q) \geqslant 1 - q$ and $P(X > a) \geqslant 1 - q$, contrary to our assumption. The second proposition can be proved similarly.

Consider now the case when $P(X \leqslant a) \geqslant q$. We have $P(X > a) \leqslant 1 - q$. Let κ_q^* be the least quantile of order q. Then

$$P(X \geqslant \kappa_q^*) \geqslant 1 - q \geqslant P(X > a).$$

It follows that either $\kappa_q^* \leqslant a$ or $P(a < X < \kappa_q^*) = 0$. In the last case every number from the interval (a, κ_q^*) is a quantile of order q of the random variable X, and κ_q^* is not the least quantile of this order. Therefore $\kappa_q^* \leqslant a$.

The fourth proposition can be proved in a similar way. □

1.3 Characteristic functions

Let X and Y be random variables with means EX and EY. If $Z = X + iY$, then we put $EZ = EX + iEY$.

Let X be a random variable. The characteristic function (c.f.) $f(t)$ of the random variable X is defined by the equality

$$f(t) = E e^{itX}$$

for every real t. Thus we have $f(t) = E \cos tX + iE \sin tX$. The functions cos u and sin u are bounded, therefore $\mathrm{E}\cos tX$ and $\mathrm{E}\sin tX$ exist for every random variable X and for every t. It follows that the c.f. is defined for every random variable.

If $F(x)$ is the distribution function of X, then

$$f(t) = \int_{-\infty}^{\infty} e^{itx}\, dF(x).$$

The following properties of characteristic functions are simple consequences of the definition: $f(0) = 1$; $|f(t)| \leq 1$ for every real t; $f(-t) = \overline{f(t)}$ where $\overline{f(t)}$ is the complex conjugate of $f(t)$. Further, $f(t)$ is uniformly continuous on the real line. If $f(t)$ is the c.f. of the random variable X, and $g(t)$ is the c.f. of the random variable $Y = aX + b$, where a and b are constants, then $g(t) = e^{ibt} f(at)$.

If the random variable X has the moment $\alpha_k = \mathrm{E}X^k$ of order k, where k is a positive integer, then the c.f. $f(t)$ of this random variable has everywhere continuous derivatives up to the order k and, moreover, $f^{(m)}(0) = i^m \alpha_m$ for $m \leq k$.

If $X_1, \ldots, X_n$ are independent random variables with the characteristic functions $f_1(t), \ldots, f_n(t)$, then the c.f. of the sum $X_1 + \cdots + X_n$ is $f_1(t) \cdots f_n(t)$.

If the random variable X with the c.f. $f(t)$ has a moment α_k of some integer order $k \geq 1$, then we have

$$f(t) = 1 + \sum_{\nu=1}^{k} \alpha_\nu \frac{(it)^\nu}{\nu!} + o(|t|^k) \qquad (t \to 0)$$

by Taylor's formula.

By definition, a random variable X and its distribution are symmetric if X and $-X$ have the same distribution. If X is a symmetric random variable with the c.f. $f(t)$, then

$$f(t) = \mathrm{E}e^{itX} = \mathrm{E}e^{-itX} = f(-t) = \overline{f(t)}.$$

Thus the c.f. of a symmetric random variable is real.

Let X be a random variable with the c.f. $f(t)$. Let us consider the random variable $X^s = X - Y$, where Y is a random variable independent of X and having the same distribution as X. The random variable X^s is called the symmetrized random variable. It has the non-negative c.f. $f(t)f(-t) = f(t)\overline{f(t)} = |f(t)|^2$. We conclude from this that if $f(t)$ is a c.f., then $|f(t)|^2$ is also a c.f.

We consider some important examples of characteristic functions. If the random variable X has a discrete distribution with the values $x_1, x_2, \ldots$ and the corresponding probabilities $p_1, p_2, \ldots$, then

$$f(t) = \mathrm{E}e^{itX} = \sum_n p_n e^{itx_n}.$$

In particular, if X takes on the value c with probability 1, then $f(t) = e^{itc}$, so that $|f(t)| \equiv 1$. The characteristic function of the binomial distribution with the parameters (n, p) is equal to $(p\,e^{it} + 1 - p)^n$. The c.f. of the Poisson distribution with the parameters (a, b, λ) has the form

$$f(t) = \exp\{iat + \lambda(e^{ibt} - 1)\}. \qquad (1.11)$$

The characteristic function of the normal distribution with the parameters (a, σ) is as follows:

$$f(t) = \exp\left\{iat - \frac{\sigma^2 t^2}{2}\right\}.$$

There are important numerical characteristics of random variables that are defined in terms of characteristic functions: these are the cumulants. If X is a random variable with the characteristic function $f(t)$, then the cumulant (or semi-invariant) of order k of this random variable is defined by the formal equality

$$\gamma_k = \frac{1}{i^k}\left[\frac{d^k}{dt^k}\log f(t)\right]_{t=0}.$$

Here and later, the expression 'log' denotes the principal value of the logarithm, so that $\log f(0) = 0$. It follows from this definition that γ_k is expressed in terms of the derivatives of the c.f. $f(t)$ up to the order k at the point zero, and, therefore, the existence of the moment α_k implies the existence of the cumulants of arbitrary order not exceeding k. In particular, $\gamma_1 = \alpha_1$, $\gamma_2 = \alpha_2 - \alpha_1^2$, $\gamma_3 = E(X - \alpha_1)^3$, if the indicated moments exist.

The characteristic function of a sum of independent random variables is equal to the product of the characteristic functions of the summands. Therefore, the cumulant of order k of the sum of independent random variables is equal to the sum of the cumulants of order k of these variables if the latter exist.

If $f(t)$ is the characteristic function of a distribution which has a moment α_k of order k for some integer k, then

$$\log f(t) = \sum_{\nu=1}^{k} \gamma_\nu \frac{(it)^\nu}{\nu!} + o(|t|^k) \qquad (t \to 0). \qquad (1.12)$$

For the normal distribution with arbitrary parameters, the cumulants of all orders beginning with the third are equal to zero.

If γ_k is the cumulant of order k of the random variable X, and γ'_k the cumulant of the same order of the random variable $Y = aX + b$, where a and b are constants, then $\gamma'_1 = a\gamma_1 + b$ and $\gamma'_k = a^k \gamma_k$ for every $k \geq 2$.

The formal identity

$$1 + \sum_{v=1}^{\infty} \frac{\alpha_v}{v!}(it)^v = \exp\left\{\sum_{v=1}^{\infty} \frac{\gamma_v}{v!}(it)^v\right\}$$

allows us to express the cumulant γ_k of arbitrary order k in terms of the moments $\alpha_1, \ldots, \alpha_k$ by the following formula:

$$\gamma_k = k! \sum (-1)^{r-1}(r-1)! \prod_{l=1}^{k} \frac{1}{m_l!}\left(\frac{\alpha_l}{l!}\right)^{m_l}, \qquad (1.13)$$

where the summation is carried out over all non-negative integer solutions of the equation $m_1 + 2m_2 + \cdots + km_k = k$, and $r = m_1 + \cdots + m_k$.

If the distribution $F(x)$ is absolutely continuous, then the Riemann–Lebesgue lemma implies that the corresponding characteristic function $f(t)$ satisfies the equality $\lim_{|t| \to \infty} f(t) = 0$. If the absolutely continuous component in eqn (1.1) is different from zero, then $\lim\sup_{|t| \to \infty} |f(t)| < 1$ (this last condition is usually called the Cramér condition (C)).

There is a useful characterization of a lattice distribution in terms of characteristic functions.

Theorem 1.3 *A distribution with the c.f. $f(t)$ is a lattice distribution if and only if there exists a number $t_0 \neq 0$ such that $|f(t_0)| = 1$.*

Proof Let $f(t)$ be the c.f. of a lattice distribution with possible values of the form $a + kh$, $k \in \mathbb{Z}$, where $h > 0$ and a are fixed numbers, and the corresponding probabilities p_k. Then

$$f(t) = \sum_{k=-\infty}^{\infty} \exp\{it(a+kh)\} p_k.$$

It follows that $|f(t + 2\pi m/h)| = |f(t)|$ for every real t and every integer m. If we here put $t = 0$ and $m = 1$, we obtain $|f(2\pi/h)| = |f(0)| = 1$. As a by-product we have proved that if $f(t)$ is the c.f. of a lattice distribution with a span h, then $|f(t)|$ is a periodic function with period $2\pi/h$.

Now suppose that there exists $t_0 \neq 0$ such that $|f(t_0)| = 1$. Then $f(t_0) = e^{ib}$, where b is a real number. Denoting the d.f. that corresponds to the c.f. $f(t)$ by $F(x)$, we obtain

$$\int_{-\infty}^{\infty} e^{it_0 x - ib} \, dF(x) = 1$$

or

$$\int_{-\infty}^{\infty} \cos(t_0 x - b) \, dF(x) = 1.$$

Therefore, the set of all points of growth of the distribution function $F(x)$

consists of the points x_k for which $t_0 x_k - b = 2\pi k$ or $x_k = 2\pi k/t_0 + b/t_0$, $k \in \mathbb{Z}$. Thus the distribution under consideration is a lattice distribution with a span $2\pi/|t_0|$. □

Using some considerations in the course of the proof of Theorem 1.3, we can prove the following proposition.

Lemma 1.2 *Let $f(t)$ be the characteristic function of a lattice distribution with a span h. The span h is maximal if and only if $|f(2\pi/h)| = 1$ and $|f(t)| < 1$ in the interval $0 < t < 2\pi/h$.*

It follows from Lemma 1.2 that if $f(t)$ is the c.f. of a lattice distribution with the maximal span h, then for every $\varepsilon > 0$ there exists a positive number c such that $|f(t)| \leqslant e^{-c}$ in the domain $\varepsilon \leqslant |t| \leqslant 2\pi/h - \varepsilon$.

Lemma 1.3 *For an arbitrary c.f. $f(t)$ and every real number t, the following inequality holds:*

$$1 - |f(2t)|^2 \leqslant 4(1 - |f(t)|^2).$$

Proof Let $G(x)$ be an arbitrary d.f. and let $g(t)$ be the corresponding c.f. Then

$$\operatorname{Re}(1 - g(t)) = \int_{-\infty}^{\infty} (1 - \cos tx)\, dG(x),$$

where Re denotes the real part. Using the elementary relations

$$1 - \cos tx = 2\sin^2 \frac{tx}{2} \geqslant \tfrac{1}{4}(1 - \cos 2tx),$$

we obtain the inequality

$$\operatorname{Re}(1 - g(2t)) \leqslant 4\operatorname{Re}(1 - g(t)) \qquad (1.14)$$

for every real t. (This inequality is of interest in itself.) If we here put $g(t) = |f(t)|^2$, we arrive at the assertion of Lemma 1.3. □

Lemma 1.4 *Let $f(t)$ be a c.f. and let b and $c < 1$ be positive constants. If $|f(t)| \leqslant c$ for $|t| \geqslant b$, then*

$$|f(t)| \leqslant 1 - \frac{1 - c^2}{8b^2} t^2$$

for $|t| < b$.

Proof It follows from Lemma 1.3 that

$$1 - |f(2^n t)|^2 \leq 4^n(1 - |f(t)|^2)$$

for every n. For $t = 0$ the inequality to be proved is obvious. Let $t \neq 0$, $|t| < b$. We choose n so that $2^{-n}b \leq |t| < 2^{-n+1}b$. Then $|f(2^n t)| \leq c$ and $1 - |f(t)|^2 > [(1 - c^2)/4b^2]t^2$, or

$$|f(t)| < 1 - \frac{1-c^2}{8b^2}t^2. \quad \square$$

The following proposition is a consequence of Lemma 1.4. If a c.f. $f(t)$ satisfies the Cramér condition (C) $\lim\sup_{|t|\to\infty} |f(t)| < 1$, then for every $\varepsilon > 0$ there exists a positive number $c < 1$ such that $|f(t)| \leq c$ for $|t| \geq \varepsilon$.

In order to prove this proposition we note that the Cramér condition (C) implies the existence of positive constants $c_0 < 1$ and b such that $|f(t)| \leq c_0$ for $|t| \geq b$. By Lemma 1.4 we have

$$|f(t)| \leq 1 - \frac{1-c_0^2}{8b^2}t^2 \leq 1 - \frac{1-c_0^2}{8b^2}\varepsilon^2$$

in the domain $\varepsilon \leq |t| < b$ for every $\varepsilon > 0$. Putting $c = \max\{c_0, 1 - [(1-c_0^2)/8b^2]\varepsilon^2\}$, we complete the proof. $\square$

Lemma 1.5 *Let $f(t)$ be the c.f. of a non-degenerate distribution. Then there exist positive constants δ and ε such that $|f(t)| \leq 1 - \varepsilon t^2$ for $|t| \leq \delta$.*

Proof Let X be a random variable with the d.f. $F(x)$ and the c.f. $f(t)$. The symmetrized random variable X^s has the d.f. $F^s(x)$ and the c.f. $|f(t)|^2$. Therefore,

$$1 - |f(t)|^2 = \int_{-\infty}^{\infty} (1 - \cos tx) \, dF^s(x).$$

From the inequality $1 - \cos x \geq \frac{11}{24}x^2$ for $|x| \leq 1$, we have

$$1 - |f(t)|^2 \geq \tfrac{11}{24}t^2 \int_{|k| \leq 1/|t|} x^2 \, dF^s(x).$$

for every $t \neq 0$. The random variable X^s has a non-degenerate distribution because the distribution of X is non-degenerate. Therefore there exists $\delta > 0$ such that $\int_{|x| \leq 1/\delta} x^2 \, dF^s(x) > 0$. Let us denote the last integral by c. If $|t| \leq \delta$ we obtain

$$1 - |f(t)|^2 \geq \tfrac{11}{24}t^2 \int_{|x| \leq 1/\delta} x^2 \, dF^s(x) = \tfrac{11}{24}ct^2.$$

The assertion of Lemma 1.5 follows from here. $\square$

We shall formulate a result containing an inversion formula.

Theorem 1.4 *Let $F(x)$ be a d.f. and $f(t)$ the corresponding c.f. If x_1 and x_2 are points of continuity of $F(x)$, then*

$$F(x_2) - F(x_1) = \frac{1}{2\pi} \lim_{T \to \infty} \int_{-T}^{T} \frac{e^{-itx_2} - e^{-itx_1}}{-it} f(t)\, dt.$$

The following uniqueness theorem is an easy consequence.

Theorem 1.5 *Two distribution functions having the same characteristic function are identical.*

There is an elementary consequence of Theorem 1.5. A random variable is symmetric if and only if its characteristic function is real. The necessity has already been proved. The sufficiency follows from the equalities

$$f(t) = \overline{f(t)} = f(-t) = E e^{-itX},$$

which are true because the c.f. $f(t)$ is real. The identity of the c.f. of the random variables X and $-X$ proves the identity of the distribution functions of the corresponding random variables.

Theorem 1.6 *If the c.f. $f(t)$ is absolutely integrable on the real line, then the c.f. corresponding d.f. $F(x)$ has an everywhere continuous derivative $p(x) = (d/dx)\, F(x)$, and moreover,*

$$p(x) = \frac{1}{2\pi} \int_{-\infty}^{\infty} e^{-itx} f(t)\, dt \qquad (1.15)$$

for every x.

The proof of Theorems 1.4–1.6 is given in, for example, the text of Gnedenko [146]. The inversion formula for a lattice distribution, analogous to (1.15), is as follows.

Theorem 1.7 *Let the random variable X have a lattice distribution with possible values of the form $a + kh$ ($k = 0, \pm 1, \pm 2, \ldots$). Let $p_k = P(X = a + kh)$. Then*

$$p_k = \frac{h}{2\pi} \int_{|t| < \pi/h} e^{-it(a+kh)} f(t)\, dt \qquad (1.16)$$

for every integer k, where $f(t)$ is the c.f. of the random variable X.

Proof We have

$$f(t) e^{-ita} = \sum_{m=-\infty}^{\infty} e^{itmh} p_m.$$

Let k be an arbitrary integer. We multiply both sides of the above equality by e^{-itkh} and integrate over the interval $|t| < \pi/h$. Then we obtain eqn (1.16). □

1.4 Convergence of distributions

Let $F(x)$, $F_1(x)$, $F_2(x)$, ... be bounded non-decreasing functions. The sequence $\{F_n(x)\}$ converges weakly to $F(x)$ if $F_n(x) \to F(x)$ at every point of continuity of $F(x)$. To indicate that the sequence $\{F_n(x)\}$ converges weakly to $F(x)$, we will use the notation $F_n \to F$. If $F_n \to F$ and $F_n(-\infty) \to F(-\infty)$, $F_n(+\infty) \to F(+\infty)$, we shall say that $F_n(x)$ converges completely to $F(x)$, and we write $F_n \rightrightarrows F$.

Later we shall need the following variant of a theorem by Helly.

Theorem 1.8 *Let the function $g(x)$ be continuous and bounded on the real line. Let $F(x)$, $F_1(x)$, $F_2(x)$, ... be bounded, non-decreasing functions, and let $F_n \rightrightarrows F$. Then*

$$\int_{-\infty}^{\infty} g(x) \, dF_n(x) \to \int_{-\infty}^{\infty} g(x) \, dF(x).$$

The following proposition is easy to prove.

Lemma 1.6 *If the sequence of characteristic functions $\{f_n(t)\}$ converges to the c.f. $f(t)$ for every t, the convergence is uniform in t in an arbitrary finite interval.*

An immediate consequence of Theorem 1.8 and Lemma 1.6 is:

Theorem 1.9 *Let $F(x)$, $F_1(x)$, $F_2(x)$, ... be distribution functions, and let $f(t)$, $f_1(t)$, $f_2(t)$, ... be the corresponding characteristic functions. If $F_n \to F$, then $f_n(t) \to f(t)$ uniformly in t in an arbitrary finite interval.*

The following inverse limit theorem for c.f. is important.

Theorem 1.10 *Let $\{f_n(t)\}$ be a sequence of c.f., $\{F_n(x)\}$ the corresponding sequence of d.f. If $f_n(t) \to f(t)$ for every t and if $f(t)$ is continuous at the point $t = 0$, there exists a d.f. $F(x)$ such that $F_n \to F$. For this d.f.,*

$$f(t) = \int_{-\infty}^{\infty} e^{itx} \, dF(x).$$

Theorem 1.11 *If the sequence of d.f. $\{F_n(x)\}$ converges to a continuous d.f. $F(x)$, the convergence is uniform in x ($-\infty < x < \infty$).*

Proof Let ε be an arbitrary positive number. The continuity of $F(x)$ implies that there exist points $x_1, \ldots x_m$ satisfying the conditions

$$F(x_1) < \frac{\varepsilon}{2}, \quad F(x_{k+1}) - F(x_k) < \frac{\varepsilon}{2} \quad (k = 1, \ldots, m-1),$$

$$1 - F(x_m) < \frac{\varepsilon}{2}.$$

Further, there exists a number n_0 such that for $n > n_0$ we have

$$|F_n(x_k) - F(x_k)| < \frac{\varepsilon}{2} \quad (k = 1, \ldots, m).$$

If $x_k \leqslant x < x_{k+1}$ ($k = 1, \ldots, m-1$), then for $n > n_0$ we get

$$F_n(x) - F(x) \leqslant F_n(x_{k+1}) - F(x_{k+1}) + F(x_{k+1}) - F(x_k) < \varepsilon$$

and

$$F_n(x) - F(x) \geqslant F_n(x_k) - F(x_{k+1}) > -\varepsilon.$$

If $x < x_1$,

$$F_n(x) - F(x) \leqslant F_n(x_1) - F(x_1) + F(x_1) < \varepsilon$$

and

$$F_n(x) - F(x) \geqslant -F(x) \geqslant -F(x_1) > -\frac{\varepsilon}{2}$$

for $n > n_0$. The case when $x \geqslant x_m$ is similarly handled. Thus $|F_n(x) - F(x)| < \varepsilon$ for all x and $n > n_0$. □

Theorem 1.12 *Let $\{F_n(x)\}$ be a sequence of d.f. that converges weakly to a d.f. $F(x)$. If*

$$\limsup \int_{-\infty}^{\infty} |x|^p \, dF_n(x) < \infty$$

for some $p > 0$, then

$$\lim \int_{-\infty}^{\infty} (1 + |x|^q)|F_n(x) - F(x)|^r \, dx = 0$$

for every $q \geqslant 0$ and r such that $pr > 1 + q$.

Proof Put

$$C = \limsup \int_{-\infty}^{\infty} |x|^p \, dF_n(x).$$

By Theorem 1.8 we have for every positive N,

$$\int_{-N}^{N} |x|^p \, dF(x) \leq \limsup \int_{-N}^{N} |x|^p \, dF_n(x) \leq C.$$

Therefore,

$$\int_{-\infty}^{\infty} |x|^p \, dF(x) = \lim_{N \to \infty} \int_{-N}^{N} |x|^p \, dF(x) \leq C.$$

If $x < -1$, then

$$|x|^p F(x) = |x|^p \int_{-\infty}^{x} dF(y) \leq \int_{-\infty}^{x} |y|^p \, dF(y) \leq C.$$

If $x > 1$, then

$$|x|^p (1 - F(x)) = x^p \int_{x}^{\infty} dF(y) \leq \int_{x}^{\infty} y^p \, dF(y) \leq C.$$

These estimates for $|x| > 1$ hold true if we replace F by F_n and C by $C + 1$, and if n is sufficiently large. Thus $|F_n(x) - F(x)| \leq C_0 |x|^{-p}$ for $|x| > 1$ and sufficiently large n where $C_0 = C + 1$. It follows that

$$(1 + |x|^q) |F_n(x) - F(x)|^r \leq G(x)$$

for every $q \geq 0$ and all sufficiently large n. Here $G(x) = 2$ for $|x| \leq 1$ and $G(x) = C_0^r (1 + |x|^q) |x|^{-pr}$ for $|x| > 1$. If $q \geq 0$ and $pr - q > 1$, the function $G(x)$ is absolutely integrable on the real line. The sequence $\{F_n(x)\}$ converges to $F(x)$ everywhere except for a finite or countable set of the points of the real line. Making the admissible passage to the limit under the sign of the integral, we obtain

$$\lim \int_{-\infty}^{\infty} (1 + |x|^q) |F_n(x) - F(x)|^r \, dx$$

$$= \int_{-\infty}^{\infty} \lim \{(1 + |x|^q) |F_n(x) - F(x)|^r \, dx = 0. \quad \square$$

Theorem 1.13 *Let $F(x), F_1(x), F_2(x), \ldots$ be a sequence of d.f. If*

$$\int_{-\infty}^{\infty} |F_n(x) - F(x)|^r \, dx \to 0$$

for some $r > 0$, then $\{F_n(x)\}$ converges weakly to $F(x)$.

Proof Suppose that there exists a point of continuity y of the function F such that $F_n(y)$ does not converge to $F(y)$. Then there exist a positive number ε and a sequence $\{n_k\}$ satisfying the conditions $|F_{n_k}(y) - F(y)| > \varepsilon$ for every k. The point y is a point of continuity of F, therefore there exist y_1 and y_2 such that $y_1 < y < y_2$ and $|F(y_i) - F(y)| < \varepsilon/2$ ($i = 1, 2$). The functions $F_n(y)$ do not decrease. This implies the inequality $|F_{n_k}(x) - F(x)| > \varepsilon/2$ either for all $x \in [y_1, y)$ or for all $x \in (y, y_2]$. It follows that

$$\int_{-\infty}^{\infty} |F_{n_k}(x) - F(x)|^r \, dx \geqslant \min\{y - y_1, y_2 - y\} \left(\frac{\varepsilon}{2}\right)^r > 0,$$

contrary to the condition of Theorem 1.13. □

Lemma 1.7 *Let X and Y be random variables defined on a common probability space. Then*

$$P(X < u - v) - P(Y > v) \leqslant P(X + Y < u) \leqslant P(X < u + v) + P(Y < -v)$$
(1.17)

for every real u, v, and

$$P(X < x - \varepsilon) - P(|Y| > \varepsilon) \leqslant P(X + Y < x) \leqslant P(X < x + \varepsilon) + P(|Y| > \varepsilon)$$
(1.18)

for every real x and every $\varepsilon > 0$.

Proof We shall prove the inequality (1.17). Obviously,

$$\{X + Y < u\} \subset \{X < u + v\} \cup \{Y < -v\}.$$

It follows that

$$P(X + Y < u) \leqslant P(X < u + v) + P(Y < -v).$$

Further,

$$\{X < u - v\} \subset \{X + Y < u\} \cup \{Y > v\},$$
$$P(X < u - v) \leqslant P(X + Y < u) + P(Y > v).$$

Inequality (1.17) is proved. We note that (1.17) holds true if we replace its left-hand side by the difference $P(X \leqslant u - v) - P(Y \geqslant v)$. Inequality (1.18) follows from (1.17). □

Lemma 1.8 *Let X and Y be arbitrary random variables, and let $F(x)$ and $H(x)$ be the distribution functions of X and $X + Y$, correspondingly. If $T(x)$*

is an arbitrary function defined on the real line, then

$$|H(x) - T(x)| \leq K + L + P(|Y| > \varepsilon) \tag{1.19}$$

for every real x and every positive ε, where

$$K = \max\{|F(x + \varepsilon) - T(x + \varepsilon)|, |F(x - \varepsilon) - T(x - \varepsilon)|\},$$
$$L = \max\{|T(x + \varepsilon) - T(x)|, |T(x - \varepsilon) - T(x)|\}.$$

Proof By (1.18) we have

$$H(x) - T(x) \leq F(x + \varepsilon) - T(x) + P(|Y| > \varepsilon).$$

Hence

$$H(x) - T(x) \leq K + L + P(|Y| > \varepsilon).$$

Similarly, we find that

$$H(x) - T(x) \geq -K - L - P(|Y| > \varepsilon). \quad \square$$

Lemma 1.8 implies the inequality

$$|H(x) - T(x)| \leq \sup_y |F(y) - T(y)| + L + P(|Y| > \varepsilon). \tag{1.20}$$

for every real x and every $\varepsilon > 0$.

Lemma 1.9 *Let X and Y be arbitrary random variables, and let F(x) and H(x) be the distribution functions of X and X + Y, correspondingly. Let $\Phi(x)$ be the standard normal distribution function. If*

$$\sup_x |F(x) - \Phi(x)| \leq M,$$

then

$$\sup_x |H(x) - \Phi(x)| \leq M + P(|Y| > \varepsilon) + \frac{\varepsilon}{\sqrt{2\pi}}$$

for every $\varepsilon > 0$.

This lemma is an immediate consequence of Lemma 1.8 if we put $T(x) \equiv \Phi(x)$ in (1.20); then we have $L \leq \varepsilon/\sqrt{2\pi}$ for every x.

Let $Y, Y_1, Y_2, \ldots$ be a sequence of random variables defined on a common probability space $(\Omega, \mathscr{A}, P)$. We say that the sequence $\{Y_n\}$ converges to Y in probability, and we write $Y_n \xrightarrow{P} Y$, if $P(|Y_n - Y| < \varepsilon) \to 1$ for every fixed $\varepsilon > 0$.

Lemma 1.10 *Let $\{X_n\}$ and $\{Y_n\}$ be sequences of random variables defined on a common probability space. If $P(X_n < x)$ converges weakly to a d.f. $F(x)$ and if $Y_n \xrightarrow{P} 0$, then $P(X_n + Y_n < x)$ converges weakly to $F(x)$.*

Proof Let x be an arbitrary point of continuity of $F(x)$. By (1.18) we have

$$P(X_n < x - \varepsilon) - F(x) - P(|Y_n| > \varepsilon)$$
$$\leqslant P(X_n + Y_n < x) - F(x) \leqslant P(X_n < x + \varepsilon) - F(x) + P(|Y_n| > \varepsilon)$$

for every $\varepsilon > 0$. The assertion of the lemma follows from here. $\square$

Theorem 1.14 *Let $\{a_n\}$ and $\{b_n\}$ be sequences of constants, in which $a_n > 0$. Let the sequence of d.f. $\{F_n(x)\}$ converge weakly to a non-degenerate d.f. $F(x)$. Then the following assertions hold:*

(A) If $F_n(a_n x + b_n) \to G(x)$, where $G(x)$ is a non-degenerate d.f., then $G(x) = F(ax + b)$, $a_n \to a$ and $b_n \to b$. In particular, if $F_n(a_n x + b_n) \to F(x)$, then $a_n \to 1$ and $b_n \to 0$.

(B) If $a_n \to a$ and $b_n \to b$, then $F_n(a_n x + b_n) \to F(ax + b)$.

Proof We shall first prove assertion (A). Let $f_n(t)$, $f(t)$ and $g(t)$ denote the c.f. of the distributions $F_n(x)$, $F(x)$ and $G(x)$ respectively. The sequence of positive constants $\{a_n\}$ contains a subsequence $\{a_{n'}\}$ such that $a_{n'} \to a$. If $a = +\infty$, then for every t we have

$$|g(t)| = \lim_{n' \to \infty} \left| f_{n'}\left(\frac{t}{a_{n'}}\right) \right| = |f(0)| = 1,$$

i.e. $g(t)$ is the c.f. of a degenerate distribution, which contradicts the hypothesis. If $a = 0$, then, writing $g_n(t) = \exp\{-it(b_n/a_n)\} f_n(t/a_n)$, we obtain $|f(t)| = \lim_{n' \to \infty} |f_{n'}(t)| = \lim_{n' \to \infty} |g_{n'}(a_{n'} t)| = |g(0)| = 1$ for every t, contradicting the assumption that $F(x)$ is non-degenerate. Thus $0 < a < \infty$. The functions $g(t)$ and $f(t/a)$ are different from zero for sufficiently small t. Therefore as $n' \to \infty$,

$$\exp\left\{-it\frac{b_{n'}}{a_{n'}}\right\} = \frac{\exp\{-it(b_{n'}/a_{n'})\} f_{n'}(t/a_{n'})}{f_{n'}(t/a_{n'})} \to \frac{g(t)}{f(t/a)}$$

and

$$b_{n'} \to b = -\frac{a}{it} \log \frac{g(t)}{f(t/a)}, \quad g(t) = \exp\left\{-it\frac{b}{a}\right\} f\left(\frac{t}{a}\right).$$

The last equality implies that $G(x) = F(ax + b)$. Suppose that there exists a subsequence $\{a_{n''}\}$ of the sequence $\{a_n\}$ satisfying the condition $a_{n''} \to a_0 \neq a$. Then $b_{n''} \to b_0$ and

$$\exp\left\{-\frac{itb}{a}\right\} f\left(\frac{t}{a}\right) = \exp\left\{\frac{-itb_0}{a_0}\right\} f\left(\frac{t}{a_0}\right).$$

Therefore $|f(t)| = |f(ct)|$ for every t and some positive $c < 1$. Hence

$$|f(t)| = |f(ct)| = |f(c^2t)| = \cdots = \lim |f(c^n t)|$$

for every t, which contradicts the assumption that $F(x)$ is non-degenerate. Thus $a_n \to a$ and accordingly $b_n \to b$.

We now prove assertion (B). Let $\varepsilon > 0$ and x be such that the function $F(x)$ is continuous at the points $ax + b$, $ax + b - \varepsilon$ and $ax + b + \varepsilon$. Since $a_n x + b_n \to ax + b$ we have $ax + b - \varepsilon \leq a_n x + b_n \leq ax + b + \varepsilon$ for sufficiently large n. Hence

$$F_n(ax + b - \varepsilon) \leq F_n(a_n x + b_n) \leq F_n(ax + b + \varepsilon)$$

and

$$F(ax + b - \varepsilon) \leq \liminf F_n(a_n x + b_n)$$

$$\leq \limsup F_n(a_n x + b_n) \leq F(ax + b + \varepsilon).$$

The number $\varepsilon > 0$ can be chosen as small as we like, and we arrive at the relation

$$F_n(a_n x + b_n) \to F(ax + b). \quad \square$$

1.5 Concentration functions

There are several different definitions of the concentration functions. This section is devoted to the Lévy concentration functions.

The Lévy concentration function $Q(X; \lambda)$ of a random variable X is defined by the equality

$$Q(X; \lambda) = \sup_x P(x \leq X \leq x + \lambda)$$

for every $\lambda \geq 0$. It is clear that $Q(X; \lambda)$ is a non-decreasing function of λ satisfying the inequalities $0 \leq Q(X; \lambda) \leq 1$ for every $\lambda \geq 0$.

We shall prove some propositions about the concentration function which will be useful later.

Lemma 1.11 *If X and Y are independent random variables, then $Q(X + Y; \lambda) \leq \min\{Q(X; \lambda), Q(Y; \lambda)\}$ for every $\lambda \geq 0$.*

Proof Let $F_U(x)$ denote the distribution function of the random variable U. Writing $Z = X + Y$, we have

$$F_Z(x + \lambda) - F_Z(x) = \int_{-\infty}^{\infty} (F_X(x + \lambda - y) - F_X(x - y)) \, dF_Y(y)$$

and
$$Q(Z; \lambda) \leqslant Q(X; \lambda) \int_{-\infty}^{\infty} dF_Y(y) = Q(X; \lambda).$$

Similarly,
$$Q(Z; \lambda) \leqslant Q(Y; \lambda). \qquad \square$$

Lemma 1.12 *For every non-negative α and λ the inequality $Q(X; \alpha\lambda) \leqslant ([\alpha] + 1)Q(X; \lambda)$ holds, where $[\alpha]$ is the largest integer not exceeding α.*

This statement is obvious.

Lemma 1.13 *Let X be a random variable with the characteristic function $f(t)$ and the concentration function $Q(X; \lambda)$. Let $h(t)$ be a real c.f. that is integrable on the real line, and let $p(x)$ be the corresponding density of the distribution. For every $\lambda \geqslant 0$, $b > 0$, and $\beta \geqslant \lambda/b$ the following inequality holds:*

$$Q(X; \lambda) \leqslant \frac{\beta}{2\pi c} \int_{-\infty}^{\infty} |f(t)h(\beta t)|\, dt \qquad (1.21)$$

where

$$c = \min_{0 \leqslant x \leqslant b/2} p(x). \qquad (1.22)$$

Proof By Theorem 1.6 the density $p(x)$ is everywhere continuous. By an assumption, the c.f. $h(t)$ is real; it implies that the corresponding distribution is symmetric and the equality $p(x) = p(-x)$ holds for every real x.

Let $F(x)$ be the distribution function of the random variable. Applying the inversion formula for the density $p(x)$ (Theorem 1.6), we obtain

$$\int_{-\infty}^{\infty} p(a(x-\gamma))\, dF(x) = \frac{1}{2\pi a} \int_{-\infty}^{\infty} e^{i\gamma u} h\left(\frac{u}{a}\right) \left\{ \int_{-\infty}^{\infty} e^{-iux}\, dF(x) \right\} du$$

for every $a > 0$ and $\gamma \in \mathbb{R}$. Taking into account the equality $f(-t) = \overline{f(t)}$ for every real t, we find that

$$\int_{-\infty}^{\infty} p(a(x-\gamma))\, dF(x) \leqslant \frac{1}{2\pi a} \int_{-\infty}^{\infty} \left| f(t) h\left(\frac{t}{a}\right) \right| dt.$$

If $0 < a\lambda \leqslant b$, then

$$\min_{-\lambda/2 \leqslant x \leqslant \lambda/2} p(ax) \geqslant \min_{-b/2 \leqslant x \leqslant b/2} p(x) = \min_{0 \leqslant x \leqslant b/2} p(x) = c$$

by eqn (1.22), since the function $p(x)$ is even. Denoting

$$I = \left\{ x \in \mathbb{R} : \gamma - \frac{\lambda}{2} \leqslant x \leqslant \gamma + \frac{\lambda}{2} \right\},$$

we have
$$\int_{-\infty}^{\infty} p(a(x-\gamma))\,dF(x) \geq cP(X \in I)$$
and
$$P(X \in I) \leq \frac{1}{2\pi ac}\int_{-\infty}^{\infty}\left|f(t)h\left(\frac{t}{a}\right)\right|dt.$$

The number γ is arbitrary, hence
$$Q(X;\lambda) \leq \frac{1}{2\pi ac}\int_{-\infty}^{\infty}\left|f(t)h\left(\frac{t}{a}\right)\right|dt \tag{1.23}$$

for every positive a and λ such that $a\lambda \leq b$. We put $\beta = 1/a$ in inequality (1.23). Inequality (1.21) follows for $\beta \geq \lambda/b$. $\square$

Let us indicate some simple consequences of Lemma 1.13.

Lemma 1.14 *Let X be a random variable with the c.f. $f(t)$ and the concentration function $Q(X;\lambda)$. Let $h(t)$ be a real c.f. such that $h(t) = 0$ for $|t| \geq 1$. Let $p(x)$ be the density of the distribution with the c.f. $h(t)$. If $\lambda \geq 0$, $a > 0$, and $b > 0$, then*
$$Q(X;\lambda) \leq \frac{\beta}{2\pi c}\int_{-a}^{a}|f(t)h(\beta t)|\,dt \tag{1.24}$$

where
$$\beta = \max(1/a, \lambda/b) \tag{1.25}$$

and the constant c is defined by the equality (1.22).

Proof Lemma 1.13 implies that
$$Q(X;\lambda) \leq \frac{\beta}{2\pi c}\int_{|t| \leq 1/\beta}|f(t)h(\beta t)|\,dt$$

for $\beta \geq \lambda/b$. Let $a > 0$. If $1/a > \lambda/b$, we put $\beta = 1/a$. Then we shall obtain (1.24). If $1/a < \lambda/b$, then we put $\beta = \lambda/b$. In this case, $a \geq 1/\beta$ and the following inequality holds:
$$\int_{|t| \leq 1/\beta}|f(t)h(\beta t)|\,dt \leq \int_{-a}^{a}|f(t)h(\beta t)|\,dt.$$

Hence the inequality (1.24) is valid with the constant β defined by eqn (1.25). $\square$

We shall formulate some other consequences of Lemma 1.13. To this end, we need the following proposition that is easy to prove.

Lemma 1.15 *The function*

$$p_0(x) = \frac{1 - \cos x}{\pi x^2} \qquad (1.26)$$

is the density of the distribution with the c.f.

$$h_0(t) = \begin{cases} 1 - |t| & \text{if } |t| \leq 1, \\ 0 & \text{if } |t| > 1. \end{cases} \qquad (1.27)$$

In Lemma 1.14 we can put $h(t) = h_0(t)$, $p(x) = p_0(x) = 1/2\pi[\sin^2(x/2)/(x/2)^2]$, and $b = 1$. From the equality $\sin x = x - (x^3/6)\cos\theta x$ for every real x where $|\theta| \leq 1$, we infer that

$$\left|\frac{\sin x}{x}\right| \geq \frac{95}{96} \quad \text{for } |x| \leq \tfrac{1}{4}. \qquad (1.28)$$

Therefore,

$$c = \min_{0 \leq x \leq 1/2} p_0(x) \geq \frac{1}{2\pi}\left(\frac{95}{96}\right)^2,$$

and (1.24) implies that

$$Q(X;\lambda) \leq \left(\tfrac{96}{95}\right)^2 \max\left(\lambda, \frac{1}{a}\right) \int_{-a}^{a} |f(t)|\left(1 - \frac{|t|}{a}\right) dt \qquad (1.29)$$

for every $\lambda \geq 0$ and $a > 0$. In turn, (1.29) implies the following proposition.

Lemma 1.16 *If $Q(X;\lambda)$ is the concentration function of a random variable X with the c.f. $f(t)$, then*

$$Q(X;\lambda) \leq \left(\tfrac{96}{95}\right)^2 \max(\lambda, 1/a) \int_{-a}^{a} |f(t)| \, dt \qquad (1.30)$$

for every $\lambda \geq 0$ and $a > 0$.

If we put

$$h(t) = \begin{cases} 0 & \text{if } |t| \geq 1, \\ 2(1 - |t|)^3 & \text{if } 1/2 \leq |t| \leq 1, \\ 1 - 6t^2 + 6|t|^3 & \text{if } |t| \leq 1/2, \end{cases}$$

then $p(x) = 3/8\pi[\sin^4(x/4)/(x/4)^4]$ (see, for example, Gnedenko and Kolmogorov [147], §39). Using inequalities (1.28) and (1.22) for $b = 2$, we get

$$c = \min_{0 \leq x \leq 1} p(x) \geq \frac{3}{8\pi}\left(\frac{95}{96}\right)^4.$$

It follows from the inequalities (1.24) and (1.29) that

$$Q(X;\lambda) \leq \frac{4}{3}\left(\frac{96}{95}\right)^4 \max\left(\frac{\lambda}{2},\frac{1}{a}\right) \int_{-a}^{a} |f(t)|\,dt \qquad (1.31)$$

for every $\lambda \geq 0$ and $a > 0$. If $a\lambda \geq 2$, the right-hand side of (1.31) is smaller than the right-hand side of (1.30), which is equal to $(\frac{96}{95})^2 \lambda \int_{-a}^{a} |f(t)|\,dt$.

Lemma 1.14 implies the following simple inequality:

$$Q(X;\lambda) \leq \frac{1}{2\pi c} \max\left(\frac{1}{a},\frac{\lambda}{b}\right) \int_{-a}^{a} |f(t)|\,dt, \qquad (1.32)$$

where c is defined by eqn (1.22).

We note a consequence of Lemma 1.16 that corresponds to the value $\lambda = 0$. If X is a random variable with the c.f. $f(t)$, then

$$\sup_{x} P(X = x) \leq \left(\frac{96}{95}\right)^2 a^{-1} \int_{-a}^{a} |f(t)|\,dt \qquad (1.33)$$

for every $a > 0$.

Later we shall need an estimate for the concentration function $Q(X;\lambda)$ from below.

Lemma 1.17 *Let X be a random variable with the c.f. $f(t)$ and the concentration function $Q(X;\lambda)$. Then*

$$Q(X;\lambda) \geq \frac{95\lambda}{256\pi(1 + 2a\lambda)} \int_{-a}^{a} |f(t)|^2\,dt \qquad (1.34)$$

for every non-negative λ and a.

Proof We consider the symmetrized random variable $X^s = X - Y$, where Y is a random variable independent of X and having the same distribution as X. Let U be a random variable independent of X and having the c.f. $h_0(t/4a)$, where $h_0(t)$ is the function defined by (1.27). The random variable $V = X^s + U$ has a continuous distribution with the c.f. $|f(t)|^2 h_0(t/4a)$. It follows from Lemma 1.11 that $Q(V;\lambda) \leq Q(X;\lambda)$. In view of Theorem 1.4 we have

$$P\left(|V| \leq \frac{1}{4a}\right) = \frac{1}{\pi}\int_{-4a}^{4a} |f(t)|^2\left(1 - \frac{|t|}{4a}\right)\frac{\sin\frac{t}{4a}}{t}\,dt.$$

Using the inequality (1.28), we find that

$$Q\left(V;\frac{1}{2a}\right) \geq P\left(|V| \leq \frac{1}{4a}\right) \geq \frac{3}{16\pi a}\cdot\frac{95}{96}\int_{-a}^{a} |f(t)|^2\,dt.$$

By Lemma 1.12 we have

$$Q\left(V; \frac{1}{2a}\right) \leq \left(\frac{1}{2a\lambda} + 1\right) Q(V; \lambda)$$

for every $\lambda > 0$. Therefore,

$$Q(X; \lambda) \geq Q(V; \lambda) \geq \frac{95\lambda}{256\pi(1 + 2a\lambda)} \int_{-\infty}^{\infty} |f(t)|^2 \, dt$$

for every positive λ and a. In the case where $\lambda = 0$ or $a = 0$, the inequality (1.33) is satisfied in an obvious way. □

We shall show that under some additional condition it is possible to replace $|f(t)|^2$ in the inequality (1.34) by $|f(t)|$.

Lemma 1.18 *If X is a random variable with a non-negative c.f. $f(t)$, then*

$$Q(X; \lambda) \geq \frac{95\lambda}{256\pi(1 + 2a\lambda)} \int_{-a}^{a} f(t) \, dt \qquad (1.35)$$

for every non-negative λ and a.

This lemma can be proved in the same way as Lemma 1.17. Instead of the random variable $V = X^s + U$, we consider the random variable $V = X + U$ which has a continuous distribution with the c.f. $f(t) h_0(t/4a)$. There is no necessity for other changes in the proof.

It is impossible to obtain an analogue to Lemma 1.18 without any additional conditions. To be more precise, in the inequality

$$Q(X; \lambda) \geq \frac{C_1 \lambda}{1 + 2a\lambda} \int_{-a}^{a} |f(t)|^2 \, dt, \qquad (1.36)$$

where C_1 is some positive constant, it is impossible to replace $|f(t)|^2$ by $|f(t)|$ with some other positive constant C_1 without introducing additional conditions.

To prove this proposition, consider a random variable X having the uniform distribution on the interval $(-1, 1)$. We have $Q(X; \lambda) = \lambda/2$ if $0 \leq \lambda \leq 2$, and $f(t) = \sin t/t$. Suppose that the inequality (1.36) holds with the replacement of $|f(t)|^2$ by $|f(t)|$ and with some positive constant C_1. Putting $a = 1/\lambda$, we obtain

$$\frac{\lambda}{2} \geq \frac{C_1 \lambda}{3} \int_{|t| \leq 1/\lambda} \left|\frac{\sin t}{t}\right| dt \qquad (1.37)$$

for $0 \leq \lambda \leq 2$. It is known that $\int_0^\infty |\sin t|(dt/|t|) = \infty$. If we divide both parts of the inequality (1.37) by λ and make the limit passage as $\lambda \downarrow 0$, we get an obviously false statement. This contradiction proves the proposition. □

We have found that the lower estimate for the concentration function in Lemma 1.17 is unimprovable in some sense. We shall prove an analogous assertion connected with Lemma 1.16. We shall show that in the inequality

$$Q(X;\lambda) \leq C_2 \max(\lambda, 1/a) \int_{-a}^{a} |f(t)|\, dt, \tag{1.38}$$

where C_2 is a positive constant, it is impossible to replace $|f(t)|$ by $|f(t)|^2$ without introducing additional conditions (even with the replacement of C_2 by any other positive constant).

To this end, consider a random variable X having the density $p(x)$ such that $\int_{-\infty}^{\infty} p^2(x)\, dx < \infty$ and $\lim_{x \to 0+} p(x) = \infty$ (an example is the function $p(x) = \frac{3}{4} x^{-1/4}$ if $0 < x < 1$ and $p(x) = 0$ if $x \leq 0$ or $x \geq 1$). For this random variable we have

$$\frac{1}{\lambda} Q(X;\lambda) \geq \frac{1}{\lambda} \int_0^{\lambda} p(x)\, dx \to \infty \tag{1.39}$$

as $\lambda \to 0$. It is known that the condition $\int_{-\infty}^{\infty} p^2(x)\, dx < \infty$ implies the relation $\int_{-\infty}^{\infty} |f(t)|^2\, dt < \infty$. Suppose that the inequality (1.38) is valid with the replacement of $|f(t)|$ by $|f(t)|^2$ for some positive constant C_2. Then

$$Q(X;\lambda) \leq C_2 \lambda \int_{|t| < 1/\lambda} |f(t)|^2\, dt.$$

After dividing by λ and the limit passage as $\lambda \downarrow 0$, we obtain

$$\limsup_{\lambda \downarrow 0} \frac{1}{\lambda} Q(X;\lambda) < \infty.$$

This contradicts (1.39). □

1.6 Infinitely divisible distributions

A distribution function $F(x)$ and the corresponding characteristic function $f(t)$ are called infinitely divisible if for every positive integer n there exists a characteristic function $f_n(t)$ such that

$$f(t) = (f_n(t))^n. \tag{1.40}$$

In other words, the distribution F is infinitely divisible if for every positive integer n there exists a distribution function $F_n(x)$ such that $F = F_n^{*n}$. Here F_n^{*n} is the n-fold convolution of the function F_n.

The normal distribution with arbitrary parameters is infinitely divisible. In fact, the c.f. of this distribution has the form $f(t) = \exp\{iat - \frac{1}{2}\sigma^2 t^2\}$, so that the equality (1.40) is satisfied for every n and $f_n(t) = \exp\{i(a/n)t - (1/2n)\sigma^2 t^2\}$; the latter function is the c.f. of the normal distribution with the parameters $(a/n, \sigma/\sqrt{n})$. The Poisson distribution with the parameters (a, b, λ) has the c.f. shown in eqn (1.11) (Section 1.3). It is clear that for every n, the equality (1.40) holds where $f_n(t)$ is the c.f. of the Poisson distribution with the parameters $(a/n, b, \lambda/n)$. The degenerate distribution with the unique growth point c has the c.f. $f(t) = e^{ict}$ and obviously is infinitely divisible.

Theorem 1.15 *Let $f(t)$ be an infinitely divisible c.f. Then $f(t) \neq 0$ for every t.*

Proof Every c.f. $f(t)$ is continuous and satisfies the equality $f(0) = 1$. Therefore, there exists a positive number a such that $f(t) \neq 0$ for $|t| \leq a$. In the same interval $|t| \leq a$ we have $f_n(t) \neq 0$, where $f_n(t)$ satisfies the equality (1.40). Let ε be an arbitrary positive number. If $|t| \leq a$, then

$$|f_n(t)| = |f(t)|^{1/n} = \left|\exp\left\{\frac{1}{n}\log f(t)\right\}\right| > 1 - \varepsilon$$

for all sufficiently large n.

By Lemma 1.3 we have

$$1 - |f_n(2t)|^2 \leq 4(1 - |f_n(t)|^2)$$

for every t. If $|t| \leq a$, then we obtain

$$1 - |f_n(2t)| \leq 1 - |f_n(2t)|^2 \leq 4(1 - |f_n(t)|^2) \leq 4(2\varepsilon - \varepsilon^2) < 8\varepsilon$$

for all sufficiently large n. Therefore, $f_n(t) \neq 0$ in the interval $|t| \leq 2a$ for all sufficiently large n, and $f(t) \neq 0$ in the same interval $|t| \leq 2a$. We have shown that the inequality $f(t) \neq 0$ for $|t| \leq a$ implies the same inequality for $|t| \leq 2a$. It follows that $f(t) \neq 0$ for any t. □

Theorem 1.15 provides a broad class of non-infinitely-divisible distributions. In fact, any c.f. $f(t)$ such that $f(t) = 0$ at some point t is non-infinitely-divisible. Hence the distribution of a random variable having two values -1 and 1 with the corresponding probabilities $\frac{1}{2}$ and $\frac{1}{2}$ is non-infinitely-divisible, since the c.f. of this random variable equals $\cos t$. Other examples are uniform distributions and the distribution from Lemma 1.15.

Lemma 1.19 *Let $f(t)$ and $g(t)$ be infinitely divisible c.f. Then $f(t)g(t)$ is also an infinitely divisible c.f.*

Proof For every n there exist c.f. $f_n(t)$ and $g_n(t)$ such that $f(t) = (f_n(t))^n$ and $g(t) = (g_n(t))^n$. Therefore, $f(t)g(t) = (f_n(t)g_n(t))^n$, and $f_n(t)g_n(t)$ is a c.f. □

Lemma 1.20 Let $\{f^{(m)}(t); m = 1, 2, \ldots\}$ be a sequence of infinitely divisible characteristic functions converging to some characteristic function $f(t)$. Then $f(t)$ is infinitely divisible.

Proof We have $f^{(m)}(t) = (f_n^{(m)}(t))^n$ for every m and n where $f_n^{(m)}(t)$ is a c.f. The conditions of the lemma imply that

$$f_n^{(m)}(t) = (f^{(m)}(t))^{1/n} \to (f(t))^{1/n}$$

as $m \to \infty$ for every n. The limit is a function which is continuous at the point $t = 0$. By Theorem 1.10 this function is a c.f. Therefore, $f(t) = ((f(t))^{1/n})^n$ is an infinitely divisible c.f. □

Now we shall derive canonical representations of infinitely divisible characteristic functions.

Theorem 1.16 *A function $f(t)$ is an infinitely divisible c.f. if and only if it admits the representation*

$$f(t) = \exp\left\{i\gamma t + \int_{-\infty}^{\infty}\left(e^{itx} - 1 - \frac{itx}{1+x^2}\right)\frac{1+x^2}{x^2}\,dG(x)\right\}, \quad (1.41)$$

where γ is a real constant, $G(x)$ is a bounded non-decreasing function, and the function under the integral sign is equal to $-t^2/2$ at the point $x = 0$.

The representation (1.41) is called the Lévy–Khintchine formula.

Note that $-t^2/2$ is equal to the limit of the function

$$\left(e^{itx} - 1 - \frac{itx}{1+x^2}\right)\frac{1+x^2}{x^2}$$

as $x \to 0$. We also note that the values of the function $G(x)$ at points of discontinuity do not influence the value of the integral on the right-hand side of (1.41). We shall suppose, for definiteness, that the function $G(x)$ is continuous on the left.

In order to prove sufficiency, we shall use the equality

$$\int_{\varepsilon}^{1/\varepsilon}\left(e^{itx} - 1 - \frac{itx}{1+x^2}\right)\frac{1+x^2}{x^2}\,dG(x) = \lim \sum_{k=0}^{n-1} T_{nk} \quad (1.42)$$

for every positive $\varepsilon < 1$, where

$$T_{nk} = \left(e^{it\xi_k} - 1 - \frac{it\xi_k}{1+\xi_k^2}\right)\frac{1+\xi_k^2}{\xi_k^2}[G(x_{k+1}) - G(x_k)]$$

$$\varepsilon = x_0 < x_1 < \cdots < x_n = 1/\varepsilon, \; x_k \leqslant \xi_k < x_{k+1}$$

$$(k = 0, 1, \ldots, n-1)$$

and the limit is taken under the condition $\max_k(x_{k+1} - x_k) \to 0$. Every summand T_{nk} can be written in the form $ita_{nk} + \lambda_{nk}(e^{itb_{nk}} - 1)$, where

$$\lambda_{nk} = \frac{1 + \xi_k^2}{\xi_k^2}[G(x_{k+1}) - G(x_k)], \quad b_{nk} = \xi_k, \quad a_{nk} = -\frac{\lambda_{nk}\xi_k}{1 + \xi_k^2}.$$

Thus $e^{T_{nk}}$ is the c.f. of a Poisson distribution. The limit of the product $\prod_{k=0}^{n-1} e^{T_{nk}}$ is a continuous function, by eqn (1.42). By Theorem 1.10 the function

$$\exp\left\{\int_{\varepsilon}^{1/\varepsilon}\left(e^{itx} - 1 - \frac{itx}{1 + x^2}\right)\frac{1 + x^2}{x^2}dG(x)\right\}$$

is the c.f. of some distribution. It follows from Lemma 1.20 that this distribution is infinitely divisible.

Passing to the limit as $\varepsilon \downarrow 0$, we obtain the same assertion for the function e^{I_+}, where

$$I_+ = \int_{x>0}\left(e^{itx} - 1 - \frac{itx}{1 + x^2}\right)\frac{1 + x^2}{x^2}dG(x). \tag{1.43}$$

Let us define I_- as the right-hand side of the equality (1.43) with replacement of the area of integration $x > 0$ by the area $x < 0$, and I as the right-hand side of (1.43) with replacement of the area of integration by the real line $\mathbb{R}$. Obviously,

$$I = I_+ + I_- - \frac{t^2}{2}(G(+0) - G(-0)).$$

The function e^{I_-} is an infinitely divisible function; this can be proved in the same fashion as a similar assertion for the function e^{I_+}. The function $\exp\{-t^2/2(G(+0) - G(-0))\}$ is the c.f. of a normal distribution; therefore, it is infinitely divisible. By Lemma 1.19 the function e^I is an infinitely divisible c.f. It remains to note that $e^{i\gamma t}$ is the c.f. of a degenerate distribution and to apply Lemma 1.19 once more.

Sufficiency is proved. Let us turn to the proof of necessity in Theorem 1.16.

We have to show that an arbitrary infinitely divisible c.f. $f(t)$ can be written in the form (1.41). By Theorem 1.15, $f(t) \neq 0$ for every t. We consider $\log f(t)$, where 'log' denotes the principal value of the logarithm. We have $f(t) = (f_n(t))^n$, for every n, where $f_n(t)$ is a c.f. Therefore,

$$\log f(t) = n \log f_n(t) = n \log[1 + (f_n(t) - 1)] = \lim[n(f_n(t) - 1)],$$

since $\log(1 + \alpha) = \alpha[1 + O(|\alpha|)]$ as $\alpha \to 0$. Denoting by $F_n(x)$ the d.f.

corresponding to the c.f. $f_n(t)$, we find that

$$\log f(t) = \lim \int_{-\infty}^{\infty} n(e^{itx} - 1)\, dF_n(x)$$

$$= \lim\left\{ it \int_{-\infty}^{\infty} \frac{nx}{1+x^2}\, dF_n(x) + \int_{-\infty}^{\infty} n\left(e^{itx} - 1 - \frac{itx}{1+x^2}\right) dF_n(x) \right\}$$

$$= \lim \psi_n(t) \qquad (1.44)$$

for every t, where $\psi_n(t)$ is defined by the equality

$$\psi_n(t) = i\gamma_n t + \int_{-\infty}^{\infty}\left(e^{itx} - 1 - \frac{itx}{1+x^2}\right)\frac{1+x^2}{x^2}\, dG_n(x) \qquad (1.45)$$

and

$$\gamma_n = n\int_{-\infty}^{\infty} \frac{x}{1+x^2}\, dF_n(x), \quad G_n(x) = n\int_{-\infty}^{x} \frac{y^2}{1+y^2}\, dF_n(y)$$

In order to derive the desired assertion from the equalities (1.44) we need some preliminaries. To any bounded non-decreasing function $G(x)$ and any real constant γ we associate the functions

$$\psi(t) = i\gamma t + \int_{-\infty}^{\infty}\left(e^{itx} - 1 - \frac{itx}{1+x^2}\right)\frac{1+x^2}{x^2}\, dG(x), \qquad (1.46)$$

$$\Lambda(x) = \int_{-\infty}^{x}\left(1 - \frac{\sin y}{y}\right)\frac{1+y^2}{y^2}\, dG(y), \qquad (1.47)$$

and

$$\lambda(t) = \psi(t) - \tfrac{1}{2}\int_0^1 (\psi(t+h) + \psi(t-h))\, dh. \qquad (1.48)$$

We have

$$\lambda(t) = \int_{-\infty}^{\infty}\left(e^{itx} - 1 - \frac{itx}{1+x^2}\right)\frac{1+x^2}{x^2}\, dG(x)$$

$$- \int_{-\infty}^{\infty}\left\{\int_0^1 \left(e^{itx}\cos hx - 1 - \frac{itx}{1+x^2}\right) dh\right\}\frac{1+x^2}{x^2}\, dG(x)$$

$$= \int_{-\infty}^{\infty} e^{itx}\left(1 - \frac{\sin x}{x}\right)\frac{1+x^2}{x^2}\, dG(x).$$

Hence

$$\lambda(t) = \int_{-\infty}^{\infty} e^{itx}\, d\Lambda(x). \qquad (1.49)$$

It is easy to prove that

$$0 < c_1 \leq \left(1 - \frac{\sin x}{x}\right)\frac{1 + x^2}{x^2} \leq c_2 \qquad (1.50)$$

for all x. Therefore, the function $\Lambda(x)$ does not decrease and is bounded. Further, the equality (1.49) implies that up to a constant multiplier, $\lambda(t)$ is a c.f.

We shall suppose from now on that the function $G(x)$ satisfies the supplementary condition $G(-\infty) = 0$.

Lemma 1.21 *There is one-to-one correspondence between the functions ψ defined by eqn (1.46) and the pairs (γ, G), where γ is a real constant and G is a non-decreasing bounded function with $G(-\infty) = 0$.*

Proof An arbitrary pair (γ, G) uniquely defines a function ψ by eqn (1.46). An arbitrary function ψ uniquely defines the function $\lambda(t)$, which is a c.f. up to a constant multiplier. It follows from eqn (1.49) and Theorem 1.5 that $\lambda(t)$ uniquely defines the function $\Lambda(x)$. In turn, $\Lambda(x)$ uniquely defines the function

$$G(x) = \int_{-\infty}^{x} \frac{y^2 \, d\Lambda(y)}{(1 + y^2)\left(1 - \dfrac{\sin y}{y}\right)}. \qquad (1.51)$$

Further, ψ and G together uniquely define the constant γ. $\square$

Using Lemma 1.21, we shall use the notation $\psi = (\gamma, G)$.

Lemma 1.22 *Let $\psi_n(t)$ be the function defined by eqn (1.45) where γ_n is a real number and $G_n(x)$ is a bounded non-decreasing function satisfying the condition $G_n(-\infty) = 0$ $(n = 1, 2, \ldots)$. If $\gamma_n \to \gamma$ and $G_n \rightrightarrows G$, where γ is a constant and $G(x)$ is a bounded non-decreasing function, then $\psi_n(t) \to \psi(t)$.*

If $\psi_n(t) \to \psi(t)$, where $\psi(t)$ is a function which is continuous at the point $t = 0$, then there exist a real number γ and a bounded non-decreasing function $G(x)$ such that $\gamma_n \to \gamma$, $G_n \rightrightarrows G$, and $\psi = (\gamma, G)$.

Proof The first assertion of the lemma follows from Theorem 1.8. We shall prove the second assertion. The function $e^{\psi(t)}$ is continuous at the point $t = 0$; it is the limit of a sequence of infinitely divisible c.f. By Theorem 1.10 and Lemma 1.20 this function is an infinitely divisible c.f. By Theorem 1.15 we have $e^{\psi(t)} \neq 0$ for every t. Therefore, the function $\psi(t)$ is finite and $\psi_n(t) \to \psi(t)$ uniformly in an arbitrary finite interval. If we introduce the functions $\lambda_n(t)$ by the equality

$$\lambda_n(t) = \psi_n(t) - \tfrac{1}{2}\int_0^1 (\psi_n(t+h) + \psi_n(t-h))\,dh,$$

then $\lambda_n(t) \to \lambda(t)$, where $\lambda(t)$ is defined by (1.48). We shall associate with $\lambda(t)$ and $\lambda_n(t)$ the functions $\Lambda(x)$ and $\Lambda_n(x)$ through the equality (1.49). Making use of the continuity of $\lambda(t)$ and applying Theorem 1.10, we obtain $\Lambda_n \to \Lambda$. Since $\lambda_n(0) \to \lambda(0)$ and since

$$\lambda_n(0) = \int_{-\infty}^{\infty} d\Lambda_n(x), \quad \lambda(0) = \int_{-\infty}^{\infty} d\Lambda(x),$$

we have $\Lambda_n(-\infty) \to \Lambda(-\infty)$ and $\Lambda_n(+\infty) \to \Lambda(+\infty)$. Thus $\Lambda_n \rightrightarrows \Lambda$. It follows from (1.50), (1.51) and Theorem 1.8 that $G_n \rightrightarrows G$. By the same theorem,

$$i\gamma_n t \to \psi(t) - \int_{-\infty}^{\infty}\left(e^{itx} - 1 - \frac{itx}{1+x^2}\right)\frac{1+x^2}{x^2}\,dG(x)$$

for every t. Therefore, there exists the limit $\lim \gamma_n = \gamma$. By the first assertion of the lemma we have $\psi = (\gamma, G)$. $\square$

Let us return to relations (1.44). By Lemma 1.22 it follows from the equality $\lim \psi_n(t) = \log f(t)$, and from continuity of $\log f(t)$ that there exist a constant γ and a bounded non-decreasing function $G(x)$, for which $\gamma_n \to \gamma$, $G_n \rightrightarrows G$ and $\log f(t) = (\gamma, G)$. Theorem 1.16 is proved. $\square$

It follows from Lemma 1.21 and Theorem 1.16 that if $G(-\infty) = 0$ the representation of an infinitely divisible c.f. $f(t)$ in the form (1.41) is unique: equality (1.41) uniquely determines the constant γ and the function $G(x)$ by means of the function $f(t)$.

The function $G(x)$ is called the Lévy–Khintchine spectral function.

We now present expressions for the constant γ and the Lévy–Khintchine spectral function for certain important infinitely divisible distributions. We recall that we have agreed to assume that the function $G(x)$ is continuous on the left and that it satisfies the condition $G(-\infty) = 0$.

For the normal distribution with parameters (a, σ) we have $\gamma = a$, $G(x) = 0$ if $x \leq 0$, and $G(x) = \sigma^2$ if $x > 0$. For the Poisson distribution with parameters (a, b, λ) we find that $\gamma = a + (b\lambda/1 + b^2)$, $G(x) = 0$ if $x \leq b$, and $G(x) = b^2\lambda/(1 + b^2)$ if $x > b$. For the degenerate distribution with the growth point c we have $\gamma = c$ and $G(x) \equiv 0$.

Formula (1.41) can be written in another way. We write

$$\sigma^2 = G(+0) - G(-0),$$

$$L(x) = \begin{cases} \int_{-\infty}^{x} \frac{1+y^2}{y^2} \, dG(y) & \text{if } x < 0, \\ -\int_{x}^{\infty} \frac{1+y^2}{y^2} \, dG(y) & \text{if } x > 0. \end{cases} \quad (1.52)$$

The function $L(x)$, defined on the real line, except at the point $x = 0$, is non-decreasing on $(-\infty, 0)$ and on $(0, \infty)$ and it satisfies the conditions $\lim_{x \to -\infty} L(x) = 0$, $\lim_{x \to +\infty} L(x) = 0$. It is continuous at precisely those points of its domain of definition at which $G(x)$ is continuous. For every finite $\delta > 0$ we have $\int_{0 < |x| < \delta} x^2 \, dL(x) < \infty$.

If γ is fixed, an arbitrary non-negative constant σ and an arbitrary function $L(x)$, satisfying the prescribed conditions, together uniquely define an infinitely divisible c.f. by means of (1.52) and (1.41). We can formulate the following proposition.

Theorem 1.17 *A function $f(t)$ is an infinitely divisible c.f. if and only if it admits the representation*

$$f(t) = \exp\left\{i\gamma t - \frac{\sigma^2 t^2}{2} + \int_{|x| > 0} \left(e^{itx} - 1 - \frac{itx}{1+x^2}\right) dL(x)\right\}, \quad (1.53)$$

where γ is a real constant, σ^2 is a non-negative constant, and the function $L(x)$ is non-decreasing on the intervals $(-\infty, 0)$ and $(0, \infty)$ and satisfies the conditions $\lim_{x \to -\infty} L(x) = 0$, $\lim_{x \to +\infty} L(x) = 0$, $\int_{0 < |x| < \delta} x^2 \, dL(x) < \infty$ for every finite $\delta > 0$.

Equality (1.53) is called the Lévy formula. The function $L(x)$ is called the Lévy spectral function.

The representation of an infinitely divisible c.f. by eqn (1.53) is unique.

We recall that if a random variable with the c.f. $f(t)$ has a finite variance, the second-order derivative $f''(0)$ exists. It is possible to show that the converse assertion is also true. From this fact and from Theorem 1.16 we can derive the following result.

Theorem 1.18 *A function $f(t)$ is the c.f. of an infinitely divisible distribution with a finite variance if and only if it admits the representation*

$$f(t) = \exp\left\{i\alpha t + \int_{-\infty}^{\infty} \frac{e^{itx} - 1 - itx}{x^2} \, dK(x)\right\}, \quad (1.54)$$

where α is a real constant, $K(x)$ is a bounded non-decreasing function, and the function under the integral sign is equal to $-t^2/2$ at the point $x = 0$.

Equality (1.54) is called the Kolmogorov formula. The function $K(x)$ is called the Kolmogorov spectral function.

1.7 Bibliographical notes

The manner of exposition of basic concepts of probability theory used in Section 1.1 is due to Kolmogorov [231].

Theorem 1.1 is the probabilistic form of a result due to Chebyshev (see, for instance, [63], pp. 128-31). The proof of this theorem given in Section 1.2 is due to Kingman [223]. Theorem 1.2 was obtained by Petrov [341].

Characteristic functions were systematically used in probability theory by Lyapunov in his papers dated 1900-1901, although sporadic uses of characteristic functions had been made before then. The basic theorems on characteristic functions (including Theorems 1.4, 1.5, 1.9, and 1.10) were obtained by Lévy [251]. Theorem 1.3 is due to Wintner [466], Lemma 1.4 to Cramér [85], and Lemma 1.5 to Hoeffding [194] and Rosén [392]. The proof of Lemma 1.5 given in section 1.3 was communicated by J. Enger (Uppsala).

Theorem 1.11 is due to Pólya [360]. Theorem 1.12 is a generalization of Laube's theorem [247] which corresponds to the case when $q = 0$. Lemma 1.9 was obtained by Petrov [328]. Lemma 1.10 is a consequence of the Slutsky-Cramér theorem (see, for example, Cramér [84]). Theorem 1.14 is due to Khintchine [222].

The concentration function $Q(X, \lambda)$ was introduced by Lévy. Lemmas 1.13 and 1.14 were obtained by Daugavet and Petrov [95, 96]. Lemma 1.16 is a modification of Esseen's result [122] given in [335]. Lemma 1.17 is due to Esseen [122].

Basic results on infinitely divisible distributions are due to Lévy and Khintchine; these results are presented in their books [222, 252].

Surveys of inequalities of Chebyshev's type can be found in the books by Godwin [148] and Karlin and Studden [210], and in the articles by Savage [412] and Nagaev [311]. Numerous inequalities for absolute moments of random variables are given in Beesack [19]. Wolfe [471] developed representations for absolute moments of arbitrary order in terms of the derivatives of fractional orders of the corresponding c.f. Characteristic functions are the principal subject matter of the books by Lukacs [257, 259] and Ramachandran [377]; other useful references are the books by Loève [256] and Linnik and Ostrovskii [255]. Surveys of investigations in infinite divisibility can be found in Fisz [136] and Steutel [437, 438]. The book by Hengartner and Theodorescu [180] is devoted to the concentration functions.

1.8 Addenda

1.8.1 Let X be a random variable with the standard normal distribution, and let $g(x)$ be an absolutely continuous function. If $g(X)$ has finite variance, then $\operatorname{Var} g(X) \leq \operatorname{E}(g'(X))^2$; equality holds if and only if $g(x) = ax + b$ for some constants a and b (Chernoff [66]). A simpler proof was given by Chen [64]. If the additional condition $\operatorname{E}|g'(X)| < \infty$ is satisfied, then $\operatorname{Var} g(X) \geq (\operatorname{E} g'(X))^2$; equality holds if and only if $g(x)$ is linear (Cacoullos [59]). For the case when X is not necessarily normal, some generalizations have been obtained by Cacoullos [59] and Klaassen [224].

1.8.2 Let X be a random variable with a discrete distribution and the mean a. We set $v_k = \operatorname{E}|X - a|^k$. Then $v_k \leq 2v_{k+1}/c$ for every integer $k \geq 1$, where c is the length of the least interval between two consecutive values of X (von Mises [465]).

1.8.3 Let X and Y be random variables defined on a common probability space. If $r > 1$ and $1/r + 1/s = 1$, then

$$\operatorname{E}|XY| \leq (\operatorname{E}|X|^r I(Y \neq 0))^{1/r} \cdot (\operatorname{E}|Y|^s I(X \neq 0))^{1/s}, \qquad (1.55)$$

where $I(B)$ is the indicator of the event B. If $r \geq 1$, then

$$(\operatorname{E}|X + Y|^r)^{1/r} \leq (\operatorname{E}|X|^r I(X + Y \neq 0))^{1/r} + (\operatorname{E}|Y|^r I(X + Y \neq 0))^{1/r} \qquad (1.56)$$

(Petrov [350]). Immediate consequences of (1.55) and (1.56) are Hölder's and Minkowski's inequalities correspondingly. The inequalities are sharp in the sense that there exist random vectors (X, Y) such that (1.55) and (1.56) become the equalities.

1.8.4 Let $g(x)$ be a function defined on an open (finite or infinite) interval I that contains the point $x = 0$. Let $g(x)$ be convex and twice differentiable on the interval I with $g(0) \geq 0$. Let X be a random variable with mean $\operatorname{E} X = m$ taking on the values from the interval I. Then

$$\operatorname{E} g(X) \geq g(m) P(X \neq 0) + g'(m) P(X = 0) m \qquad (1.57)$$

(Petrov [350]). If $P(X = 0) = 0$, (1.57) reduces to Jensen's inequality.

1.8.5 Let X be a random variable, and let $\beta_r = \operatorname{E}|X|^r$. If B is a Borel set in $\mathbb{R}$, then for $0 < r < s$,

$$\beta_r \leq \operatorname{E}|X|^r I(X \in B) + (P(X \notin B))^{1-r/s} \cdot (\operatorname{E}|X|^s I(X \notin B))^{r/s}$$

(Arnold [13]).

In subsections 1.8.6–1.8.16 we shall use the following notation: $F(x)$ is a d.f.; $f(t)$ is the corresponding c.f.; n is a positive integer; and $\beta_k = \int_{-\infty}^{\infty} |x|^k \, dF(x)$.

1.8.6 If $2n > p > 0$, then

$$\beta_{2n-p} = \frac{(-1)^n 2^{1-p}}{\sqrt{\pi}\,\Gamma\left(\frac{p}{2}\right)} \int_0^{\infty} y^{-2n+p-1} \, dy \int_{-\infty}^{\infty} f(yt) \frac{d^{2n}}{dt^{2n}} e^{-t^2} \, dt$$

(Hsu [196]).

1.8.7 If $k > 0$ is not an even number, $k = m + \delta$, where $0 < \delta \leq 1$, m is an integer, and if $\beta_k < \infty$, then

$$\beta_k = -\frac{2}{\pi} \Gamma(1+\delta) \sin \frac{k\pi}{2} \lim_{\substack{\varepsilon \to 0 \\ T \to \infty}} \int_{\varepsilon}^{T} \operatorname{Re}(f^{(m)}(t) - f^{(m)}(0)) t^{-1-\delta} \, dt.$$

If $f(t)$ is differentiable $m + 1$ times, then

$$\beta_k = -\frac{2}{\pi} \Gamma(\delta) \sin \frac{k\pi}{2} \lim_{\substack{\varepsilon \to 0 \\ T \to \infty}} \int_{\varepsilon}^{T} \operatorname{Re} f^{(m+1)}(t) t^{-\delta} \, dt$$

(Brown [54]).

1.8.8 If $0 < k < 2$, then the condition $\beta_k < \infty$ is equivalent to the condition

$$\int_{-\infty}^{\infty} (1 - \operatorname{Re} f(t)) |t|^{-k-1} \, dt < \infty.$$

The following equality holds:

$$\beta_k = c_k \int_0^{\infty} (1 - \operatorname{Re} f(t)) t^{-k-1} \, dt \qquad (0 < k < 2),$$

where

$$c_k = \frac{k(1-k)}{\Gamma(2-k)} \cdot \frac{1}{\sin\left((1-k)\frac{\pi}{2}\right)} \quad \text{for } k \neq 1 \text{ and } c_1 = \frac{2}{\pi}.$$

If $2m < k < 2m + 2$, where m is a positive integer, and if $\beta_k < \infty$, then

$$\beta_k = \frac{2}{b_k} \int_0^{\infty} \left(1 - \operatorname{Re} f(t) + \sum_{j=1}^{m} f^{(2j)}(0) \frac{t^{2j}}{(2j)!}\right) t^{-k-1} \, dt,$$

where

$$b_k = \int_0^{\infty} \left(1 - \cos t - \sum_{j=1}^{m} (-1)^j \frac{t^{2j}}{(2j)!}\right) t^{-k-1} \, dt$$

(see, for example, Kawata [215], pp. 429–31).

1.8.9 If $f(t) \in L^p(\mathbb{R})$ (i.e. $\int_{-\infty}^{\infty} |f(t)|^p \, dt < \infty$) for some p from the interval $1 \leq p \leq 2$, then the corresponding d.f. $F(x)$ is absolutely continuous and the density $p(x) = (d/dx) F(x)$ satisfies the condition $p(x) \in L^r(\mathbb{R})$ for every r from the interval $1 \leq r \leq p/(p-1)$. If $p > 2$, then there exists a c.f. $f(t) \in L^p(\mathbb{R})$ such that $F(x)$ is singular (see, for example, Kawata [215], pp. 437–8).

1.8.10 Let $0 < \alpha < 1$. The relation
$$1 - F(x) + F(-x) = O(x^{-\alpha})$$
as $x \to +\infty$ is equivalent to the condition $f(t) \in \text{Lip } \alpha$ (i.e. $f(t)$ satisfies the Lipschitz condition with the exponent α). In order that
$$1 - F(x) + F(-x) = O(1/x)$$
as $x \to +\infty$, it is necessary that the following condition be satisfied:
$$f(t+h) + f(t-h) - 2f(t) = O(h)$$
as $h \to 0$ uniformly in t, and sufficient that this condition holds for $t = 0$. The latter proposition remains true if O is everywhere replaced by o (Boas [44]).

1.8.11 Let $1 < \alpha < 2$. The condition $\int_{-\infty}^{\infty} |x|^{-\alpha-1} \, dF(x) < \infty$ is equivalent to the convergence of the integral
$$\int_b^{b+1} (t-b)^{-\alpha} |f(t) - f(b)| \, dt \tag{1.58}$$
for every real b. This proposition remains true for $\alpha = 1$ if $|x|^{-\alpha-1}$ is replaced by $\max\{0, \log|x|\}$, and for $\alpha = 2$ if the integral (1.58) is replaced by the integral
$$\int_0^1 t^{-2} |f(b+t) + f(b-t) - 2f(b)| \, dt$$
(Boas [44]).

1.8.12 Let k be an odd integer. The derivative $f^{(k)}(0)$ exists if and only if $\lim_{x \to +\infty} x^k (1 - F(x) + F(-x)) = 0$ and there exists the limit $\lim_{a \to +\infty} \int_{-a}^{a} x^k \, dF(x) = l_k$. If these conditions are satisfied, then $f^{(k)}(0) = i^k l_k$ (Pitman [359]).

1.8.13 If k is an odd integer, then it is possible to construct a c.f. $f(t)$ such that $f^{(k)}(0)$ exists but $f^{(k)}(t)$ fails to exist almost everywhere (Wolfe [472]).

1.8.14 If $0 < \alpha < 2$, the relation
$$1 - F(x) + F(-x) = o(x^{-\alpha}) \qquad (x \to +\infty) \tag{1.59}$$

is equivalent to the condition

$$1 - \operatorname{Re} f(t) = o(t^\alpha) \qquad (t \to 0^+).$$

This proposition remains true when we replace everywhere o by O. If $\alpha = 2n + \beta$, where n is a positive integer and $0 < \beta < 2$, then (1.59) is equivalent to

$$\operatorname{Re} f^{(2n)}(0) - \operatorname{Re} f^{(2n)}(t) = o(t^\beta) \qquad (t \to 0+)$$

(Binmore and Stratton [42]).

1.8.15 Let $0 < \alpha < 1$. If $F(x) \in \operatorname{Lip} \alpha$, then

$$\frac{1}{T} \int_{-T}^{T} |f(t)|^2 \, dt = O(T^{-\alpha}) \quad (T \to +\infty). \tag{1.60}$$

If condition (1.60) is satisfied, then $F(x) \in \operatorname{Lip} \alpha/2$. If

$$\int_{1}^{\infty} t^{\alpha-1} |f(t)| \, dt < \infty,$$

then $F(x) \in \operatorname{Lip} \alpha$ (Makabe [270]).

1.8.16 For every real x there exists the limit

$$\lim_{T \to +\infty} \frac{1}{2T} \int_{-T}^{T} e^{-itx} f(t) \, dt;$$

it is equal to $F(x+0) - F(x-0)$ (see, for instance, Cramér [85]).

1.8.17 Let $f(t)$ be the c.f. of a discrete distribution with the possible values $x_1, x_2, \ldots$ and the corresponding probabilities $p_1, p_2, \ldots$. Then the limit

$$\lim_{T \to +\infty} \frac{1}{2T} \int_{-T}^{T} |f(t)|^2 \, dt$$

exists and is equal to $\sum_k p_k^2$ (Lévy [251]; see also [85]).

1.8.18 Let $f(t)$ be the c.f. of a random variable with finite variance. Then

$$\sup_x \left| \int_x^{x+c} f^n(t) \, dt \right| = O(n^{-1/2})$$

for every $c > 0$ (Hall [162]).

1.8.19 For every c.f. $f(t)$, every real t, and every positive integer n, the

following inequalities hold:
$$1 - \operatorname{Re} f(nt) \leq n(1 - (\operatorname{Re} f(t))^n) \leq n^2(1 - \operatorname{Re} f(t)),$$
$$1 - |f(nt)| \leq n(1 - |f(t)|^n) \leq n^2(1 - |f(t)|)$$
(Heathcote and Pitman [179]).

1.8.20 Let $0 \leq a < 1$, $0 < b \leq B$, and let the c.f. $f(t)$ satisfy the condition $|f(t)| \leq a$ for $B \leq t \leq B + b$. Then
$$|f(t)| \leq \left(\frac{B + at}{B + t}\right)^{t/(B+t)} \leq 1 - \frac{(1-a)t^2}{(B+t)^2}$$
for $0 < t < b$. It is possible to replace $|f(t)|$ here by $\operatorname{Re} f(t)$ (Heathcote and Pitman [179]).

1.8.21 Let X be a random variable with the c.f. $f(t)$, zero mean, and finite variance σ^2. Then $|f(t)| < 1$ in the area $0 < |t| < \pi \beta_1/\sigma^2$, where $\beta_1 = \mathrm{E}|X|$. If additionally $\beta_3 = \mathrm{E}|X|^3 < \infty$, then $|f(t)|^2 \leq 1 - \sigma^2 t^2 + (\beta_3 + \sigma^2 \beta_1)|t|^3/5$ for all real t (Prawitz [363]).

1.8.22 Let X be a bounded random variable, $|X| \leq C$, and let $f(t)$ be its c.f., and σ^2 its variance. Then
$$e^{-\sigma^2 t^2} \leq |f(t)| \leq e^{-\sigma^2 t^2/3}$$
in the interval $|t| \leq 1/(4C)$ (Doob [104]).

1.8.23 Let $f(t)$ be a continuous non-negative even function convex in the interval $t > 0$ and satisfying the conditions $f(0) = 1$ and $\lim_{t \to \infty} f(t) = 0$. Then $f(t)$ is a c.f. (Pólya [361]; see also Feller [133] and Lukacs [257]).

1.8.24 If $f(t)$ is the c.f. of a non-negative random variable, then $f(t) \neq 0$ in an arbitrary interval (Smith [430]).

1.8.25 Let $f(t)$ be a c.f., and let $F(x)$ be the corresponding d.f. Then
$$\operatorname{Im} f(t) = -\frac{1}{\pi} \int_0^\infty \frac{1}{u} \{\operatorname{Re} f(t+u) - \operatorname{Re} f(t-u)\}\, du.$$
If $F(+0) = 0$, then
$$\operatorname{Re} f(t) = \frac{1}{\pi} \int_0^\infty \frac{1}{u} \{\operatorname{Im} f(t+u) - \operatorname{Im} f(t-u)\}\, du$$
(Laue [248]).

1.8.26 If $F(x)$ is a d.f. with zero mean and unit variance, then
$$\sup_x |F(x) - \Phi(x)| \leq 0.5416,$$
where $\Phi(x)$ is the standard normal d.f. (Bhattacharya and Ranga Rao [34]).

1.8.27 Let M be the set of distribution functions $F(x)$ such that
$$\int_{-\infty}^{\infty} x \, dF(x) = 0, \quad \int_{-\infty}^{\infty} x^2 \, dF(x) = 1, \quad \int_{-\infty}^{\infty} |x|^3 \, dF(x) = \beta < \infty.$$
Then the least absolute constant A such that
$$\sup_{F \in M} \sup_x |F(x) - \Phi(x)| \leq A\beta$$
is equal to $0.370352\ldots$. The equality is achieved if and only if F is the d.f. of a random variable with two values.

Let M_s be the subset of the set M consisting of all symmetric distribution functions F. Then the least absolute constant A_s such that
$$\sup_{F \in M_s} \sup_x |F(x) - \Phi(x)| \leq A_s \beta$$
is equal to $0.341345\ldots$. Furthermore,
$$\sup_{F \in M} \sup_x |x|^3 |F(x) - \Phi(x)| \leq \beta,$$
$$\sup_{F \in M_s} \sup_x |x|^3 |F(x) - \Phi(x)| \leq \beta/2,$$
and the equality is not achieved (Bentkus and Kirsha [21]).

1.8.28 Let X and Y be arbitrary random variables defined on a common probability space. Let $F(x)$ and $H(x)$ be the d.f. of the random variables X and $X + Y$ correspondingly. If
$$|F(x) - \Phi(x)| \leq M(1 + |x|)^{-\alpha}$$
for every real x and some positive constants M and α, then
$$|H(x) - \Phi(x)| \leq C(1 + |x|)^{-\alpha}(M + \varepsilon + \varepsilon^{-\alpha} E|Y|^\alpha)$$
for every real x and every positive $\varepsilon < 1/2$, where C is a constant depending only on α (Maejima [263]).

1.8.29 Let $F_1(x), F_2(x), \ldots$ be a sequence of d.f., identically equal to zero for $x \leq 0$, and let $f_1(t), f_2(t), \ldots$ be the corresponding sequence of c.f. If $f_n(t) \to f(t)$ at every point of some interval $|t| < a$ and $f(t)$ is continuous at the point $t = 0$, there exists a d.f. $F(x)$ such that $F_n(x) \to F(x)$ at every point of continuity of $F(x)$. The assertion remains true when the condition

$F_n(x) = 0$ for $x \leq 0$ is replaced by the weaker condition $F_n(x) \leq b \, e^{-c|x|}$ for $x \leq x_0$, where $b > 0$, $c > 0$ and x_0 are some constants not depending on n (Zygmund [487]).

1.8.30 If the d.f. $F(x)$ is defined uniquely by its moments, and if $\{F_n(x)\}$ is a sequence of d.f. for which the moments of arbitrary positive integer order converge to the corresponding moments of $F(x)$, then $F_n(x) \to F(x)$ at every point of continuity of $F(x)$ (Fréchet and Shohat [138]; see also [256] and [379]).

1.8.31 Let F and G be two distribution functions. The Lévy metric $L(F, G)$ is defined as the greatest lower bound of the set of values h, for which

$$F(x - h) - h \leq G(x) \leq F(x + h) + h$$

for all real x. For the weak convergence of the distributions F_n to a distribution F it is necessary and sufficient that $L(F, G) \to 0$ (see, for instance, [147] and [258]).

1.8.32 Let F and G be two distribution functions. The uniform metric (or the Kolmogorov metric) $\rho(F, G)$ is defined by equality

$$\rho(F, G) = \sup_x |F(x) - G(x)|.$$

The following inequality holds: $L(F, G) \leq \rho(F, G)$. If $G(x)$ is absolutely continuous, then

$$\rho(F, G) \leq \left(1 + \sup_x |G'(x)|\right) L(F, G).$$

1.8.33 Let $P_1(B)$ and $P_2(B)$ be probability functions, and let $F_1(x)$ and $F_2(x)$ be the corresponding distribution functions, i.e. $F_k(x) = P_k((-\infty, x))$ ($k = 1, 2$). We write

$$\rho_{\text{var}}(P_1, P_2) = \sup_{B \in \mathscr{B}} |P_1(B) - P_2(B)|,$$

where $\mathscr{B}$ is the set of all Borel sets on the real line. Then

$$\rho_{\text{var}}(P_1, P_2) = \tfrac{1}{2} \text{Var}(F_1(x) - F_2(x)).$$

If the distribution functions $F_1(x)$ and $F_2(x)$ are absolutely continuous with the corresponding densities $p_1(x)$ and $p_2(x)$, then

$$\rho_{\text{var}}(P_1, P_2) = \tfrac{1}{2} \int_{-\infty}^{\infty} |p_1(x) - p_2(x)| \, dx.$$

We write $\rho(P_1, P_2) = \rho(F_1, F_2)$. The following inequality is obvious: $\rho(P_1, P_2) \leq \rho_{\text{var}}(P_1, P_2)$.

If the distributions $P, P_1, P_2, \ldots$ are such that $\rho_{\text{var}}(P_n, P) \to 0$, we say that P_n converges in variation to P.

The Lévy–Prohorov metric $\pi(P_1, P_2)$ is defined by the equality

$$\pi(P_1, P_2) = \inf\{\varepsilon > 0\colon P_1(B) \leq P_2(B^\varepsilon) + \varepsilon,\ P_2(B) \leq P_1(B^\varepsilon) + \varepsilon$$

$$\text{for every } B \in \mathscr{A}\},$$

where B^ε is the ε-neighbourhood of the set B which is defined by the equality

$$B^\varepsilon = \left\{ x \in \mathbb{R}\colon \inf_{y \in B} |x - y| < \varepsilon \right\},$$

and $\mathscr{A}$ is the set of all closed sets on the real line. The weak convergence of the distributions P_n to a distribution P is equivalent to the relation $\pi(P_n, P) \to 0$. (Different kinds of convergence of distributions are considered, for instance, in the books of Billingsley [38], Lukacs [258], Rachev [375], and Zolotarev [486], and the articles of Kolmogorov [232] and Prohorov [366].)

1.8.34 Let $p(x), p_1(x), p_2(x), \ldots$ be densities of distributions. Let $p_n(x) \to p(x)$ for all real x, with the exception of a set of values of Lebesgue measure zero. Then

$$\sup_{B \in \mathscr{B}} \left| \int_B p_n(x)\, dx - \int_B p(x)\, dx \right| \to 0,$$

where $\mathscr{B}$ is the set of all Borel sets of the real time (Scheffé [418]).

1.8.35 Let $F(x), F_1(x), F_2(x), \ldots$ be a sequence of d.f., and $f(t), f_1(t), f_2(t), \ldots$ be the corresponding sequence of c.f. If $f_n(t) \to f(t)$ uniformly in t on the real line, then $F_n(x) \to F(x)$ uniformly in x on the real line (Dyson [105]).

1.8.36 We write

$$\|g\| = \lim_{T \to +\infty} \frac{1}{2T} \int_{-T}^{T} |g(t)|^2\, dt$$

for those functions g for which this limit exists. Let $F(x), F_1(x), F_2(x), \ldots$ be distribution functions, and let $f(t), f_1(t), f_2(t), \ldots$ be the corresponding characteristic functions. The following conditions are equivalent: (1) $F_n(x) \to F(x)$ uniformly in x on the real line; (2) $f_n(t) \to f(t)$ for every real t and $\|f_n - f\| \to 0$ (Eisenberg and Gan Shixin [112]). A consequence of this proposition is the previous result of Dyson. The converse of the latter result does not hold.

Let X be a random variable with the d.f. $F(x)$. The Kawata concentration function $C(X; \lambda)$ is defined by the equality

$$C(X; \lambda) = \frac{1}{\lambda} \int_{-\infty}^{\infty} (F(x + \lambda) - F(x))^2 \, dx$$

for every $\lambda > 0$. The Kunisawa concentration functions $\Psi_1(X; \lambda)$ and $\Psi_2(X; \lambda)$ are defined by the equalities

$$\Psi_1(X; \lambda) = \int_{-\infty}^{\infty} \frac{\lambda^2}{x^2 + \lambda^2} \, dF^s(x)$$

and

$$\Psi_2(X; \lambda) = \int_{-\infty}^{\infty} \frac{\lambda^2}{x^2 + \lambda^2} \, dF(x)$$

for every $\lambda \geq 0$, where $F^s(x)$ is the d.f. of the symmetrized random variable X^s.

1.8.37 Let X be a random variable with the c.f. $f(t)$. Then

$$C(X; \lambda) = \frac{2\lambda}{\pi} \int_{-\infty}^{\infty} \left(\frac{\sin \frac{\lambda t}{2}}{\lambda t} \right)^2 |f(t)|^2 \, dt,$$

$$\lim_{\lambda \to 0} C(X; \lambda) = \lim_{T \to +\infty} \frac{1}{T} \int_0^T |f(t)|^2 \, dt$$

(Kawata [214], [215]).

1.8.38 Let $f(t)$ be the c.f. of the random variable X. If $0 \leq \alpha < 1$, then

$$\frac{1}{T} \int_0^T |f(t)|^2 \, dt = O(T^{-\alpha}) \quad \text{as } T \to \infty \tag{1.61}$$

is equivalent to the condition

$$C(X; \lambda) = O(\lambda^\alpha) \quad \text{as } \lambda \to 0. \tag{1.62}$$

If (1.61) holds for $\alpha = 1$, then (1.62) holds for $\alpha = 1$. If (1.62) holds for $\alpha = 1$, then

$$\frac{1}{T} \int_0^T |f(t)|^2 \, dt = O\left(\frac{\log T}{T}\right) \quad \text{as } T \to \infty$$

(Kawata [215]).

1.8.39 If $\lambda > 0$, then

$$Q^2(X; \lambda/2)/2 \leq C(X; \lambda) \leq Q(X; \lambda)$$

(Kawata [214], [215]). Here Q is Lévy's concentration function.

1.8.40 If $\lambda > 0$, then $C(X; \lambda) \geq [(2 - \alpha)/\alpha]Q^2(X; \alpha\lambda)$ for every positive $\alpha \leq 1$ (Ananjevsky [6]).

1.8.41 Let X be a random variable with the c.f. $f(t)$. Then

$$\Psi_1(X; \lambda) = \lambda \int_0^\infty e^{-\lambda t} |f(t)|^2 \, dt$$

and

$$\Psi_2(X; \lambda) = \lambda \int_0^\infty e^{-\lambda t} \operatorname{Re} f(t) \, dt$$

(Kunisawa [245]).

Let $\mathscr{P}$ be the set of non-negative even functions $p(x)$ that are integrable on $\mathbb{R}$, non-increasing on $\mathbb{R}_+$, continuous at the point $x = 0$, and satisfying the condition $p(0) = 1$. We put $\hat{p}(t) = \int_{-\infty}^\infty e^{itx} p(x) \, dx$. Let $\mathscr{P}_+$ be the set of the functions $p(x) \in \mathscr{P}$ satisfying the additional condition $\hat{p}(t) \geq 0$ for all $t \in \mathbb{R}$. Let X be a random variable with the d.f. $F(x)$. Kunisawa [245] introduced the generalized concentration function $K_0(X; p, \lambda)$ by the equality

$$K_0(X; p, \lambda) = \int_{-\infty}^\infty p(x/\lambda) \, dF^s(x) \tag{1.63}$$

for $\lambda > 0$, where $p(x) \in \mathscr{P}_+$. A broader class of the concentration functions was introduced by Ananjevsky [7] by the equality

$$K(X; p, \lambda) = \sup_{a \in \mathbb{R}} \int_{-\infty}^\infty p(x/\lambda) \, dF^s(x - a) \tag{1.64}$$

for every function $p(x) \in \mathscr{P}$ and every $\lambda > 0$.

1.8.42 If $p(x) \in \mathscr{P}_+$, then

$$K_0(X; p, \lambda) = K(X; p, \lambda) = \frac{\lambda}{2\pi} \int_{-\infty}^\infty \hat{p}(\lambda t) |f(t)|^2 \, dt.$$

If $p(x) = (1 + x^2)^{-1}$, then

$$K_0(X; p, \lambda) = K(X; p, \lambda) = \Psi_1(X; \lambda).$$

If $p(x) = 1 - |x|$ for $|x| \leq 1$ and $p(x) = 0$ for $|x| > 1$, then

$$K_0(X; p, \lambda) = K(X; p, \lambda) = C(X; \lambda).$$

If $p(x) = 1$ for $|x| \leq \frac{1}{2}$ and $p(x) = 0$ for $|x| > \frac{1}{2}$, then

$$K(X; p, \lambda) = Q(X^s; \lambda)$$

(Kunisawa [245] and Ananjevsky [7]).

1.8.43 Let $p(x) \in \mathscr{P}$. The generalized concentration function $K(X; p, \lambda)$

has the following properties:

(A) $0 < K(X; p, \lambda) \leq 1$;
(B) if $0 < \lambda_1 < \lambda_2$, then

$$K(X; p, \lambda_1) \leq K(X; p, \lambda_2);$$

(C) $\lim_{\lambda \to \infty} K(X; p, \lambda) = 1$;
(D) if X and Y are independent random variables, then

$$K(X + Y; p, \lambda) \leq \min\{K(X; p, \lambda), K(Y; p, \lambda)\};$$

(E) there exists a positive constant $C(p)$ depending only on $p(x)$ such that

$$K(X; p, \lambda) \leq C(p) Q(X^s; \lambda)$$

for every $\lambda > 0$ (Ananjevsky [7]; Kunisawa [245] proved that these conditions are satisfied by the concentration function $K_0(X; p, \lambda)$, where $p(x) \in \mathscr{P}_+$).

Let $F(x)$ be a d.f. We denote by S_F its spectrum, that is, the set of growth points of $F(x)$. Let $l_F = \inf S_F$, $u_F = \sup S_F$.

1.8.44 A non-degenerate infinitely divisible distribution has an unbounded spectrum (Chatterjee and Pakshirajan [62]).

1.8.45 Let $F(x)$ be an infinitely divisible d.f. with the c.f. $f(t)$ having the Lévy representation (1.53). Then l_F will be finite if and only if $\sigma^2 = 0$, $L(x) = 0$ for $x < 0$, and $\int_{+0}^{1} x \, dL(x) < \infty$. Also, u_F will be finite if and only if $\sigma^2 = 0$, $L(x) = 0$ for $x > 0$, and $\int_{-1}^{-0} x \, dL(x) < \infty$ (Baxter and Shapiro [18]).

If $l_F > -\infty$, then $l_F = \gamma - \int_{+0}^{\infty} [x \, dL(x)/1 + x^2]$. If $u_F < \infty$, then $u_F = \gamma - \int_{-\infty}^{-0} [x \, dL(x)/1 + x^2]$ (Tucker [456] and Esseen [121]).

1.8.46 Let W be the set of non-negative functions $w(x)$ defined on the real line and such that one of the following conditions is satisfied:

(a) $w(x + y) \leq B(w(x) + w(y))$ for every real x and y;
(b) $w(x + y) \leq Bw(x)w(y)$ for every real x and y.

Here B is a constant. Let $F(x)$ be an infinitely divisible function with the corresponding Lévy spectral function $L(x)$. If $w(x) \in W$, the condition

$$\int_{-\infty}^{\infty} w(x) \, dF(x) < \infty$$

is equivalent to the condition

$$\int_{|x| > 1} w(x) \, dL(x) < \infty$$

(Kruglov [240]). This is a generalization of Ramachandran's theorem [378]: an infinitely divisible d.f. $F(x)$ will have a moment of order α if and only if $\int_{|x|>1} |x|^\alpha \, dL(x) < \infty$.

1.8.47 Let $F(x)$ be an infinitely divisible d.f. with the Lévy–Khintchine spectral function $G(x)$. Let $p > 0$. Then $1 - F(x) = O(x^{-p})$ as $x \to +\infty$ if and only if $G(+\infty) - G(x) = O(x^{-p})$ as $x \to +\infty$; $F(-x) = O(x^{-p})$ as $x \to +\infty$ if and only if $G(-x) = O(x^{-p})$ as $x \to +\infty$. These propositions remain true when O is everywhere replaced by o (Wolfe [469]). (Grübel [152] obtained a generalization of this result; he replaced the function x^{-p} by an arbitrary continuous decreasing function $u(x) \colon \mathbb{R}_+ \to \mathbb{R}_+$ that satisfies the condition $u(x) \to 0$ as $x \to +\infty$ and some additional condition excluding the possibility of too rapid a decrease.)

1.8.48 Let $F(x)$ be an infinitely divisible d.f. with the c.f. $f(t)$ having the Lévy representation (1.53). The following conditions are equivalent:

(a) $\lim_{x \to +\infty} \{-x^{-2} \log F(-x)\} > 0$,
(b) $\lim_{x \to +\infty} \{-(x \log x)^{-1} \log F(-x)\} = \infty$,
(c) $L(x) = 0$ for $x < 0$.

Moreover, the following are equivalent:

(d) $\lim_{x \to +\infty} \{-x^{-2} \log F(-x)\} = \infty$,
(e) $L(x) = 0$ for $x < 0$ and $\sigma^2 = 0$.

These propositions remain true if $F(-x)$ is replaced by $T(x) = 1 - F(x) + F(-x)$ and $L(x)$ by $L(x) + |L(-x)|$ (Ohkubo [316]).

1.8.49 Let $F(x)$ be an infinitely divisible d.f. with the c.f. $f(t)$ having the Lévy representation (1.53).
(1) If $\sigma^2 = 0$ and $L(x) = 0$ for $x < 0$, then

$$\liminf_{x \to +\infty} \{-x^{-1-a} \log F(-x)\} \geq k_a \liminf_{x \to 0+} x^{-1-a} |L(x)|^{-a}$$

and

$$\liminf_{x \to +\infty} \{-x^{-1-a} \log F(-x)\} \leq h_a \liminf_{x \to 0+} x^{-1-a} |L(x)|^{-a},$$

where $1 < a < \infty$, and k_a and h_a are positive constants that do not depend on $F(x)$ and $L(x)$;
(2) If $\sigma^2 = 0$, $L(x) = 0$ for $x < 0$ and if there exists the limit $\lim_{x \to 0+} x^{1+1/a} |L(x)|$, where $1 < a < \infty$, then

$$\liminf_{x \to +\infty} \{-x^{-1-a} \log F(-x)\} = k_a \liminf_{x \to 0+} x^{-1-a} |L(x)|^{-a}.$$

(3) If $L(x) = 0$ for $x < 0$, then

$$\lim_{x \to +\infty} \{-x^{-2} \log F(-x)\} = \frac{1}{2\sigma^2}.$$

(4) $\lim_{x \to +\infty} \{-(x \log x)^{-1} \log F(-x)\} = (-\inf[\{x < 0: L(x) > 0\} \cup \{0\}])^{-1}$.
(5) If $L(x) > 0$ for $x < 0$, then

$$\liminf_{x \to +\infty} \{-x^{-1}(\log x)^{-a} \log F(-x)\} = c_a \liminf_{x \to +\infty} \{x^{-1}(-\log L(-x))^{1-a}\},$$

where $0 < a < 1$ and c_a is a positive constant independent of $F(x)$ and $L(x)$. (Ohkubo [316]).

1.8.50 Let $F(x)$ be an infinitely divisible d.f. with the Lévy–Khintchine spectral function $G(x)$.
(a) $F(x)$ is continuous if and only if $\int_{-\infty}^{\infty} x^{-2} dG(x) = \infty$ (Doeblin [102]; Hartman and Wintner [177]).
(b) $F(x)$ is discrete if and only if $G(x)$ is a step function and $\int_{-\infty}^{\infty} x^{-2} dG(x) < \infty$ (Blum and Rosenblatt [43]).

1.8.51 An arbitrary absolutely continuous and infinitely divisible distribution has the density such that its set of zeroes either has Lebesgue measure zero or coincides with some infinite interval except for a set of Lebesgue measure zero (Hudson and Tucker [201]).

1.8.52 Let $F(x)$ be an infinitely divisible d.f. Let $\Phi(x)$ be the standard normal d.f. The following propositions hold (they are arranged in accordance with an increase in generality):
(1) if $F(x) = \Phi(x)$ for $x < 0$, then $F(x) \equiv \Phi(x)$ (Rossberg [395]);
(2) if $F(x)/\Phi(x) \to 1$ as $x \to -\infty$, then $F(x) \equiv \Phi(x)$ (Riedel [384]);
(3) if there exist numerical sequences $\{x_n\}$ and $\{y_n\}$ such that $x_n \to -\infty$, $y_n \to -\infty$, $F(x_n) \geqslant \Phi(x_n)(1 + o(1))$, and

$$\frac{F(y_n)}{\Phi(y_n)} = o(e^{c|y_n|})$$

as $n \to \infty$ for every $c > 0$, then $F(x) \equiv \Phi(x)$ (Rossberg and Siegel [398]).

1.8.53 Let the c.f. $f(t)$ of an infinitely divisible d.f. $F(x)$ admit an analytical continuation to the upper (lower) half-plane of a complex variable $z = t + is$. If an infinitely divisible d.f. $G(x)$ coincides with $F(x)$ on the half-line $(-\infty, a)$ (on the half-line (a, ∞)), then either $F(x) = 0$ (correspondingly, $F(x) = 1$) on some half-line or $F(x) \equiv G(x)$ (Ibragimov [204]).

2

Probability inequalities for sums of independent random variables

2.1 Inequalities for the maximum of sums of independent random variables

Let $X_1, X_2, \ldots, X_n$ be independent random variables, with $S_k = \sum_{i=1}^{k} X_i$, and let $\kappa_q(X)$ denote a quantile of order q, $0 < q < 1$, for a random variable X.

Theorem 2.1 *For every q from the interval $0 < q < 1$ and for every real x, the following inequality holds:*

$$P\left(\max_{1 \leq k \leq n} \{S_k - \kappa_q(S_k - S_n)\} \geq x\right) \leq \frac{1}{q} P(S_n \geq x). \quad (2.1)$$

Proof We write

$$\bar{S}_k = \max_{1 \leq l \leq k} \{S_l - \kappa_q(S_l - S_n)\} \quad (k = 1, \ldots, n),$$

$$D_1 = [S_1 - \kappa_q(S_1 - S_n) \geq x],$$

$$D_k = [\bar{S}_{k-1} < x, S_k - \kappa_q(S_k - S_n) \geq x] \quad (k = 2, \ldots, n).$$

$$E_k = [S_n - S_k - \kappa_{1-q}(S_n - S_k) \geq 0] \quad (k = 1, \ldots, n).$$

We have $[\bar{S}_n \geq x] = \bigcup_{k=1}^{n} D_k$ and

$$P(\bar{S}_n \geq x) = \sum_{k=1}^{n} P(D_k), \quad (2.2)$$

since $P(D_k D_j) = 0$ for $k \neq j$. Furthermore,

$$P(E_k) \geq q \quad (k = 1, \ldots, n). \quad (2.3)$$

It is easy to see that if X is a random variable and if $\kappa_q(X)$ is a quantile of order q for X, then $-\kappa_q(X)$ is a quantile of order $1 - q$ for the random variable $-X$. This proposition follows from the inequalities $P(X \leq \kappa_q(X)) \geq q$ and $P(X \geq \kappa_q(X)) \geq 1 - q$, admitting the form $P(-X \geq -\kappa_q(X)) \geq q$ and $P(-X \leq -\kappa_q(X)) \geq 1 - q$.

In the definition of the event E_k we can put

$$\kappa_{1-q}(S_n - S_k) = -\kappa_q(S_k - S_n).$$

Then we have

$$\bigcup_{k=1}^{n} D_k E_k \subset [S_n \geq x]$$

and

$$P(S_n \geq x) \geq P\left(\bigcup_{k=1}^{n} D_k E_k\right) = \sum_{k=1}^{n} P(D_k E_k) = \sum_{k=1}^{n} P(D_k) P(E_k),$$

since the events D_k and E_k are independent. Taking into account (2.2) and (2.3) we conclude that

$$P(S_n \geq x) \geq q \sum_{k=1}^{n} P(D_k) = q P(\bar{S}_n \geq x). \quad \square$$

By definition, a median mX of a random variable X is $\kappa_{1/2}(X)$. The following proposition is an immediate consequence of Theorem 2.1.

Theorem 2.2 *For every real x,*

$$P\left(\max_{1 \leq k \leq n} \{S_k - \mathrm{m}(S_k - S_n)\} \geq x\right) \leq 2P(S_n \geq x). \tag{2.4}$$

For every $x \geq 0$ we have

$$P\left(\max_{1 \leq k \leq n} |S_k - \mathrm{m}(S_k - S_n)| \geq x\right) \leq 2P(|S_n| \geq x). \tag{2.5}$$

If $X_1, \ldots, X_n$ are independent symmetric random variables, then

$$P\left(\max_{1 \leq k \leq n} S_k \geq x\right) \leq 2P(S_n \geq x) \tag{2.6}$$

for every real x and

$$P\left(\max_{1 \leq k \leq n} |S_k| \geq x\right) \leq 2P(|S_n| \geq x) \tag{2.7}$$

for every $x \geq 0$.

The inequalities (2.4)–(2.7) are called Lévy's inequalities. The inequality (2.5) follows from (2.4) if we apply (2.4) to the random variables $-X_1, \ldots, -X_n$.

Theorem 2.3 *If*

$$P(S_n - S_k \geq -C) \geq q \quad (k = 1, \ldots, n-1) \tag{2.8}$$

for some constants $C \geq 0$ and $q > 0$, then

$$P\left(\max_{1 \leq k \leq n} S_k \geq x\right) \leq \frac{1}{q} P(S_n \geq x - C) \qquad (2.9)$$

for every real x.

This theorem follows from Lemma 1.1 (Chapter 1) and the following proposition:

Lemma 2.1 *Let there exist a non-negative constant C and a set of quantiles $\kappa_q(S_1 - S_n), \ldots, \kappa_q(S_{n-1} - S_n)$ of order q $(0 < q < 1)$ such that*

$$\kappa_q(S_k - S_n) \leq C \qquad \text{for } k = 1, \ldots, n-1. \qquad (2.10)$$

Then inequality (2.9) holds for every real x.

Proof By (2.10),

$$\left[\max_{1 \leq k \leq n} S_k \geq x + C\right] \subset \left[\max_{1 \leq k \leq n} \{S_k - \kappa_q(S_k - S_n)\} \geq x\right]$$

for every real x. Therefore, the probability of the first event does not exceed the probability of the second event. Applying the inequality (2.1) and replacing x by $x - C$, we arrive at the inequality (2.10). □

So far we have made no assumptions about the existence of any moments of the random variables under consideration. We shall formulate some consequences of the above-mentioned results under some additional moment conditions.

Theorem 2.4 *Let $n \geq 2$, $EX_k = 0$, $EX_k^2 < \infty$ for $k = 2, \ldots, n$. We write $b_n = \sum_{k=2}^{n} EX_k^2$. Then for every positive $q < 1$ and every real x we have*

$$P\left(\max_{1 \leq k \leq n} S_k \geq x\right) \leq \frac{1}{q} P\left(S_n \geq x - \left(\frac{b_n}{1-q}\right)^{1/2}\right). \qquad (2.11)$$

Note the absence of any condition concerning the moments of the random variable X_1. If $q = \frac{1}{2}$, (2.11) takes the form

$$P\left(\max_{1 \leq k \leq n} S_k \geq x\right) \leq 2P(S_n \geq x - (2b_n)^{1/2}). \qquad (2.12)$$

for every real x. In turn, (2.12) implies the following Kolmogorov inequality:

$$P\left(\max_{1 \leq k \leq n} S_k \geq x\right) \leq 2P\left(S_n \geq x - \left(2 \sum_{k=1}^{n} EX_k^2\right)^{1/2}\right) \qquad (2.13)$$

for every real x. (This inequality holds for independent random variables $X_1, \ldots, X_n$ having finite variances and zero means.) Inequality (2.12) is a strengthening of (2.13), because $b_n \leqslant \sum_{k=1}^n EX_k^2$.

For a proof of Theorem 2.4 we note that

$$E(S_n - S_k)^2 = \sum_{i=k+1}^n EX_i^2 \leqslant b_n$$

if $1 \leqslant k \leqslant n - 1$, and by Chebyshev's inequality we have

$$P\left(|S_n - S_k| \geqslant \left(\frac{b_n}{1-q}\right)^{1/2}\right) \leqslant \frac{1-q}{b_n} E(S_n - S_k)^2 \leqslant 1 - q.$$

Hence

$$P\left(S_n - S_k \geqslant -\left(\frac{b_n}{1-q}\right)^{1/2}\right) \geqslant q \qquad (k = 1, \ldots, n-1).$$

Taking into account Theorem 2.3, we obtain the inequality (2.11). $\square$

Theorem 2.5 *Let* $EX_k = 0$, $EX_k^2 < \infty$ $(k = 1, \ldots, n)$, *and let* $0 < c_n \leqslant c_{n-1} \leqslant \cdots \leqslant c_1$. *Then*

$$P\left(\max_{m \leqslant k \leqslant n} c_k |S_k| \geqslant x\right) \leqslant x^{-2}\left(c_m^2 \sum_{k=1}^m EX_k^2 + \sum_{k=m+1}^n c_k^2 EX_k^2\right) \qquad (2.14)$$

for every $x > 0$ *and every positive integer* $m < n$.

Proof We define

$$Y = \sum_{k=m}^{n-1} (c_k^2 - c_{k+1}^2) S_k^2 + c_n^2 S_n^2.$$

It is easy to show that

$$EY = c_m^2 \sum_{k=1}^m EX_k^2 + \sum_{k=m+1}^n c_k^2 EX_k^2. \qquad (2.15)$$

We also define $B_m = \{c_m|S_m| \geqslant x\}$ and

$$B_k = \left\{\max_{m \leqslant r \leqslant k-1} c_r |S_r| < x, \; c_k|S_k| \geqslant x\right\} \qquad (k = m+1, \ldots, n).$$

It is clear that

$$P\left(\max_{m \leqslant k \leqslant n} c_k |S_k| \geqslant x\right) = \sum_{k=m}^n P(B_k) \qquad (2.16)$$

If X is a random variable defined on a probability space $(\Omega, \mathscr{A}, P)$, and if $B \in \mathscr{A}$, $P(B) > 0$, then for every x we write

$$F(x \mid B) = \frac{1}{P(B)} P(\{X < x\} \cap B)$$

and

$$E(X \mid B) = \int_{-\infty}^{\infty} x \, dF(x \mid B).$$

We have $EY \geq \sum_{k=m}^{n} P(B_k) E(Y \mid B_k)$. If $j > k$, the independence of the random variables $X_1, \ldots, X_n$ implies that

$$E(X_j S_k \mid B_k) = E(X_j \mid B_k) E(S_k \mid B_k) = 0.$$

Hence $E(S_j^2 \mid B_k) \geq E(S_k^2 \mid B_k) \geq x^2/c_k^2$. For $k \geq m$ we have

$$E(Y \mid B_k) = \sum_{l=m}^{n-1} (c_l^2 - c_{l+1}^2) E(S_l^2 \mid B_k) + c_n^2 E(S_n^2 \mid B_k)$$

$$\geq \sum_{l=k}^{n-1} (c_l^2 - c_{l+1}^2) E(S_l^2 \mid B_k) + c_n^2 E(S_n^2 \mid B_k)$$

$$\geq \frac{x^2}{c_k^2} \left\{ \sum_{l=k}^{n-1} (c_l^2 - c_{l+1}^2) + c_n^2 \right\} = x^2.$$

Therefore, $EY \geq x^2 \sum_{k=m}^{n} P(B_k)$. This inequality together with (2.15) and (2.16) implies (2.14). □

Inequality (2.14) is called the Hàjek–Rényi inequality. Its important consequence is the Kolmogorov inequality

$$P\left(\max_{1 \leq k \leq n} |S_k| \geq x \right) \leq x^{-2} \sum_{k=1}^{n} EX_k^2, \qquad (2.17)$$

which holds for every positive x and for independent random variables $X_1, \ldots, X_n$ with finite variances and zero means.

2.2 Exponential bounds

We consider independent random variables $X_1, \ldots, X_n$ and we write $S_n = \sum_{k=1}^{n} X_k$.

Theorem 2.6 *Suppose there exist positive constants $g_1, \ldots, g_n$ and T such that*

$$E e^{tX_k} \leq e^{g_k t^2/2} \qquad (k = 1, \ldots, n) \qquad (2.18)$$

for $0 \leqslant t \leqslant T$. We write $G_n = \sum_{k=1}^{n} g_k$. Then

$$P(S_n \geqslant x) \leqslant e^{-x^2/(2G_n)} \quad \text{if } 0 \leqslant x \leqslant G_n T, \qquad (2.19)$$

$$P(S_n \geqslant x) \leqslant e^{-Tx/2} \quad \text{if } x \geqslant G_n T. \qquad (2.20)$$

Proof If $0 < t \leqslant T$, we have

$$P(S_n \geqslant x) = P(e^{tS_n} \geqslant e^{tx}) \leqslant e^{-tx} \, \mathrm{E} e^{tS_n}$$

for every x, because e^{tS_n} is a non-negative random variable. The random variables $X_1, \ldots, X_n$ are independent, therefore,

$$\mathrm{E} e^{tS_n} = \prod_{k=1}^{n} \mathrm{E} e^{tX_k} \leqslant e^{G_n t^2/2}$$

by condition (2.18). It follows that

$$P(S_n \geqslant x) \leqslant e^{-tx + G_n t^2/2} \qquad (2.21)$$

for every x and $0 < t \leqslant T$. For a fixed value of x we write $f(t) = G_n t^2/2 - tx$ and we shall minimize the function $e^{f(t)}$.

If $x = 0$, the assertion of the theorem is obvious. Suppose $0 < x \leqslant G_n T$. The equation $f'(t) = 0$ has the unique solution $t = x/G_n$ which yields the minimum of the function $f(t)$. This solution satisfies the condition $0 < t \leqslant T$. Therefore, in (2.21) we may put $t = x/G_n$. Then we obtain (2.19).

Now suppose $x \geqslant G_n T$. Then $f'(t) = G_n t - x \leqslant 0$ and $f(t)$ is a non-increasing function. Putting $t = T$ in (2.21), we have

$$P(S_n \geqslant x) \leqslant e^{-Tx + G_n T^2/2} \leqslant e^{-Tx/2}.$$

The inequality (2.20) is proved. □

The following proposition is a left-sided analogue of Theorem 2.6. If condition (2.18) is satisfied for $-T \leqslant t \leqslant 0$ and some positive constants T, $g_1, \ldots, g_n$, then

$$P(S_n \leqslant -x) \leqslant e^{-x^2/(2G_n)} \quad \text{if } 0 \leqslant x \leqslant G_n T, \qquad (2.22)$$

$$P(S_n \leqslant -x) \leqslant e^{-Tx/2} \quad \text{if } x \geqslant G_n T. \qquad (2.23)$$

This result follows from Theorem 2.6. In fact, if the random variables $X_1, \ldots, X_n$ satisfy condition (2.18) for $-T \leqslant t \leqslant 0$, the random variables $-X_1, \ldots, -X_n$ satisfy it for $0 \leqslant t \leqslant T$. Therefore, the inequalities (2.19) and (2.20) are true when $-S_n$ is substituted for S_n, and are then equivalent to the inequalities (2.22) and (2.23).

Combining all these results, we arrive at the following theorem.

Theorem 2.7 *Suppose that condition (2.18) is satisfied for $|t| \leqslant T$ and for some positive constants $g_1, \ldots, g_n$ and T. Then the assertions (2.19), (2.20), (2.22) and (2.23) are true.*

Let us clarify the probabilistic meaning of the conditions of this theorem.

Lemma 2.2 *Let X be a random variable. The following assertions are equivalent:*
(I) *There exists a positive constant H such that*
$$\mathrm{E} e^{tX} < \infty \quad \text{for } |t| < H.$$
(II) *There exists a positive constant a such that*
$$\mathrm{E} e^{a|X|} < \infty.$$
(III) *There exist positive constants b and c such that*
$$P(|X| \geq x) \leq b e^{-cx} \quad \text{for all } x > 0.$$
If $\mathrm{E} X = 0$, the above assertions are each equivalent to the assertion:
(IV) *There exist positive constants g and T such that*
$$\mathrm{E} e^{tX} \leq e^{gt^2} \quad \text{for } |t| \leq T.$$

Proof The inequality $e^{tX} \leq e^{|tX|}$ implies that if (II) is fulfilled, $\mathrm{E} e^{tX} < \infty$ for $|t| < a$; that is, (I) is true. It is easy to show that (I) implies (II).

If (II) is satisfied, then
$$P(|X| \geq x) = P(e^{a|X|} \geq e^{ax}) \leq e^{-ax} \mathrm{E} e^{a|X|}$$

for every positive x, so that (III) is also satisfied. We shall show that (III) implies (II). Let $V(x)$ be the distribution function of the random variable X. For $0 \leq a < c$ we have

$$\mathrm{E} e^{a|X|} = \int_{-\infty}^{\infty} e^{a|x|} \, dV(x) = \int_{-\infty}^{0} e^{-ax} d\left(\int_{-\infty}^{x} dV(u) \right)$$
$$+ \int_{0}^{\infty} e^{ax} d\left(-\int_{x}^{\infty} dV(u) \right) \leq 2 + ab \int_{-\infty}^{0} e^{(c-a)x} \, dx$$
$$+ ab \int_{0}^{\infty} e^{-(c-a)x} \, dx < \infty.$$

The equivalence of (I), (II), and (III) is proved.

Now suppose $\mathrm{E} X = 0$. If (I) is fulfilled, then the random variable X has the moments of all orders, and the following relation holds:
$$\log \mathrm{E} e^{tX} = \tfrac{1}{2} \sigma^2 t^2 + o(t^2)$$

as $t \to 0$, where $\sigma^2 = \mathrm{E} X^2$. For any constant $g > \sigma^2/2$ the inequalities $\log \mathrm{E} e^{tX} \leq gt^2$ and $\mathrm{E} e^{tX} \leq e^{gt^2}$ hold for all sufficiently small t; that is, (IV) is true. It is clear that (IV) implies (I). □

Note that (I) is the Cramér condition.

An insight into the exactness of the estimates in Theorems 2.6 and 2.7 is afforded by the following example. Suppose that X_k ($k = 1, \ldots, n$) has the normal $(0, \sigma_k)$ distribution. Then

$$\log E e^{tX_k} = \tfrac{1}{2}\sigma_k^2 t^2, \quad E e^{tX_k} = e^{\sigma_k^2 t^2 / 2},$$

and condition (2.18) is satisfied for $g_k = \sigma_k^2$ and every positive T. In this case $G_n = \sum_{k=1}^n \sigma_k^2$, and by Theorem 2.6 we have $P(S_n \geq x) \leq e^{-x^2/(2G_n)}$ for every $x \geq 0$. The factor $\tfrac{1}{2}$ in the exponent cannot be replaced by any larger number, since the sum $S_n = X_1 + \cdots + X_n$ has the normal distribution with the parameters $(0, G_n^{1/2})$ and we have

$$P(S_n \geq x) = (2\pi)^{-1/2} \int_{xG_n^{-1/2}}^{\infty} e^{-t^2/2}\, dt$$

$$= x^{-1} G_n^{1/2} (2\pi)^{-1/2} e^{-x^2/(2G_n)} (1 + O(x^{-2}))$$

as $x \to +\infty$. (Here we have used the relation

$$\int_x^{\infty} e^{-t^2/2}\, dt = x^{-1} e^{-x^2/2} (1 + O(x^{-2}))$$

as $x \to +\infty$, which is easy to prove by integrating by parts in the integral $\int_x^\infty t^{-1}\, d(-e^{-t^2/2})$.)

Theorem 2.8 Suppose that $EX_k = 0$, $\sigma_k^2 = EX_k^2 < \infty$ ($k = 1, \ldots, n$), $B_n = \sum_{k=1}^n \sigma_k^2$. Suppose there exists a positive constant H such that

$$|EX_k^m| \leq \tfrac{1}{2} m! \sigma_k^2 H^{m-2} \quad (k = 1, \ldots, n) \qquad (2.24)$$

for all integers $m \geq 2$. Then

$$P(S_n \geq x) \leq e^{-x^2/(4B_n)} \quad \text{if } 0 \leq x \leq B/H, \qquad (2.25)$$

$$P(S_n \geq x) \leq e^{-x/(4H)} \quad \text{if } x \geq B/H, \qquad (2.26)$$

$$P(S_n \leq -x) \leq e^{-x^2/(4B_n)} \quad \text{if } 0 \leq x \leq B/H, \qquad (2.27)$$

$$P(S_n \leq -x) \leq e^{-x/(4H)} \quad \text{if } x \geq B/H. \qquad (2.28)$$

Proof Consider the formal equality

$$E e^{tX_k} = 1 + \tfrac{1}{2} t^2 \sigma_k^2 + \tfrac{1}{6} t^3\, EX_k^3 + \cdots \quad (k = 1, \ldots, n).$$

Condition (2.24) implies that the series on the right is majorized by the series

$$1 + \tfrac{1}{2} t^2 \sigma_k^2 (1 + H|t| + H^2 t^2 + \cdots).$$

If $|t| \leq 1/(2H)$, the sum of the latter series does not exceed

$$1 + \frac{t^2}{2(1 - H|t|)} \sigma_k^2 \leq 1 + t^2 \sigma_k^2 \leq e^{t^2 \sigma_k^2}.$$

Thus condition (2.18) is satisfied for $|t| \leq 1/(2H)$ and $g_k = 2\sigma_k^2$. Accordingly, $G_n = \sum_{k=1}^n g_k = 2B_n$. Applying Theorem 2.7 for $T = 1/(2H)$, we obtain the assertion of Theorem 2.8. □

Inequalities (2.25) to (2.28) are called the Bernstein inequalities.

Let us indicate some sufficient conditions for the Bernstein condition (2.24). If the random variables $X_1, \ldots, X_n$ with zero means are uniformly bounded (i.e. if there exists a positive constant C such that $P(|X_k| \leq C) = 1$ for all k), then

$$|EX_k^m| = \left| \int_{-C}^{C} x^m \, dV_k(x) \right| \leq C^{m-2} \int_{-C}^{C} x^2 \, dV_k(x) = C^{m-2} \sigma_k^2$$

for all integers $m \geq 2$, where $V_k(x)$ is the d.f. of X_k. Thus condition (2.24) is satisfied with $H = C$.

It is not hard to prove that if $X_1, \ldots, X_n$ are independent identically distributed random variables satisfying the Cramér condition, then condition (2.24) is also satisfied.

Note that Theorem 2.7 represents not only a more general but also a more exact result than Theorem 2.8.

2.3 Inequalities for moments of sums of independent random variables

If $X_1, \ldots, X_n$ are arbitrary random variables (not necessarily independent), then

$$E|S_n|^p \leq \sum_{k=1}^n E|X_k|^p \quad \text{if } 0 < p \leq 1 \tag{2.29}$$

and

$$E|S_n|^p \leq n^{p-1} \sum_{k=1}^n E|X_k|^p \quad \text{if } p > 1. \tag{2.30}$$

Here $S_n = \sum_{k=1}^n X_k$. (We have made no assumptions about the finiteness of the moments $E|X_k|^p$, because these inequalities are obviously true if at least one moment $E|X_k|^p$ is infinite.)

Inequalities (2.29) and (2.30) follow from the elementary inequalities

$$\left| \sum_{k=1}^n a_k \right|^p \leq \sum_{k=1}^n |a_k|^p \quad (0 < p \leq 1)$$

and
$$\left|\sum_{k=1}^{n} a_k\right|^p \le n^{p-1} \sum_{k=1}^{n} |a_k|^p \quad (p > 1)$$

for every positive integer n and arbitrary real numbers $a_1, \ldots, a_n$. Replacing here a_k by X_k we obtain (2.29) and (2.30) after integration.

Under additional assumptions it is possible to strengthen the inequality (2.30).

In what follows, we suppose that $X_1, \ldots, X_n$ are independent random variables.

Theorem 2.9 *Let $EX_k = 0$ ($k = 1, \ldots, n$), and let $p \ge 2$. We write*

$$M_{p,n} = \sum_{k=1}^{n} E|X_k|^p, \quad B_n = \sum_{k=1}^{n} EX_k^2.$$

Then

$$E|S_n|^p \le c(p)(M_{p,n} + B_n^{p/2}), \tag{2.31}$$

where $c(p)$ is a positive constant depending only on p.

Inequality (2.31) is called the Rosenthal inequality.
We shall prove this theorem by means of the following lemmas.

Lemma 2.3 *Let $X_1, \ldots, X_n$ be independent random variables with the distribution functions $V_1(x), \ldots, V_n(x)$, and let y be a positive number. We write*

$$\mu(-\infty, y) = \sum_{k=1}^{n} \int_{x<y} x \, dV_k(x),$$

$$B(-\infty, y) = \sum_{k=1}^{n} \int_{x<y} x^2 \, dV_k(x).$$

Then

$$P(S_n \ge x) \le \sum_{k=1}^{n} P(X_k \ge y)$$

$$+ \exp\left\{\frac{x}{y} - \left(\frac{x - \mu(-\infty, y)}{y} + \frac{B(-\infty, y)}{y^2}\right) \log\left(1 + \frac{xy}{B(-\infty, y)}\right)\right\}$$

for every positive x.

Proof We introduce the random variables

$$Z_k = \begin{cases} X_k & \text{if } X_k < y \\ 0 & \text{if } X_k \ge y \end{cases}$$

and $T_n = \sum_{k=1}^{n} Z_k$. It is clear that
$$\{S_n \geqslant x\} \subset \{T_n \geqslant x\} \cup \{S_n \neq T_n\}$$
for every real x. Hence
$$P(S_n \geqslant x) \leqslant P(T_n \geqslant x) + P(S_n \neq T_n).$$
Further, $P(T_n \geqslant x) \leqslant e^{-hx} E e^{hT_n}$ for every $h > 0$, therefore,
$$P(S_n \geqslant x) \leqslant e^{-hx} E e^{hT_n} + \sum_{k=1}^{n} P(X_k \geqslant y). \tag{2.32}$$

It is easy to find that the function $(e^x - 1 - x)/x^2$ does not decrease on the real line. It follows that the function $(e^{hx} - 1 - hx)/x^2$, where $h > 0$, has the same property. We have

$$E e^{hZ_k} = \int_{-\infty}^{y} e^{hx} \, dV_k(x) + \int_{y}^{\infty} dV_k(x)$$

$$\leqslant 1 + h \int_{-\infty}^{y} x \, dV_k(x) + \int_{-\infty}^{y} (e^{hx} - 1 - hx) \, dV_k(x)$$

$$\leqslant 1 + h \int_{-\infty}^{y} x \, dV_k(x) + (e^{hy} - 1 - hy) y^{-2} \int_{-\infty}^{y} x^2 \, dV_k(x).$$

The random variables $Z_1, \ldots, Z_n$ are independent because the variables $X_1, \ldots, X_n$ are independent. Therefore,

$$e^{-hx} E e^{hT_n} = e^{-hx} \prod_{k=1}^{n} E e^{hZ_k}$$

$$\leqslant \exp\{-hx + h\mu(-\infty, y) + (e^{hy} - 1 - hy) y^{-2} B(-\infty, y)\}.$$

We put here
$$h = \frac{1}{y} \log\left(1 + \frac{xy}{B(-\infty, y)}\right).$$
Then
$$e^{hy} - 1 = \frac{xy}{B(-\infty, y)},$$

$$(e^{hy} - 1 - hy) y^{-2} B(-\infty, y) = \frac{x}{y} - \frac{B(-\infty, y)}{y^2} \log\left(1 + \frac{xy}{B(-\infty, y)}\right),$$
and
$$e^{-hx} E e^{hT_n} \leqslant \exp\left\{\frac{x}{y} - \left[\frac{B(-\infty, y)}{y^2} - \frac{\mu(-\infty, y)}{y} + \frac{x}{y}\right] \log\left(1 + \frac{xy}{B(-\infty, y)}\right)\right\}.$$

Taking into account (2.32), we arrive at the assertion of the lemma. $\square$

We note that an analogous estimate holds for $P(S_n \leq -x)$, with the replacement of $P(X_k \geq y)$ by $P(X_k \leq -y)$, $\mu(-\infty, y)$ by

$$-\mu(-y; \infty) = -\sum_{k=1}^{n} \int_{x > -y} x \, dV_k(x),$$

and $B(-\infty, y)$ by

$$B(-y, \infty) = \sum_{k=1}^{n} \int_{x > -y} x^2 \, dV_k(x).$$

Let us continue the proof of Theorem 2.9. Since $EX_k = 0$ for all k, we have $\mu(-\infty, y) \leq 0$. The definitions of $B(-\infty, y)$ and B_n imply that $B(-\infty, y) \leq B_n$. By Lemma 2.3 and the accompanying note, we obtain

$$P(|S_n| \geq x) \leq \sum_{k=1}^{n} P(|X_k| \geq y) + 2 \exp\left\{\frac{x}{y} - \frac{x}{y} \log\left(1 + \frac{xy}{B_n}\right)\right\} \quad (2.33)$$

for every positive x and y.

Lemma 2.4 *Let $g(x)$ be a non-negative even function, non-decreasing on the positive half-line, and such that $g(0) = 0$. Let X be an arbitrary random variable. If $Eg(X) < \infty$, then*

$$Eg(X) = \int_0^{\infty} P(|X| \geq x) \, dg(x).$$

It is easy to prove this lemma by integrating by parts the expression appearing on the right-hand side. We take into account the equalities $g(0) = 0$ and $\lim_{x \to +\infty} g(x) P(|X| \geq x) = 0$; the latter follows from the relations

$$g(x) P(X \geq x) = g(x)(1 - F(x)) \leq \int_x^{\infty} g(y) \, dF(y)$$

for $x > 0$, where $F(y)$ stands for the distribution function of X.

The function $g(x) = |x|^p$, where $p > 0$, satisfies the conditions of Lemma 2.4. Therefore,

$$E|X|^p = p \int_0^{\infty} P(|X| \geq x) x^{p-1} \, dx \quad (2.34)$$

for an arbitrary random variable X.

We put $y = x/r$, where $r > p/2$, in (2.33), then multiply both sides of this inequality by px^{p-1} and integrate on the positive half-line. Using the equality (2.34), we find that

$$E|S_n|^p \leq r^p \sum_{k=1}^{n} E|X_k|^p + 2p \, e^r \int_0^{\infty} x^{p-1} \left(1 + \frac{x^2}{rB_n}\right)^{-r} dx.$$

The integral is equal to $\frac{1}{2}r^{p/2}B_n^{p/2}B(p/2, r - p/2)$ (see, for instance, Gradshteyn and Ryzhik [151]), where $B(a, b) = \int_0^1 x^{a-1}(1 - x)^{b-1} \, dx$ is the beta function. Thus we have

$$E|S_n|^p \leqslant r^p M_{p,n} + pr^{p/2} e^r B\left(\frac{p}{2}, r - \frac{p}{2}\right) B_n^{p/2} \tag{2.35}$$

for every $r > p/2$. Inequality (2.31) follows from (2.35). □

Theorem 2.10 *Let $X_1, \ldots, X_n$ be independent random variables with zero means, and let $p \geqslant 2$. Then*

$$E|S_n|^p \leqslant C(p) n^{p/2-1} \sum_{k=1}^n E|X_k|^p, \tag{2.36}$$

where $C(p)$ is a positive constant depending only on p.

Proof If Y is an arbitrary random variable and $0 < r < s$, then $(E|Y|^r)^{1/r} \leqslant (E|Y|^s)^{1/s}$ (Lyapunov's inequality). We shall apply this inequality to the random variable Y having the distribution function $1/n \sum_{k=1}^n V_k(x)$, where $V_k(x)$ is the d.f. of X_k. We find that $(B_n/n)^{1/2} \leqslant (M_{p,n}/n)^{1/p}$ for every $p \geqslant 2$, or $B_n^{p/2} \leqslant n^{p/2-1} M_{p,n}$. Taking into account Rosenthal's inequality (2.31), we obtain (2.36). □

Inequality (2.36) is stronger than (2.30), which was established without the independence condition.

It is possible to obtain some strengthening of (2.36), using an improvement of Lyapunov's inequality for moments of a random variable. By Theorem 1.2 (Chapter 1), we have

$$(E|Y|^r)^{1/r} \leqslant \gamma^{1/r - 1/s} (E|Y|^s)^{1/s}, \tag{2.37}$$

where Y is an arbitrary random variable, $0 < r \leqslant s$, and $\gamma = P(Y \neq 0)$. If Y is a random variable with the d.f. $1/n \sum_{k=1}^n P(X_k < x)$, then $E|Y|^r = 1/n \sum_{k=1}^n E|X_k|^r$ for any $r > 0$ and

$$P(Y \neq 0) = \frac{1}{n} \sum_{k=1}^n P(X_k \neq 0).$$

Applying inequality (2.37) to Y with $r = 2$ and $s = p \geqslant 2$, we conclude that

$$B_n^{1/2} \leqslant \left(\sum_{k=1}^n P(X_k \neq 0)\right)^{1/2 - 1/p} M_{p,n}^{1/p}.$$

The following result is a consequence of this last inequality and (2.31).

Theorem 2.11 *Suppose conditions of Theorem 2.9 are satisfied. Then*

$$E|S_n|^p \leq c(p)\left[1 + \left(\sum_{k=1}^{n} P(X_k \neq 0)\right)^{p/2-1}\right]M_{p,n} \qquad (2.38)$$

If the sum $\sum_{k=1}^{n} P(X_k \neq 0)$ grows slower than n (for example, slower than cn^q with some $q < 1$), then (2.38) gives a better estimate than (2.36) does.

So far we have considered independent random variables $X_1, \ldots, X_n$ having moments of order $p \geq 2$. We turn now to the case $p > 1$.

Theorem 2.12 *Suppose $p > 1$. Then*

$$E|S_n|^p \leq c(p)(M_{p,n} + D_n^p), \qquad (2.39)$$

where

$$M_{p,n} = \sum_{k=1}^{n} E|X_k|^p, \quad D_n = \sum_{k=1}^{n} E|X_k|,$$

and $c(p)$ is a positive constant depending only on p.

Inequality (2.39) is also due to Rosenthal.

We can prove this theorem by following the lines of the proof of Theorem 2.9. Instead of (2.33) it is possible to use the inequality

$$P(|S_n| \geq x) \leq \sum_{k=1}^{n} P(|X_k| \geq y) + 2\exp\left\{\frac{x}{y} - \frac{x}{y}\log\left(1 + \frac{x}{D_n}\right)\right\}$$

for every positive x and y.

Theorem 2.11 has the following analogue for the case $p > 1$.

Theorem 2.13 *Suppose $X_1, \ldots, X_n$ are independent random variables and $p > 1$. Then*

$$E|S_n|^p \leq c(p)\left[1 + \left(\sum_{k=1}^{n} P(X_k \neq 0)\right)^{p-1}\right]M_{p,n}.$$

Note that the condition $EX_k = 0$ $(k = 1, \ldots, n)$ is absent in Theorems 2.12 and 2.13.

2.4 Inequalities for the concentration functions of sums of independent random variables

Consider a random variable X with the distribution function $F(x)$. For every $\lambda > 0$ we write

$$D(X; \lambda) = \lambda^{-2}\int_{|x|<\lambda} x^2\, dF(x) + \int_{|x|\geq\lambda} dF(x). \qquad (2.40)$$

Note that
$$D(X; \lambda) \leq 1 \quad \text{for every } \lambda > 0. \tag{2.41}$$
If $0 < \lambda_0 < \lambda$, then
$$\lambda^{-2} \int_{|x|<\lambda} x^2\, dF(x) \leq \lambda_0^{-2} \int_{|x|<\lambda_0} x^2\, dF(x) + \lambda^{-2} \int_{\lambda_0 \leq |x| < \lambda} x^2\, dF(x)$$
$$\leq \lambda_0^{-2} \int_{|x|<\lambda_0} x^2\, dF(x) + \int_{\lambda_0 \leq |x| < \lambda} dF(x).$$

It follows that the function $D(X; \lambda)$ does not increase on the positive half-line $\lambda > 0$.

We put $D(X; 0) = P(X \neq 0)$. It is clear that $D(X; \lambda) = 0$ for some $\lambda \geq 0$ if and only if $P(X = 0) = 1$. If $u \geq \lambda$, then
$$D(X; \lambda) \geq u^{-2} \int_{|x|<\lambda} x^2\, dF(x) + \int_{\lambda \leq |x| < u} dF(x).$$

Therefore,
$$D(X; \lambda) \geq u^{-2} \int_{|x| \leq u} x^2\, dF(x) \quad \text{for } u \geq \lambda. \tag{2.42}$$

If X is a random variable, then, as in Chapter 1, we denote by X^s the corresponding symmetrized random variable.

Theorem 2.14 Let $X_1, \ldots, X_n$ be independent random variables, $S_n = \sum_{k=1}^n X_k$. Let $\lambda_1, \ldots, \lambda_n$ be positive numbers, $\lambda_k \leq \lambda$ $(k = 1, \ldots, n)$. Then there exists an absolute positive constant A, such that
$$Q(S_n; \lambda) \leq A\lambda \left(\sum_{k=1}^n \lambda_k^2 D(X_k^s; \lambda_k) \right)^{-1/2}. \tag{2.43}$$

Proof Let $V_k(x)$ and $v_k(t)$ denote respectively the d.f. and the c.f. of the random variable X_k. We shall apply Lemma 1.16 (Chapter 1) to the sum S_n. Setting $a = 1/\lambda$ in the inequality (1.30) (Chapter 1), we find that
$$Q(S_n; \lambda) \leq A_1 \lambda \int_{|t| \leq 1/\lambda} \prod_{k=1}^n |v_k(t)|\, dt.$$
The inequality $1 + x \leq e^x$ for every real x implies that
$$|v_k(t)|^2 \leq \exp\{-(1 - |v_k(t)|^2)\}.$$
If $V_k^s(x)$ is the d.f. of the symmetrized random variable X_k^s, then
$$1 - |v_k(t)|^2 = \int_{-\infty}^{\infty} (1 - \cos tx)\, dV_k^s(x).$$

Therefore,

$$Q(S_n; \lambda) \leq A_1 \lambda \int_{|x| \leq 1/\lambda} \exp\left\{-\frac{1}{2} \sum_{k=1}^{n} \int_{-\infty}^{\infty} (1 - \cos tx) \, dV_k^s(x)\right\} dt. \quad (2.44)$$

The function $L_k(x)$, such that $L_k(x) = V_k^s(x) - 1$ for $x > 0$ and $L_k(x) = V_k^s(x)$ for $x < 0$, is a Lévy spectral function. In order to estimate the integral in (2.44) from above we need the following lemma.

Lemma 2.5 *Let $L_k(x)$ be a Lévy spectral function for $k = 1, \ldots, n$. Let δ be a positive number, and let $0 < \lambda_k \leq \lambda$ $(k = 1, \ldots, n)$. Then*

$$\int_{|t| \leq 1/\lambda} \exp\left\{-\delta \sum_{k=1}^{n} \int_{|x| > 0} (1 - \cos tx) \, dL_k(x)\right\} dt$$

$$\leq A\delta^{-1/2} \left\{\sum_{k=1}^{n} \left[\int_{0 < |x| < \lambda_k} x^2 \, dL_k(x) + \lambda_k^2 \int_{|x| \geq \lambda_k} dL_k(x)\right]\right\}^{-1/2}, \quad (2.45)$$

where A is an absolute positive constant.

Proof If $|x| \leq 1$, then $1 - \cos x \geq \frac{11}{24} x^2$. It follows that

$$\int_{|x| > 0} (1 - \cos tx) \, dL_k(x)$$

$$= \int_{0 < |x| < \lambda_k} (1 - \cos tx) \, dL_k(x) + \int_{|x| \geq \lambda_k} (1 - \cos tx) \, dL_k(x)$$

$$\geq \tfrac{11}{24} t^2 \int_{0 < |x| < \lambda_k} x^2 \, dL_k(x) + \int_{|x| \geq \lambda_k} (1 - \cos tx) \, dL_k(x)$$

for $|t| \leq 1/\lambda$. We denote the left-hand side of the inequality (2.45) by I and we write

$$\beta_0 = \delta \sum_{k=1}^{n} \int_{0 < |x| < \lambda_k} x^2 \, dL_k(x).$$

Then

$$I \leq \int_{|t| \leq 1/\lambda} \{-\tfrac{11}{24}\beta_0 t^2\} \prod_{k=1}^{n} \exp\left\{-\delta \int_{|x| \geq \lambda_k} (1 - \cos tx) \, dL_k(x)\right\} dt. \quad (2.46)$$

We put

$$\beta_k = \delta \lambda_k^2 \int_{|x| \geq \lambda_k} dL_k(x) \quad (1 \leq k \leq n),$$

$$B = \sum_{k=0}^{n} \beta_k, \quad \alpha_k = \beta_k / B \quad (0 \leq k \leq n).$$

Obviously,

$$B = \delta \sum_{k=1}^{n} \left\{ \int_{0 < |x| < \lambda_k} x^2 \, dL_k(x) + \lambda_k^2 \int_{|x| \geq \lambda_k} dL_k(x) \right\}. \tag{2.47}$$

Without loss of generality we may assume that $\alpha_k > 0$ for all k. Applying Hölder's inequality to the right-hand side of (2.46), we obtain

$$I \leq \prod_{k=0}^{n} I_k^{\alpha_k}, \tag{2.48}$$

where

$$I_0 = \int_{|t| \leq 1/\lambda} \exp\{-\tfrac{11}{24} B t^2\} \, dt,$$

$$I_k = \int_{|t| \leq 1/\lambda} \exp\left\{ -\frac{B}{\lambda_k^2} \int_{-\infty}^{\infty} (1 - \cos tx) \, dM_k(x) \right\} dt \qquad (k = 1, \ldots, n).$$

$M_k(x)$ is a d.f. such that

$$dM_k(x) = \begin{cases} \dfrac{1}{p_k} dL_k(x) & \text{if } |x| \geq \lambda_k \\ 0 & \text{if } |x| < \lambda_k \end{cases}$$

and

$$p_k = \int_{|x| \geq \lambda_k} dL_k(x).$$

It is clear that the integral I_0 has the following upper bound:

$$I_0 \leq \int_{-\infty}^{\infty} \exp\{-\tfrac{11}{24} B t^2\} \, dt \leq A B^{-1/2}, \tag{2.49}$$

where A is an absolute positive constant. We can estimate the integral I_k ($k = 1, \ldots, n$) using inequality (1.7) (Chapter 1) (Jensen's inequality) with $g(x) = e^{-x}$. Then we obtain

$$\exp\left\{ -\frac{B}{\lambda_k^2} \int_{-\infty}^{\infty} (1 - \cos tx) \, dM_k(x) \right\} \leq \int_{-\infty}^{\infty} \exp\left\{ -\frac{B}{\lambda_k^2} (1 - \cos tx) \right\} dM_k(x)$$

and

$$I_k \leq \int_{|x| \geq \lambda_k} \left(\int_{|t| \leq 1/\lambda} \exp\left\{ -\frac{B}{\lambda_k^2} (1 - \cos tx) \right\} dt \right) dM_k(x).$$

We set

$$J_k = \int_{|t| \leq 1/\lambda} \exp\left\{ -\frac{B}{\lambda_k^2} (1 - \cos tx) \right\} dt.$$

To prove (2.45) it is sufficient to show that

$$J_k \leq AB^{-1/2} \quad \text{if } |x| \geq \lambda_k \text{ and } 1 \leq k \leq n. \tag{2.50}$$

In fact, (2.50) implies that $I_k \leq AB^{-1/2}$ ($k = 1, \ldots, n$). In turn, these inequalities together with (2.48), (2.49), and the equality $\alpha_0 + \alpha_1 + \cdots + \alpha_n = 1$ imply that $I \leq AB^{-1/2}$. Then (2.45) follows from the latter inequality and from (2.47).

If $\lambda_k \leq |x| \leq \pi\lambda$, then we have $|tx| \leq \pi$ for $|t| \leq 1/\lambda$. Using the inequality $\sin u/u \geq 2/\pi$ for $|u| \leq \pi/2$, we obtain

$$1 - \cos tx = 2\sin^2\frac{tx}{2} \geq \frac{2}{\pi^2} t^2 x^2 \geq \frac{2}{\pi^2} t^2 \lambda_k^2$$

and

$$J_k \leq \int_{|t| \leq 1/\lambda} \exp\left\{-\frac{2}{\pi^2} Bt^2\right\} dt \leq AB^{-1/2}.$$

If, however, $|x| > \pi\lambda$, it is sufficient to consider the case when $x > \pi\lambda$. Then

$$\lambda J_k = \frac{\lambda}{x}\int_{|u| \leq x/\lambda} \exp\left\{-\frac{B}{\lambda^2}(1 - \cos u)\right\} du$$

$$\leq \frac{2\lambda}{x}\left(\left[\frac{x}{2\pi\lambda}\right] + 1\right)\int_{|u| \leq \pi} \exp\left\{-\frac{B}{\lambda^2}(1 - \cos u)\right\} du$$

$$\leq \frac{3}{\pi}\int_{|u| \leq \pi} \exp\left\{-\frac{B}{\lambda^2}(1 - \cos u)\right\} du,$$

because the function under the integral sign has the period 2π. Here $[y]$ denotes the largest integer not exceeding y. If $|u| \leq 1$, we have $1 - \cos u \geq u^2/3$. Therefore,

$$\int_{|u| \leq 1} \exp\left\{-\frac{B}{\lambda^2}(1 - \cos u)\right\} du \leq \int_{|u| \leq 1} \exp\left\{-\frac{B}{3\lambda^2} u^2\right\} du \leq A\lambda B^{-1/2}.$$

From the inequality $1 - \cos u > 1/3$ for $1 \leq |u| \leq \pi$, we infer that

$$\int_{1 \leq |u| \leq \pi} \exp\left\{-\frac{B}{\lambda^2}(1 - \cos u)\right\} du \leq 2\pi\exp\left\{-\frac{B}{3\lambda^2}\right\} \leq A\lambda B^{-1/2}.$$

These estimates imply the desired result (2.50). □

To complete the proof of Theorem 2.14 we apply Lemma 2.5 to the integral on the right-hand side of (2.44). We obtain

$$Q(S_n; \lambda) \leq A\lambda\left\{\sum_{k=1}^{n}\left(\int_{|x| < \lambda_k} x^2\, dV_k^s(x) + \lambda_k^2\int_{|x| \geq \lambda_k} dV_k^s(x)\right)\right\}^{-1/2}. \quad \square$$

Note that for an arbitrary random variable X and every $\lambda > 0$ the following inequalities are true:

$$\lambda^2 D(X;\lambda) = \int_{|x|<\lambda} x^2\, dF(x) + \lambda^2 \int_{|x|\geq\lambda} dF(x)$$

$$\geq \frac{\lambda^2}{4}\int_{\lambda/2\leq|x|<\lambda} dF(x) + \lambda^2 \int_{|x|\geq\lambda} dF(x) \geq \frac{\lambda^2}{4} P\!\left(|X|\geq\frac{\lambda}{2}\right). \quad (2.51)$$

By Lemma 1.11 (Chapter 1) we have

$$P\!\left(|X^s|\geq\frac{\lambda}{2}\right) \geq 1 - Q(X^s;\lambda) \geq 1 - Q(X;\lambda) \quad (2.52)$$

It follows from (2.51) and (2.52) that

$$D(X^s;\lambda) \geq \tfrac{1}{4}(1 - Q(X;\lambda)). \quad (2.53)$$

A consequence of Theorem 2.14 and the inequalities (2.51) and (2.53) is the following.

Theorem 2.15 Let $X_1,\ldots,X_n$ be independent random variables, $S_n = \sum_{k=1}^{n} X_k$. Let $\lambda_1,\ldots,\lambda_n$ be positive numbers such that $\lambda_k \leq \lambda$ $(k=1,\ldots,n)$. Then

$$Q(S_n;\lambda) \leq A\lambda\left\{\sum_{k=1}^{n} \lambda_k^2 P\!\left(|X_k^s|\geq\frac{\lambda_k}{2}\right)\right\}^{-1/2}, \quad (2.54)$$

$$Q(S_n;\lambda) \leq A\lambda\left\{\sum_{k=1}^{n} \lambda_k^2 (1 - Q(X_k;\lambda_k))\right\}^{-1/2}, \quad (2.55)$$

We recall that the symbols $A, A_1, A_2,\ldots$ denote absolute positive constants.

Inequality (2.55) is called the Kolmogorov–Rogozin inequality.

We shall formulate some other consequences of Theorem 2.14. If we put $\lambda_k = \lambda$ for all k in (2.43), we obtain

$$Q(S_n;\lambda) \leq A\left(\sum_{k=1}^{n} D(X_k^s;\lambda)\right)^{-1/2}. \quad (2.56)$$

If independent random variables $X_1,\ldots,X_n$ have a common non-degenerate distribution, then $D(X_1^s;\lambda) > 0$ for every $\lambda \geq 0$ and

$$Q(S_n;\lambda) \leq An^{-1/2}(D(X_1^s;\lambda))^{-1/2}. \quad (2.57)$$

The inequalities (2.57) and (2.53) imply that for these random variables we have

$$Q(S_n;\lambda) \leq An^{-1/2}(1 - Q(X_1;\lambda))^{-1/2} \quad (2.58)$$

for every $\lambda \geqslant 0$. In the general case of non-identical distributions it follows from (2.56) and (2.42) that

$$Q(S_n; \lambda) \leqslant A \left\{ \sup_{u \geqslant \lambda} u^{-2} \sum_{k=1}^{n} \int_{|x| < u} x^2 \, dV_k^s(x) \right\}^{-1/2}. \qquad (2.59)$$

Using the same method as was used for the proof of (2.44), we can obtain from Lemma 1.16 (Chapter 1) the inequality

$$Q(S_n; \lambda) \leqslant \frac{A}{a} \int_{-a}^{a} \exp\left\{ -\frac{1}{2} \sum_{k=1}^{n} \int_{-\infty}^{\infty} (1 - \cos tx) \, dV_k^s(x) \right\} dt,$$

if $0 < a\lambda \leqslant 1$. Hence, and from the obvious estimates

$$\int_{-\infty}^{\infty} (1 - \cos tx) \, d\tilde{V}_k(x) \geqslant \int_{|x| \leqslant 1/|t|} (1 - \cos tx) \, dV_k^s(x) \geqslant \tfrac{11}{24} t^2 \varphi_k(t),$$

where

$$\varphi_k(t) = \int_{|x| \leqslant 1/|t|} x^2 \, dV_k^s(x),$$

we conclude that

$$Q(S_n; \lambda) \leqslant \frac{A}{a} \int_{-a}^{a} \exp\left\{ -\tfrac{11}{48} t^2 \sum_{k=1}^{n} \varphi_k(t) \right\} dt \qquad (2.60)$$

if $0 < a\lambda \leqslant 1$.

The Kolmogorov–Rogozin inequality (2.55) allows us to estimate the concentration function of a sum of independent random variables by an upper bound depending on the concentration functions of the summands. It is possible to improve this inequality.

Theorem 2.16 *Let $X_1, \ldots, X_n$ be independent random variables. Let $0 < \lambda_k \leqslant \lambda$ ($k = 1, \ldots, n$). Then*

$$Q(S_n; \lambda) \leqslant A\lambda \left\{ \sum_{k=1}^{n} \lambda_k^2 D(X_k^s; \lambda_k) Q^{-2}(X_k; \lambda_k) \right\}^{-1/2}. \qquad (2.61)$$

To prove this theorem, we need two lemmas.

Lemma 2.6 *If independent random variables $X_1, \ldots, X_n$ have a common non-degenerate distribution, then*

$$Q(S_n; \lambda) \leqslant An^{-1/2} Q(X_1; \lambda)(D(X_1^s; \lambda))^{-1/2} \qquad (2.62)$$

for every $\lambda > 0$.

Proof In the case when $Q(X_1; \lambda) \geq \frac{1}{2}$, (2.62) follows from (2.57). Consider the case when $Q(X_1; \lambda) < \frac{1}{2}$. Moreover, we shall assume that $n \geq 2$, since for $n = 1$ the inequality (2.62) follows immediately from (2.41).

Let a be an arbitrary real number, $I = [a, a + \lambda] = \{x: a \leq x \leq a + \lambda\}$, and let the segment $J = [b, b + l]$, where $l \geq \lambda$ be such that $P(X_1 \in J) \geq \frac{1}{2}$ but $P(X_1 \in J_1) < \frac{1}{2}$ for every segment J_1 with a length smaller than l. Consider the event $B = [X_1 \in J]$ and its complement B^c. We have $P(B^c) \leq \frac{1}{2}$ and

$$P(S_n \in I) = \int_{-\infty}^{\infty} P(X_1 + x \in I) \, dF_{n-1}(x)$$

where $F_{n-1}(x) = P(S_{n-1} < x)$. Furthermore,

$$P(S_n \in I) = \int_{-\infty}^{\infty} P([X_1 + x \in I] \cap B) \, dF_{n-1}(x)$$

$$+ \int_{-\infty}^{\infty} P([X_1 + x \in I] \cap B^c) \, dF_{n-1}(x)$$

$$\leq \int_{-\infty}^{\infty} P([X_1 + x \in I] \cap B) \, dF_{n-1}(x)$$

$$+ \int_{-\infty}^{\infty} P([X_1 + x \in I] \mid B^c) P(B^c) \, dF_{n-1}(x).$$

If $x \in I - J = \{y - z: y \in I, z \in J\}$, then

$$P([X_1 + x \in I] \cap B) = P([X_1 + x \in I] \cap [X_1 \in J]) \leq Q(X_i; \lambda)$$

If, however, $x \notin I - J$, then $P([X_1 + x] \cap B) = 0$. Therefore,

$$P(S_n \in I) \leq Q(X_1; \lambda) P(S_{n-1} \in I - J) + \frac{1}{2} \int_{-\infty}^{\infty} P([X_1 + x \in I] \mid B^c) \, dF_{n-1}(x),$$

since $P(B^c) \leq \frac{1}{2}$. We have

$$P(S_{n-1} \in I - J) \leq Q(S_{n-1}; 2l)$$

$$\leq 4Q(S_{n-1}; l/2) \leq A((n-1)(1 - Q(X_1; l/2)))^{-1/2};$$

the latter inequality follows from (2.58). Since $Q(X_1; l/2) < \frac{1}{2}$, we find that $P(S_{n-1} \in I - J) \leq An^{-1/2}$ and

$$P(S_n \in I) \leq An^{-1/2} Q(X_1; \lambda) + \frac{1}{2} \int_{-\infty}^{\infty} P(S_{n-1} + x \in I) \, dP(X_1 < x \mid B^c)$$

$$\leq An^{-1/2} Q(X_1; \lambda) + \frac{1}{2} Q(S_{n-1}; \lambda).$$

In view of the arbitrariness of a in the definition of I, this implies that

$$Q(S_n; \lambda) \leq An^{-1/2}Q(X_1; \lambda) + \tfrac{1}{2}Q(S_{n-1}; \lambda)$$

$$\leq AQ(X_1; \lambda)\left\{n^{-1/2} + \sum_{k=0}^{n-1} 2^{-k}(n-k)^{-1/2}\right\}.$$

Clearly,

$$\sum_{k=0}^{n-1} 2^{-k}(n-k)^{-1/2} = \sum_{k=0}^{[n/2]} 2^{-k}(n-k)^{-1/2} + \sum_{k=[n/2]+1}^{n-1} 2^k(n-k)^{-1/2}$$

$$\leq n^{-1/2}2^{1/2} \sum_{k=0}^{[n/2]} 2^{-k} + \sum_{k=[n/2]+1}^{n-1} 2^{-k} \leq An^{-1/2}.$$

Hence $Q(S_n; \lambda) \leq An^{-1/2}Q(X_1; \lambda)$. This inequality and (2.41) imply (2.62). □

Lemma 2.7 *Let X be a random variable with the characteristic function $f(t)$ and the concentration function $Q(X; \lambda)$. Then*

$$\int_{-a}^{a} |f(t)|^p \, dt \leq \frac{A}{\lambda} Q(X; \lambda)(pD(X^s; \lambda))^{-1/2} \quad (2.63)$$

for every $a > 0$, $p \geq 2$ and $0 < \lambda \leq 1/a$.

Proof For every positive integer n we have

$$\int_{-a}^{a} |f(t)|^{2n} \, dt \leq \int_{|t| \leq 1/\lambda} |f(t)|^{2n} \, dt = \int_{|t| \leq 1/\lambda} |f^n(t)|^2 \, dt \leq \frac{A}{\lambda} Q(S_n; \lambda)$$

by Lemma 1.17 (Chapter 1). Here $S_n = \sum_{k=1}^{n} X_k$ and the random variables $X_1, \ldots, X_n$ are independent and have the same distribution as X. By Lemma 2.6,

$$\int_{-a}^{a} |f(t)|^{2n} \, dt \leq \frac{A}{\lambda} n^{-1/2} Q(X; \lambda)(D(X^s; \lambda))^{-1/2}.$$

Thus the inequality (2.63) is proved for any even p. If, however, $2n < p < 2n+2$ for some positive integer n, we note that $|f(t)|^p \leq |f(t)|^{2n}$, since $|f(t)| \leq 1$ for all t, and, therefore,

$$\int_{-a}^{a} |f(t)|^p \, dt \leq \int_{-a}^{a} |f(t)|^{2n} \, dt \leq \frac{A}{\lambda} Q(X; \lambda)(pD(X^s; \lambda))^{-1/2}. \quad \square$$

To complete the proof of Theorem 2.16 we put

$$p_k = Q^2(X_k; \lambda_k)(\lambda_k^2 D(X_k^s; \lambda_k))^{-1} \sum_{j=1}^{n} \frac{\lambda_j^2 D(X_j^s; \lambda_j)}{Q^2(X_j; \lambda_j)} \quad (2.64)$$

for $k = 1, \ldots, n$. Obviously,

$$\sum_{k=1}^{n} \frac{1}{p_k} = 1. \tag{2.65}$$

We shall first prove Theorem 2.16 under the additional assumption $p_k \geq 2$ for $k = 1, \ldots, n$. Let $f_k(t)$ be the c.f. of X_k. By Lemma 1.16 (Chapter 1), we have

$$Q(S_n; \lambda) \leq A\lambda \int_{|t| \leq 1/\lambda} \prod_{k=1}^{n} |f_k(t)| \, dt$$

for every $\lambda > 0$. By Hölder's inequality,

$$\int_{|t| \leq 1/\lambda} \prod_{k=1}^{n} |f_k(t)| \, dt = \int_{|t| \leq 1/\lambda} \prod_{k=1}^{n} (|f_k(t)|^{p_k})^{1/p_k} \, dt$$

$$\leq \prod_{k=1}^{n} \left(\int_{|t| \leq 1/\lambda} |f_k(t)|^{p_k} \, dt \right)^{1/p_k}$$

$$\leq \prod_{k=1}^{n} \left(\int_{|t| \leq 1/\lambda_k} |f_k(t)|^{p_k} \, dt \right)^{1/p_k},$$

since $\lambda_k \leq \lambda$ for all k. Lemma 2.7 implies that

$$\prod_{k=1}^{n} \left(\int_{|t| \leq 1/\lambda_k} |f_k(t)|^{p_k} \, dt \right)^{1/p_k}$$

$$\leq A \prod_{k=1}^{n} \lambda_k^{-1/p_k} Q^{1/p_k}(X_k; \lambda_k)(p_k D(X_k^s; \lambda_k))^{-1/(2p_k)}.$$

Taking into account (2.64), we obtain

$$Q(S_n; \lambda) \leq A\lambda \prod_{k=1}^{n} \left(\sum_{j=1}^{n} \frac{\lambda_j^2 D(X_j^s; \lambda_j)}{Q^2(X_j; \lambda_j)} \right)^{-1/(2p_k)}$$

$$= A\lambda \prod_{k=1}^{n} \left[\left(\sum_{j=1}^{n} \frac{\lambda_j^2 D(X_j^s; \lambda_j)}{Q^2(X_j; \lambda_j)} \right)^{-1/2} \right]^{1/p_k}$$

$$= A\lambda \left(\sum_{j=1}^{n} \frac{\lambda_j^2 D(X_j^s; \lambda_j)}{Q^2(X_j; \lambda_j)} \right)^{-1/2}.$$

by (2.65). Thus the inequality (2.61) is proved under the additional assumption $p_k \geq 2$ for every k.

It remains to consider the case when $p_k < 2$ for some k from the set $\{1, \ldots, n\}$. By (2.65), the inequality $p_k < 2$ cannot be fulfilled for two or more indices k.

Lemma 1.12 (Chapter 1) implies that $Q(X; L) \leq (2L/\lambda)Q(X; \lambda)$ for an arbitrary random variable X and $0 < \lambda \leq L$. Therefore,

$$\frac{1}{\lambda}Q(S_n; \lambda) \leq \frac{1}{\lambda}Q(X_k; \lambda) \leq \frac{2}{\lambda_k}Q(X_k; \lambda_k)$$

if $\lambda_k \leq \lambda$. It follows from the relations (2.41), (2.64) and the condition $p_k < 2$ that

$$Q(S_n; \lambda) \leq 2\lambda Q(X_k; \lambda_k)(\lambda_k^2 D(X_k^s; \lambda_k))^{-1/2}$$

$$= 2\lambda p_k^{1/2} \left(\sum_{j=1}^n \frac{\lambda_j^2 D(X_j^s; \lambda_j)}{Q^2(X_j; \lambda_j)} \right)^{-1/2}$$

$$< 2^{3/2}\lambda \left(\sum_{j=1}^n \frac{\lambda_j^2 D(X_j^s; \lambda_j)}{Q^2(X_j; \lambda_j)} \right)^{-1/2}. \quad \square$$

Theorem 2.16 has much simpler corollaries.

Theorem 2.17 *If $X_1, \ldots, X_n$ are independent random variables and $0 < \lambda_k \leq \lambda$ ($k = 1, \ldots, n$), then*

$$Q(S_n; \lambda) \leq A\lambda \left\{ \sum_{k=1}^n \lambda_k^2 (1 - Q(X_k; \lambda_k)) Q^{-2}(X_k; \lambda_k) \right\}^{-1/2}. \quad (2.66)$$

Inequality (2.66) follows from (2.61) and (2.53).

Theorem 2.18 *Let conditions of Theorem 2.17 be satisfied. Then*

$$Q(S_n; \lambda) \leq A\lambda \max_{1 \leq k \leq n} Q(X_k; \lambda_k) \left\{ \sum_{j=1}^n \lambda_j^2 (1 - Q(X_j; \lambda_j)) \right\}^{-1/2}. \quad (2.67)$$

Inequality (2.67) is called the Kesten inequality. It follows immediately from (2.66). Note that the estimates (2.66) and (2.67) are essential strengthenings of the Kolmogorov–Rogozin inequality (2.55). It is possible to show that the assertions of Theorems 2.16–2.18 hold true if the condition $0 < \lambda_k \leq \lambda$ ($k = 1, \ldots, n$) of these theorems is replaced by the condition $0 < \lambda_k \leq 2\lambda$ ($k = 1, \ldots, n$).

In the case when $\lambda_k = \lambda$, the above-mentioned results imply that

$$Q(S_n; \lambda) \leq A \left\{ \sum_{k=1}^n (1 - Q(X_k; \lambda)) Q^{-2}(X_k; \lambda) \right\}^{-1/2} \quad (2.68)$$

and

$$Q(S_n; \lambda) \leq A \max_{1 \leq k \leq n} Q(X_k; \lambda) \left\{ \sum_{j=1}^n (1 - Q(X_j; \lambda)) \right\}^{-1/2}. \quad (2.69)$$

If $X_1, \ldots, X_n$ are independent random variables with a common nondegenerate distribution, then we have the following simple consequences of

(2.61) and (2.68):

$$Q(S_n; \lambda) \leq An^{-1/2}Q(X_1; \lambda)(D(X_1^s; \lambda))^{-1/2} \qquad (2.70)$$

and

$$Q(S_n; \lambda) \leq An^{-1/2}Q(X_1; \lambda)(1 - Q(X_1; \lambda))^{-1/2} \qquad (2.71)$$

for every $\lambda > 0$. The latter inequality is a sharpening of (2.58).

So far we have dealt with some upper estimates for the concentration function of a sum of n independent random variables. These estimates are valid for an arbitrary n and for arbitrary distributions of the summands. By simpler means it is possible to obtain some asymptotic estimates for $Q(S_n; \lambda)$ under additional assumptions about the distributions of the summands.

Theorem 2.19 *Let $\{X_n; n = 1, 2, \ldots\}$ be a sequence of independent random variables having a common distribution. If $EX_1^2 = \infty$, then*

$$Q(S_n; \lambda) = o(n^{-1/2}) \qquad (2.72)$$

for every fixed $\lambda \geq 0$.

Proof Let $V^s(x)$ denote the d.f. of the symmetrized random variable X_1^s. The function

$$\varphi(t) = \int_{|x| \leq 1/|t|} x^2 \, dV_1^s(x)$$

does not increase on the half-line $t > 0$ and is an even function. It is easy to see that $EX_1^2 < \infty$ if and only if $\lim_{t \to 0} \varphi(t) < \infty$. By the hypothesis, $EX_1^2 = \infty$, therefore, $\lim_{t \to 0} \varphi(t) = \infty$.

Let λ be an arbitrary fixed number. Let a be such that $0 < a\lambda \leq 1$ and $\varphi(a) > 0$. (The existence of such an a is guaranteed by non-degeneracy of the random variable X_1 as a consequence of the hypothesis $EX_1^2 = \infty$.) Inequality (2.60) implies that for the special case of identical distributions the following estimates are true for every positive $\varepsilon < a$:

$$Q(S_n; \lambda) \leq \frac{A}{a} \int_{-\varepsilon}^{\varepsilon} \exp\{-\tfrac{11}{48}t^2 n\varphi(\varepsilon)\} \, dt + \frac{2A}{a} \int_{\varepsilon}^{a} \exp\{-\tfrac{11}{48}t^2 n\varphi(a)\} \, dt$$

$$\leq C_1(n\varphi(\varepsilon))^{-1/2} + C_2 n^{-1/2} \int_{\varepsilon\sqrt{n\varphi(a)}}^{\infty} \exp\{-\tfrac{11}{48}u^2\} \, du,$$

where C_1 and C_2 are constants not depending on n and ε. We now put $\varepsilon = n^{-1/4}$. Since $\lim \varphi(n^{-1/4}) = \infty$, the relation (2.72) is proved for every fixed $\lambda > 0$. The concentration function does not decrease, and, therefore, this relation holds for $\lambda = 0$. □

Let us turn to the general case of not necessarily identical distributions.

Consider a sequence of independent random variables $\{X_n; n = 1, 2, \ldots\}$ with the c.f. $\{v_n(t)\}$. We shall say that the sequence $\{X_n\}$ satisfies the condition (Δ), if there exist a constant $\delta > 0$ and a function $\psi(n)$ such that $\psi(n) \to \infty$ as $n \to \infty$ and

$$\int_{|t| \leq \delta} \prod_{k=1}^{n} |v_k(t)| \, dt \leq \frac{K}{\psi(n)}$$

for $n \geq N$ and some constants K and N.

The condition (Δ) is satisfied with $\psi(n) = n^{1/2}$ for an arbitrary sequence of independent random variables having a common non-degenerate distribution. In fact, if $v(t)$ is the c.f. of a non-degenerate distribution, then by Lemma 1.5 (Chapter 1) there exist positive constants δ and ε such that

$$|v(t)| \leq e^{-\varepsilon t^2} \qquad \text{for } |t| \leq \delta. \tag{2.73}$$

Therefore,

$$\int_{|t| \leq \delta} |v(t)|^n \, dt \leq \int_{|t| \leq \delta} e^{-n\varepsilon t^2} \, dt \leq (n\varepsilon)^{-1/2} \int_{-\infty}^{\infty} e^{-u^2} \, du = \left(\frac{\pi}{n\varepsilon}\right)^{1/2}$$

for all n.

We now state another more general sufficient condition for the condition (Δ). Let the sequence $\{v_n(t)\}$ contain a sub-sequence $\{v_{n_m}(t); m = 1, 2, \ldots\}$, such that
(A) there exist a c.f. $v(t)$ of a non-degenerate distribution and a constant $\gamma > 0$, such that $|v_{n_m}(t)| \leq |v(t)|$ for $|t| \leq \gamma$ ($m = 1, 2, \ldots$);
(B) if $\mu(n)$ is the number of terms of the sub-sequence $\{v_{n_m}(t)\}$ in the set $v_1(t), v_2(t), \ldots, v_n(t)$, then $\lim \mu(n) = \infty$. Then the condition (Δ) will be satisfied for $\psi(n) = (\mu(n))^{1/2}$.

For the proof of this assertion we note that for the c.f. $v(t)$ there exist, by condition (A), positive constants σ and ε for which (2.73) holds. Putting $\delta_0 = \min(\delta, \gamma)$ and using (A) and (B), we obtain

$$\int_{|t| \leq \delta_0} \prod_{k=1}^{n} |v_k(t)| \, dt \leq \int_{|t| \leq \delta_0} |v(t)|^{\mu(n)} \, dt \leq \int_{|t| \leq \delta_0} e^{-\mu(n)\varepsilon t^2} \, dt \leq \pi^{1/2} (\varepsilon \mu(n))^{-1/2}.$$

If $\liminf \mu(n)/n > 0$, the condition (Δ) is satisfied for $\psi(n) = n^{1/2}$.

Theorem 2.20 *Let $\{X_n\}$ be a sequence of independent random variables satisfying condition (Δ) and let $S_n = \sum_{k=1}^{n} X_k$. Then*

$$Q(S_n; \lambda) \leq C \frac{\lambda + 1}{\psi(n)} \tag{2.74}$$

for all $\lambda \geq 0$ and all sufficiently large n. Here C is a constant, independent of λ and n.

Proof The sum S_n has the c.f. $\prod_{k=1}^{n} v_k(t)$, where $v_k(t)$ is the c.f. of X_k. By Lemma 1.16 (Chapter 1), we have

$$Q(S_n; \lambda) \leqslant (\tfrac{96}{95})^2 \max(\lambda, 1/\delta) \int_{|t| \leqslant \delta} \prod_{k=1}^{n} |v_k(t)| \, dt. \tag{2.75}$$

Taking into account the condition (Δ), we find that

$$Q(S_n; \lambda) \leqslant (\tfrac{96}{95})^2 \max(\lambda, 1/\delta) \frac{K}{\psi(n)} \tag{2.76}$$

for $n \geqslant N$ and every $\lambda \geqslant 0$. Obviously, (2.74) follows from (2.76). □

Consider the case when λ depends on n. Theorem 2.20 implies that if condition (Δ) is satisfied and $\lambda_n \geqslant 0$, then

$$Q(S_n; \lambda_n) = O((\lambda_n + 1)/\psi(n))$$

as $n \to \infty$. In particular, if $\lambda_n = o(\psi(n))$, then $Q(S_n; \lambda_n) \to 0$; if $\lambda_n = O(1)$, then $Q(S_n; \lambda_n) = O(1/\psi(n))$. For the special case $\psi(n) = n^{1/2}$ we can formulate the following theorem.

Theorem 2.21 *Let $\{X_n\}$ be a sequence of independent random variables satisfying the condition (Δ) with $\psi(n) = n^{1/2}$. Then the following assertions are true:*

(I) *if $\lambda_n = O(n^p)$, $0 < p < \tfrac{1}{2}$, then $Q(S_n; \lambda_n) = O(n^{p-1/2})$;*
(II) *if $\lambda_n = o(n^{1/2})$, then $Q(S_n; \lambda_n) \to 0$;*
(III) *if $\lambda_n = O(1)$, then $Q(S_n; \lambda_n) = O(n^{-1/2})$;*
(IV) $\sup_x P(S_n = x) = O(n^{-1/2})$.

As we have already remarked, the conditions of Theorem 2.21 are fulfilled in the case when the random variables $X_1, X_2, \ldots$ have a common non-degenerate distribution. For this particular case the relations (2.75) and (2.73) imply the following proposition.

Theorem 2.22 *If $\{X_n\}$ is a sequence of independent random variables with a common non-degenerate distribution, then*

$$Q(S_n; \lambda) \leqslant C(\lambda + 1)n^{-1/2} \tag{2.77}$$

for all $\lambda \geqslant 0$ and all n. Here C is a constant not depending on λ and n.

In the inequality (2.77) we may put

$$C = (\tfrac{96}{95})^2 \max(1, 1/\delta)(\pi/\varepsilon)^{1/2},$$

where δ and ε are positive constants as in condition (2.73), which is satisfied by the c.f. $v(t)$ of the random variable X_1.

2.5 Bibliographical notes

Theorems 2.1, 2.3 and 2.4 were obtained by Petrov [340]. Theorem 2.2 is due to Lévy, Theorem 2.5 to Hàjek and Rényi [154], Theorems 2.6 and 2.7 to Petrov [332], and Theorem 2.8 to Bernstein [30].

Theorems 2.9 and 2.12 were obtained by Rosenthal [393, 394] and Lemma 2.3 by Fuk and Nagaev [142]. The proof of Theorem 2.9 given in section 2.3 is due to Nagaev and Pinelis [312]. Theorem 2.10 is an immediate consequence of an inequality due to Marcinkiewicz and Zygmund (see, for instance, Subsection 2.6.18 below). The least upper bounds for the moments ES_n^{2m}, where S_n is a sum of n independent random variables and m is a positive integer, are found by Pinelis and Utev [358], Utev [459], and Bestsennaya and Utev [33].

Theorems 2.11 and 2.13 were obtained by Petrov [352].

The first results concerning estimates for the concentration functions of sums of independent random variables were obtained by Lévy [252] and Doeblin [102, 103]. Their investigations were continued by Kolmogorov [233–235], Rogozin [388, 389], Esseen [122], Kesten [216], Miroshnikov and Rogozin [305], and others.

Theorems 2.14 and 2.19 and the inequalities (2.54), (2.59), and (2.60) are due to Esseen [122]. The inequality (2.55) is a generalization of one obtained by Kolmogorov [234], and is due to Rogozin [389]. Theorems 2.16 and 2.17 and Lemmas 2.6 and 2.7 are due to Miroshnikov and Rogozin [305, 306]. A result that is close to Theorem 2.17 was obtained by Postnikova and Yudin [362]. Theorem 2.18 is due to Kesten [216]. Theorems 2.20–2.22, due to Petrov [335], are generalizations of some results of Chung and Erdös [80], Rosén [392], and Heyde [181]. Using the examples given in Rosén [392], it is easy to show that the estimates in the assertions of Theorem 2.21 are optimal.

2.6 Addenda

In Subsections 2.6.1–2.6.14 and 2.6.18–2.6.37 we consider independent random variables $X_1, \ldots, X_n$ and their sum $S_n = \sum_{k=1}^n X_k$.

2.6.1 Let $|X_k| \leqslant c$, $EX_k = 0$ ($k = 1, \ldots, n$). We put $B_n = \sum_{k=1}^n \operatorname{Var} X_k$. Then

$$P(S_n \geqslant x) \leqslant \exp\left\{-\frac{x}{2c} \operatorname{arsinh} \frac{cx}{2B_n}\right\}$$

for every real x (Prohorov [368]).

2.6.2 Let $a_k \leqslant X_k \leqslant b_k$ ($k = 1, \ldots, n$). Then

$$P(S_n - ES_n \geqslant nx) \leqslant \exp\left\{-\frac{2n^2x^2}{\sum_{k=1}^n (b_k - a_k)^2}\right\}$$

for every $x > 0$ (Hoeffding [195]).

2.6.3 Let $X_k \leqslant b$ and $EX_k = 0$ ($k = 1, \ldots, n$). Then

$$P(S_n \geqslant nx) \leqslant \left\{\left(1 + \frac{bx}{\sigma^2}\right)^{-(1+bx/\sigma^2)/(1+b^2/\sigma^2)} \left(1 - \frac{x}{b}\right)^{-(1-x/b)/(1+\sigma^2/b^2)}\right\}^n$$

$$\leqslant \exp\left\{-\frac{nx}{b}\left[\left(1 + \frac{\sigma^2}{bx}\right)\log\left(1 + \frac{bx}{\sigma^2}\right) - 1\right]\right\}$$

in the interval $0 < x < b$. Here $\sigma^2 = (1/n)\sum_{k=1}^n \text{Var } X_k$ (Hoeffding [195] and Bennett [20]).

2.6.4 Let $p \geqslant 2$, $\beta = p/(p+2)$, $\alpha = 1 - \beta$. Let $Y = \{y_1, \ldots, y_n\}$ be an arbitrary set of positive numbers, and let $y \geqslant \max(y_1, \ldots, y_n)$. We put $V_k(x) = P(X_k < x)$,

$$A(p; 0, Y) = \sum_{k=1}^n \int_0^{y_k} x^p \, dV_k(x), \quad B(-\infty; Y) = \sum_{k=1}^n \int_{-\infty}^{y_k} x^2 \, dV_k(x).$$

If $EX_k = 0$ ($k = 1, \ldots, n$), then

$$P(S_n \geqslant x) \leqslant \sum_{k=1}^n P(X_k \geqslant y_k)$$

$$+ \exp\left\{\max\left[-\beta\frac{x}{y}\log\left(1 + \frac{\beta x y^{p-1}}{A(p; 0, Y)}\right), -\frac{\alpha^2 x^2}{2e^p B(-\infty; Y)}\right]\right\}$$

$$\leqslant \sum_{k=1}^n P(X_k \geqslant y_k) + (\beta x y^{p-1}/A(p; 0, Y) + 1)^{-\beta x/y}$$

$$+ \exp\left\{-\frac{\alpha^2 x^2}{2 e^p B(-\infty; Y)}\right\} \tag{2.78}$$

for every $x > 0$ (Fuk and Nagaev [142]).

2.6.5 Let $EX_k = 0$ and $E|X_k|^p < \infty$ ($k = 1, \ldots, n$) for some $p \geqslant 2$. We put $B_n = \sum_{k=1}^n \text{Var } X_k$ and $M_{p,n} = \sum_{k=1}^n E|X_k|^p$. Then

$$P(S_n \geqslant x) \leqslant \left(1 + \frac{2}{p}\right)^p M_{p,n} x^{-p} + \exp\{-2(p+2)^{-2} e^{-p} x^2 B_n^{-1}\} \tag{2.79}$$

for every $x > 0$ (Fuk and Nagaev [142]).

2.6.6 The inequalities (2.78) and (2.79) remain true for every $x > 0$ if we replace $P(S_n \geq x)$ by $P(\max_{1 \leq k \leq n} S_k \geq x)$. The same replacement is possible in Lemma 2.3, if we replace $\mu(-\infty, y)$ by zero in the assertion of this lemma (Borovkov [51]).

2.6.7 Let $EX_k = 0$, $E|X_k|^p < \infty$ ($k = 1, \ldots, n$) for some $p \geq 2$. We put $V_k(x) = P(X_k < x)$, $Y_k = V_k^{-1}(Z_k) - \Phi_k^{-1}(Z_k)$, where Z_k are random variables uniformly distributed in the interval $(0, 1)$, $\Phi_k(x)$ is the normal d.f. with parameters $(0, (\operatorname{Var} X_k)^{1/2})$, and $g^{-1}(x)$ denotes the inverse of g. Moreover, we put

$$l_k(p) = E|Y_k|^p, \quad L_n(p) = \sum_{k=1}^n l_k(p).$$

Then

$$P\left(\max_{1 \leq k \leq n} S_k \geq xB_n^{1/2}\right)$$

$$\leq 2(1 - \Phi((1-\alpha)x)) + \exp\left\{-c_1 \frac{\alpha^2 x^2 B_n}{L_n(2)}\right\} + c_2 \frac{L_n(p)}{\alpha^p x^p B_n^{p/2}}$$

for every positive $\alpha < 1$ and every $x > 0$, where $\Phi(x)$ is the standard normal d.f., c_1 and c_2 are positive constants depending only on p, and $B_n = \sum_{k=1}^n \operatorname{Var} X_k$ (Borovkov [51]; a similar inequality was obtained by Ebralidze [106] for $P(S_n \geq x)$ instead of $P(\max_{1 \leq k \leq n} S_k \geq x)$). It was shown in [106] that $l_k(p) \leq v_k(p)$, where

$$v_k(p) = p \int_{-\infty}^{\infty} |x|^{p-1} |V_k(x) - \Phi_k(x)| \, dx.$$

2.6.8 Let $EX_k = 0$ and $E|X_k|^3 < \infty$ ($k = 1, \ldots, n$). Then

$$P(S_n \geq xB_n^{1/2}) \leq 1 - \Phi(x/2) + e^{-cx^2} + c_0 B_n^{-3/2} \sum_{k=1}^n v_k(3)$$

for every $x > 0$, where $c \leq 0.0003$ and $c_0 \leq 150$; B_n and $v_k(p)$ are the same as in Subsection 2.6.7 (Ebralidze [106]).

2.6.9 Let $X_1, \ldots, X_n$ be independent symmetric random variables and let $Y_1, \ldots, Y_n$ be independent random variables with the infinitely divisible characteristic functions

$$g_k(t) = \exp\left\{\int_{-\infty}^{\infty} (e^{itu} - 1) \, dV_k(u)\right\} \quad (k = 1, \ldots, n),$$

where $V_k(u) = P(X_k < u)$. Then

$$P\left(\sum_{k=1}^{n} X_k \geq x\right) \leq 8P\left(\sum_{k=1}^{n} Y_k \geq x/2\right)$$

for every $x > 0$. If, moreover, the random variables $X_1, \ldots, X_n$ are identically distributed, then

$$P\left(\sum_{k=1}^{n} X_k \geq x\right) \geq \tfrac{1}{8}P\left(\sum_{k=1}^{n} Y_k \geq 2x\right) - \tfrac{1}{2}e^{-n\Gamma}$$

for every $x > 0$, where $\Gamma = \log 2 - \tfrac{1}{2} = 0.1931\ldots$ (Prohorov [367]).

2.6.10 Let $Z_n = \max_{1 \leq k \leq n} |X_k - mX_k|$, where mX is a median of the random variable X. Then

$$P(|S_n - mS_n| \geq x/4) \geq \tfrac{1}{8}P(Z_n > x)$$

for every $x > 0$ (Rogozin [390]).

2.6.11 Let $EX_k \geq 0$, $Ee^{tX_k} \leq e^{g_k t^2/2}$ ($k = 1, \ldots, n$) for $0 \leq t \leq T$ and some positive constants $g_1, \ldots, g_n$ and T. Then

$$P\left(\max_{1 \leq k \leq n} S_k \geq x\right) \leq \begin{cases} e^{-x^2/(2G_n)} & \text{if } 0 \leq x \leq G_n T, \\ e^{-Tx/2} & \text{if } x \geq G_n T. \end{cases}$$

Here $G_n = \sum_{k=1}^{n} g_k$ (Petrov [332]). In [332] this result was proved under the stronger condition that $EX_k = 0$ for all k; however, the proof in [332] makes it possible to replace the latter condition by the condition $EX_k \geq 0$.

2.6.12 Let $X_1, \ldots, X_n$ be independent symmetric random variables. Let $0 \leq c_n \leq c_{n-1} \leq \cdots \leq c_1$, and let $g(x)$ be a non-negative function convex in $\mathbb{R}$. Then

$$P\left(\max_{1 \leq k \leq n} c_k g(S_k) \geq x\right) \leq 2P(G_n \geq x)$$

for every $x > 0$, where

$$G_n = \sum_{k=1}^{n-1} (c_k - c_{k+1})g(S_k) + c_n g(S_n)$$

(Bickel [35]).

2.6.13 Let $X_1, \ldots, X_n$ be independent symmetric random variables. Then

$$P\left(\max_{1 \leq k \leq n} S_k \geq x\right) \geq 2P(S_n \geq x + 2\varepsilon) - 2\sum_{k=1}^{n} P(X_k \geq \varepsilon)$$

for every $x > 0$ and $\varepsilon > 0$ (Doob [104], Chapter 3).

2.6.14 Let $X_1, \ldots, X_n$ be independent random variables satisfying the condition $E|X_k|^p < \infty$ ($k = 1, \ldots, n$) for some positive $p \leq 2$ and the additional condition $EX_k = 0$ ($k = 1, \ldots, n$) in the case when $1 < p \leq 2$. Then for every positive $q < 1$ and every $x \in \mathbb{R}$,

$$P\left(\max_{1 \leq k \leq n} S_k \geq x\right) \leq \frac{1}{q} P\left(S_n \geq x - \left(\frac{L}{1-q} \sum_{k=2}^{n} E|X_k|^p\right)^{1/p}\right),$$

where $L = 1$ if $0 < p \leq 1$ or $p = 2$, and $L = 2$ if $1 < p < 2$ (Petrov [340]).

2.6.15 Let X and Y be independent random variables, and let $g(x)$ be a non-negative even function defined on $\mathbb{R}$, strictly increasing on $\mathbb{R}_+$ and satisfying the condition $g(x) \to \infty$ as $x \to \infty$. If $Eg(Y) < \infty$, then for every $x \in \mathbb{R}$ and every positive $q < 1$,

$$P(\max(X, X+Y) \geq x) \leq \frac{1}{q} P\left(X + Y \geq x - g^{-1}\left(\frac{Eg(Y)}{1-q}\right)\right),$$

where g^{-1} is the inverse of g. In particular, if X and Y are independent random variables and if $\beta = E|Y|^p < \infty$ for some $p > 0$, then

$$P(\max(X, X+Y) \geq x) \leq \frac{1}{q} P\left(X + Y \geq x - \left(\frac{\beta}{1-q}\right)^{1/p}\right)$$

for every $x \in \mathbb{R}$ and every positive $q < 1$ (Petrov [340]).

2.6.16 Let X and Y be independent random variables with a common non-degenerate distribution. Let $EX = 0$, $\beta_1 = E|X|$, $\sigma^2 = EX^2$, $\beta_3 = E|X|^3 < \infty$. Then

$$2\beta_3 + 2\beta_1^3 \leq E|X - Y|^3 \leq 2\beta_3 + 3\beta_1 \sigma^2 - \beta_1^3 \leq 2\beta_3 + 2\sigma^3,$$

$$2\beta_3 + \tfrac{3}{2}\beta_1^3 \leq E|X + Y|^3 \leq 2\beta_3 + 3\beta_1 \sigma^2 - \beta_1^3 \leq 2\beta_3 + 2\sigma^3.$$

All constants in these inequalities are optimal (Esseen [124]). Daley [91] obtained some strengthenings of these inequalities and upper estimates for $E|X \pm Y|^3$ in the case when X and Y are independent non-identically distributed random variables.

2.6.17 Let X and Y be independent identically distributed random variables. Let $EX = 0$ and $p \geq 1$. Then

$$A_p(E|X|^p + E|Y|^p) \leq E|X + Y|^p \leq B_p(E|X|^p + E|Y|^p)$$

with the following optimal constants: $A_p = 1$ for $p \geq 3$, $A_p = 2^{p-2}$ for $1 \leq p \leq 2$, and $B_p = 2^{p-2}$ for $p \geq 2$. If $2 \leq p < 3$, then $A_p = \inf_{0 \leq x \leq 1} T(x)$, where

$$T(x) = 2^{p-1}(x + x^{p-1}) + (1-x)^p / [(1+x)(1+x^{p-1})].$$

If $1 \leq p \leq 2$, then $B_p = \sup_{0 \leq x \leq 1} T(x)$ (Cox and Kemperman [82]).

2.6.18 Let $EX_k = 0$, $E|X_k|^p < \infty$ $(k = 1, \ldots, n)$ for some $p \geq 1$. Then

$$B(p) E\left[\left(\sum_{k=1}^n X_k^2\right)^{p/2}\right] \leq E|S_n|^p \leq C(p) E\left[\left(\sum_{k=1}^n X_k^2\right)^{p/2}\right],$$

where $B(p)$ and $C(p)$ are positive constants depending only on p. If $1 \leq p \leq 2$, then $B(p) \geq A_1$ and $C(p) \leq A_2$, where A_1 and A_2 are absolute positive constants (Marcinkiewicz and Zygmund [275, 277]; see also [274] and [215]).

2.6.19 If $EX_k = 0$ and $E|X_k|^p < \infty$ $(k = 1, \ldots, n)$ for some $p \geq 1$, then

$$E\left(\max_{1 \leq k \leq n} |S_k|^p\right) \leq C(p) E|S_n|^p,$$

where $C(p)$ depends only on p (Marcinkiewicz and Zygmund [275]; see also [274]). Mogyoródi [307] obtained the following generalization of the latter inequality:

$$E\left(\max_{1 \leq k \leq n} g(|S_k|)\right) \leq A E g(|S_n|),$$

where $g(x)$ is a convex function function satisfying some additional conditions, and A is an absolute positive constant. In particular,

$$E\left(\max_{1 \leq k \leq n} |S_k|^p (\log(1 + |S_k|))^q\right) \leq A E(|S_n|^p (\log(1 + |S_n|))^q)$$

for $p \geq 1$ and $q > 1$, where A is an absolute constant.

2.6.20 Let $EX_k = 0$ and $E|X_k|^p < \infty$ $(k = 1, \ldots, n)$ for some p, $1 \leq p \leq 2$. Then

$$E|S_n|^p \leq \left(2 - \frac{1}{n}\right) \sum_{k=1}^n E|X_k|^p.$$

We put

$$D(p) = \frac{13.52}{(2.6\pi)^p} \Gamma(p) \sin \frac{\pi p}{2}.$$

If $D(p) < 1$, then

$$E|S_n|^p \leq (1 - D(p))^{-1} \sum_{k=1}^n E|X_k|^p$$

(von Bahr and Esseen [464]).

2.6.21 If the conditions of Theorem 2.10 are satisfied, then the inequality (2.36) holds with the constant $C(p)$ defined by the equality

$$C(p) = \tfrac{1}{2}p(p-1)\max(1, 2^{p-3})\left(1 + \frac{2}{p}K_{2m}^{(p-2)/(2m)}\right),$$

where m is the integer satisfying the condition $2m \leq p < 2m+2$, and $K_{2m} = \sum_{r=1}^{m} r^{2m-1}/(r-1)!$ (Dharmadhikari and Jogdeo [101]).

2.6.22 Let $EX_k = 0$ and $E|X_k|^p < \infty$ $(k = 1, \ldots, n)$ for some $p > 2$. We put

$$M_{p,n} = \sum_{k=1}^{n} E|X_k|^p, \quad B_n = \sum_{k=1}^{n} \operatorname{Var} X_k. \tag{2.80}$$

Then

$$E|S_n|^p \geq 2^{-p}\max\{M_{p,n}, B_n^{p/2}\}$$

(Rosenthal [393, 394]).

2.6.23 If $E|X_k|^p < \infty$ $(k = 1, \ldots, n)$ for some $p > 1$, then

$$E|S_n|^p \leq 2^{p^2}\max\left\{\sum_{k=1}^{n} E|X_k|^p, \left(\sum_{k=1}^{n} E|X_k|\right)^p\right\}$$

(Rosenthal [393]).

2.6.24 Let $X_1, \ldots, X_n$ be independent symmetric random variables such that $E|X_k|^p < \infty$ $(k = 1, \ldots, n)$ for some $p > 2$. Then

$$E|S_n|^p \geq \max\{M_{p,n}, B_n^{p/2}\}$$

and

$$E|S_n|^p \leq (Kp/\operatorname{Log} p)^p \max\{M_{p,n}, B_n^{p/2}\}$$

where $\operatorname{Log} p = \max\{1, \log p\}$, $M_{p,n}$ and B_n are defined by (2.80), and $K \leq 7.35$ is an absolute constant (Johnson et al. [209]).

2.6.25 If X is a random variable and $t > 0$, we put $\|X\|_t = (E|X|^t)^{1/t}$. For $p \geq 2$ let $C = C(p)$ be the smallest constant such that for all independent symmetric identically distributed random variables $X_1, X_2, \ldots$ and all $m = 1, 2, \ldots$ we have

$$\|S_m\|_p \leq C\max\{m^{1/2}\|X_1\|_2, m^{1/p}\|X_1\|_p\}.$$

Then $C \geq p/(2^{1/2} e \operatorname{Log} p)$ (Johnson et al. [209]).

2.6.26 Let G_0 be the set of non-negative even functions $g(x)$, $x \in \mathbb{R}$, non-decreasing on $\mathbb{R}_+$ and satisfying the condition $g(0) = 0$. If $EX_k = 0$

$(k = 1, \ldots, n)$,

$$Eg(X_k) < \infty \quad (k = 1, \ldots, n) \text{ for some } g \in G_0, \qquad (2.81)$$

and $0 < B_n < \infty$, where $B_n = \sum_{k=1}^n \operatorname{Var} X_k$, then

$$Eg(S_n) \leq \sum_{k=1}^n Eg(rX_k) + 2 e^r \int_0^\infty \left(1 + \frac{x^2}{rB_n}\right)^{-r} dg(x)$$

for every $r > 0$. If we replace the condition (2.81) by the condition

$$E^+ g(X_k) < \infty \quad (k = 1, \ldots, n) \text{ for some } g \in G_0, \qquad (2.82)$$

where

$$E^+ g(X) = \int_0^\infty g(x) \, dV(x)$$

for any random variable X with the d.f. $V(x)$, then

$$E^+ g(S_n) \leq \sum_{k=1}^n E^+ g(rX_k) + e^r \int_0^\infty \left(1 + \frac{x^2}{rB_n}\right)^{-r} dg(x)$$

for every $r > 0$. These propositions remain true if we replace S_n by $\max_{1 \leq k \leq n} S_k$ (Petrov [353]). In particular, putting $g(x) = |x|^p$, $p \geq 2$, and $r > p/2$, we obtain

$$E \left| \max_{1 \leq k \leq n} S_k \right|^p \leq C(p) \left(\sum_{k=1}^n E|X_k|^p + B_n^{p/2} \right),$$

where $C(p)$ is a positive constant depending only on p.

2.6.27 Suppose $0 < D_n < \infty$, where $D_n = \sum_{k=1}^n E|X_k|$. If the condition (2.81) is satisfied, then

$$Eg(S_n) \leq \sum_{k=1}^n Eg(rX_k) + 2 e^r \int_0^\infty \left(1 + \frac{x}{D_n}\right)^{-r} dg(x)$$

for every $r > 0$. If (2.82) is satisfied, then

$$E^+ g(S_n) \leq \sum_{k=1}^n E^+ g(rX_k) + e^r \int_0^\infty \left(1 + \frac{x}{D_n}\right)^{-r} dg(x)$$

for every $r > 0$. The inequalities remain true if we replace S_n by $\max_{1 \leq k \leq n} S_k$ (Petrov [353]).

2.6.28 Let G be the set of non-negative even functions $g(x)$ defined on $\mathbb{R}$ such that $g(1) = 1$ and $g(x)$, $x/g(x)$ do not decrease on $\mathbb{R}_+$. Let $X_1, \ldots, X_n$ be independent identically distributed random variables satisfying the conditions $EX_1 = 0$, $\operatorname{Var} X_1 = 1$, $E|X_1|^m g(X_1) < \infty$ for some integer $m \geq 2$ and some function $g(x) \in G$. Then

$$E|S_n|^m g(S_n) \leqslant C(m)\left\{\sum_{v=1}^{[m/2]} n^v \, E|X_1|^{m-2(v-1)} g(X_1) + n^{m/2} g(n^{1/2})\right\}.$$

In particular,

$$ES_n^2 g(S_n) \leqslant 5n \, EX_1^2 g(X_1) + 6ng(n^{1/2}),$$

$$E|S_n/n^{1/2}|^{2+\delta} \leqslant 2^{1-\delta} - 1 + (2^{1+\delta} + 1)n^{-\delta/2} \, E|X_1|^{2+\delta}$$

if $0 \leqslant \delta \leqslant 1$ (Sazonov [413]; in this paper some upper bounds are obtained for $E|S_n|^m g(S_n)$ in the case when the random variables $X_1, \ldots, X_n$ are non-identically-distributed).

2.6.29 Let $X_1, X_2, \ldots$ be a sequence of independent identically distributed random variables, $0 < EX_1 < \infty$. Let $M(x)$ be a non-decreasing function defined on $\mathbb{R}_+$ such that $M(0) > 0$ and $x/M(x)$ does not decrease in the area $x \geqslant x_0$ for some x_0. Let r be a non-negative integer and $\mu = EX_1$. If $E|X_1|^r M(|X_1|) < \infty$, then

$$E|S_n|^r M(|S_n|) \sim (n\mu)^r M(n\mu)$$

as $n \to \infty$. In particular,

$$EM(|X_1|) < \infty \Rightarrow EM(|S_n|) \sim M(n\mu).$$

If $\mu = \infty$ and $EM(|X_1|) < \infty$, then $EM(|S_n|) = 0(n)$ as $n \to \infty$ (Smith [431]).

2.6.30 Suppose $E|X_k|^p < \infty$ ($k = 1, \ldots, n$ for some positive $p < 2$. We put

$$\Lambda_n(y, p) = \left(\sum_{k=1}^n \int_{|x|<y} x^2 \, dV_k^s(x)\right)^{p/2} + \sum_{k=1}^n \int_{|x| \geqslant y} |x|^p \, dV_k^s(x)$$

and $\lambda_n(p) = \inf_{y \geqslant 0} \Lambda_n(y, p)$, where $V_k^s(x)$ is the d.f. of the symmetrized random variable X_k^s. If $1 \leqslant p < 2$ and $EX_k = 0$ for all k, then

$$c_1(p)\lambda_n(p) \leqslant E|S_n|^p \leqslant c_2(p)\lambda_n(p).$$

If $0 < p < 1$ and each random variable X_k has zero as its median, then

$$c(p)\lambda_n(p) \leqslant E|S_n|^p \leqslant 2\lambda_n(p).$$

Here $c_1(p)$, $c_2(p)$, and $c(p)$ are positive constants depending only on p (Manstavičius [273]).

2.6.31 Suppose $EX_k = 0$ ($k = 1, \ldots, n$). Let $g(x)$ be a non-negative convex function defined on $\mathbb{R}$ and satisfying the conditions

$$g(x + y) \leqslant K(g(x) + g(y)) \qquad \text{for every } x, y \in \mathbb{R} \qquad (2.83)$$

and
$$g(-x) = g(x) \quad \text{for every } x \in \mathbb{R}, \tag{2.84}$$
where K is a constant. Then
$$E \max_{1 \leq k \leq n} g(S_k) \leq 4K Eg(S_n).$$
If the random variables $X_1, \ldots, X_n$ are symmetric, then
$$E \max_{1 \leq k \leq n} g(S_k) \leq 2 Eg(S_n)$$
even without the conditions (2.83) and (2.84) (Bickel [35]).

2.6.32 Suppose that $P(|X_k| \leq 1) = 1$ $(k = 1, \ldots, n)$. If $P(|S_n| \geq a) \leq 1/(8e)$, then $E|S_n|^m \leq C(m)(a+1)^m$ for every positive integer m, where $C(m)$ is a positive constant depending only on m (Skorohod [428]).

2.6.33 Let $X_1, \ldots, X_n$ be independent random variables with a common non-degenerate d.f. $V(x)$. Suppose $E|X_1|^p < \infty$ for some positive $p \leq 2$. We set $v(x) = x^{-2} \int_0^x u P(|X_1| \geq u)\, du$ for $x > 0$. Let $y(n)$ be a solution of the equation $v(y) = 1/n$ (this solution exists and is unique for all $n \geq N$ and some N). Moreover, we set

$$\lambda_p(n) = y^p(n) + n \int_{|x| \geq y(n)} |x|^p\, dV(x),$$

$$\mu_n = n \int_{|x| < y(n)} x\, dV(x), \quad \text{if } 0 < p < 1,$$

$$\mu_n = n E X_1, \quad \text{if } 1 \leq p \leq 2.$$

Then $E|S_n - mS_n|^p \leq K_1 E|S_n - \mu_n|^p \leq K_1 K_2 \lambda_p(n)$ for all $n \geq N$, where

$$K_1 = \begin{cases} 4, & \text{if } 0 < p < 1, \\ 2^{p+1}, & \text{if } 1 \leq p \leq 2, \end{cases} \qquad K_2 = \begin{cases} 2^{p/2}, & \text{if } 0 < p < 1, \\ 2^{2p}, & \text{if } 1 \leq p \leq 2. \end{cases}$$

It follows from here (1) that in the case when $0 < p < 1$ there exists a constant $c > 0$ depending only on p such that
$$c\bar{\lambda}_p(p) \leq E|S_n|^p \leq 2^{p/2} \bar{\lambda}_n(p)$$
for $n \geq N$, where
$$\bar{\lambda}_n(p) = \lambda_n(p) + \left(n \left| \int_{|x| < y} x\, dV(x) \right| \right)^p;$$
and (2) that in the case when $0 < p \leq 2$ there exists a constant C depending

only on p such that
$$|mS_n - \mu_n| \leq C\left\{y(n) + \left(n\int_{|x|\geq y} |x|^p \, dV(x)\right)^{1/p}\right\}$$
for $n \geq N$ (Hall [159]).

2.6.34 Let $X_1, \ldots, X_n$ be independent identically distributed random variables, and let $E|X_1|^r < \infty$ for some positive $r \leq 2$. Then
$$Q(S_n; \lambda) \geq K(r)\lambda(\lambda + (nv_r(a))^{1/r})^{-1}$$
for every $\lambda \geq 0$, where Q is the Lévy concentration function, $v_r(a) = E|X_1 - a|^r$, a is an arbitrary number, and $K(r)$ is a positive number depending only on r (Esseen [122]).

2.6.35 Let $X_1, \ldots, X_n$ be independent random variables with a common non-degenerate d.f. $F(x)$. The following conditions are equivalent:
(A) $EX_1^2 < \infty$;
(B) there exist positive constants $K_1(\lambda, F)$ and $K_2(\lambda, F)$, depending only on λ and F, such that
$$K_1(\lambda, F)n^{-1/2} \leq Q(S_n; \lambda) \leq K_2(\lambda, F)n^{-1/2}$$
for every $n \geq 1$ and every $\lambda \geq 0$ (Esseen [122]).

2.6.36 Let $X_1, \ldots, X_n$ be independent random variables with a common d.f. $F(x)$, and let $0 < \alpha < 2$. If $F(x)$ is unimodal and
$$\liminf_{x \to +\infty} x^\alpha(1 - F(x) + F(-x)) > 0,$$
then $Q(S_n; \lambda) \leq C_1 n^{-1/\alpha}$ for every $n \geq 1$ and every $\lambda \geq 0$. If
$$\liminf_{x \to +\infty} x^\alpha(1 - F(x) + F(-x)) < \infty,$$
then $Q(S_n; \lambda) \geq C_2 n^{-1/\alpha}$ for every $n \geq 1$ and every $\lambda \geq 0$. Here C_1 and C_2 are positive constants not depending on n but possibly depending on λ and F (Suchkov and Ushakov [445]).

2.6.37 Let the generalized concentration function $K(X; p, \lambda)$ be defined by the equality (1.64) (Chapter 1), where $p(x) \in \mathscr{P}_+$. Suppose that $\inf_{|t|\leq 1} \hat{p}(t) > 0$ and $p(x) \geq 1 - |x|^r$ for $|x| \leq 1$ and some positive $r \leq 2$. If $0 < \lambda_k \leq \lambda$ ($k = 1, \ldots, n$), then
$$K(S_n; p, \lambda) \leq C(p)\lambda\left\{\sum_{k=1}^{n} \lambda_k^2(1 - K(X_k; p, \lambda_k))^{2/r}(K(X_k; p, \lambda_k))^{-2}\right\}^{-1/2}$$
for every $n \geq 1$ and every $\lambda > 0$, where $C(p)$ is a positive constant depending only on $p(x)$ (Ananjevsky [7]).

3

Weak limit theorems: convergence to infinitely divisible distributions

3.1 The condition of infinite smallness

Consider a sequence of series of random variables

$$X_{11}, X_{12}, \ldots, X_{1k_1},$$
$$X_{21}, X_{22}, \ldots, X_{2k_2},$$
$$\ldots\ldots\ldots\ldots\ldots$$
$$X_{n1}, X_{n2}, \ldots, X_{nk_n},$$
$$\ldots\ldots\ldots\ldots\ldots$$

that are independent within each series. Suppose $k_n \to \infty$ as $n \to \infty$. We set ourselves the task of finding all the limit distributions for sums of the form

$$\sum_{k=1}^{k_n} X_{nk} \tag{3.1}$$

as $n \to \infty$. In the absence of any additional restrictions, the solution is obvious. Namely, any distribution function can serve as a limit of this kind. In fact, if the random variable X_{n1} has the distribution function $F(x)$ for every n, and if $X_{nk} \equiv 0$ for all n and for $k > 1$, then the sum (3.1) has the distribution function $F(x)$ for all n.

In order to exclude such cases, it seems natural to introduce some restrictions to make the role of any individual term in (3.1) become infinitely small as $n \to \infty$. As a suitable convention we shall accept the following condition:

$$\max_{1 \leq k \leq k_n} P(|X_{nk}| \geq \varepsilon) \to 0 \quad \text{as } n \to \infty \text{ for every fixed } \varepsilon > 0. \tag{3.2}$$

This condition is called the condition of infinite smallness.

It is possible to reformulate this condition in different terms.

We shall denote by $F_{nk}(x)$, $f_{nk}(t)$ and mX_{nk} respectively the distribution function, the characteristic function, and a median of the random variable X_{nk}. We shall write $\max_k$ in place of $\max_{1 \leq k \leq k_n}$, and $\sum_k$ in place of $\sum_{k=1}^{k_n}$

Lemma 3.1 *The following conditions are equivalent:*

(I) *the condition of infinite smallness,*

(II)
$$\max_k \int_{-\infty}^{\infty} \frac{x^2}{1+x^2} \, dF_{nk}(x) \to 0,$$

(III) $\max_k |f_{nk}(t) - 1| \to 0$ *uniformly in t in an arbitrary finite interval.*

Proof The function $(1+x^2)/x^2$ decreases on the positive half-line, therefore,

$$\max_k \int_{|x| \geq \varepsilon} dF_{nk}(x) \leq \frac{1+\varepsilon^2}{\varepsilon^2} \max_k \int_{|x| \geq \varepsilon} \frac{x^2}{1+x^2} \, dF_{nk}(x)$$

for every $\varepsilon > 0$ and for all n. Hence (II) implies (I).

If x is an arbitrary real number, then $|e^{ix} - 1| \leq |x|$. (A more general result will be stated in Lemma 3.2.) It follows that for $|t| \leq b$ and every $b < \infty$ we have

$$\max_k |f_{nk}(t) - 1| \leq \max_k \left| \int_{|x|<\varepsilon} (e^{itx} - 1) \, dF_{nk}(x) \right|$$

$$+ \max_k \left| \int_{|x| \geq \varepsilon} (e^{itx} - 1) \, dF_{nk}(x) \right| \leq b\varepsilon + 2 \max_k \int_{|x| \geq \varepsilon} dF_{nk}(x).$$

Therefore, (I) implies (III). It remains only to show that (III) implies (II). It is not hard to find that

$$\int_0^{\infty} e^{-t}(1 - \operatorname{Re} f(t)) \, dt = \int_{-\infty}^{\infty} \frac{x^2}{1+x^2} \, dF(x)$$

for an arbitrary d.f. $F(x)$ and the corresponding c.f. $f(t)$. In fact,

$$\int_0^{\infty} e^{-t}(1 - \operatorname{Re} f(t)) \, dt = 1 - \int_{-\infty}^{\infty} \left\{ \int_0^{\infty} e^{-t} \cos tx \, dt \right\} dF(x),$$

and the inner integral can be calculated by integrating by parts. Hence

$$\int_{-\infty}^{\infty} \frac{x^2}{1+x^2} \, dF_{nk}(x) \leq \int_0^{\infty} e^{-t} |f_{nk}(t) - 1| \, dt$$

$$\leq \int_0^T \max_k |f_{nk}(t) - 1| \, dt + 2 \int_T^{\infty} e^{-t} \, dt$$

for every $T > 0$, n and k. If (III) is fulfilled, the right-hand side of this inequality can be made arbitrarily small for all sufficiently large n by a choice of T. Thus (II) follows from (III). □

Lemma 3.2 *If x is an arbitrary real number and k is any positive integer, then*

$$e^{ix} = \sum_{v=0}^{k-1} \frac{(ix)^v}{v!} + \theta \frac{x^k}{k!} \tag{3.3}$$

where $|\theta| \leqslant 1$. (We put $0^0 = 1$.)

Proof The equality

$$\int_0^x e^{it}\, dt = \frac{1}{i}(e^{ix} - 1)$$

implies that $|e^{ix} - 1| \leqslant |x|$. We write $\theta = \theta(x) = (e^{ix} - 1)/x$ for $x \neq 0$ and $\theta(0) = i$. Then $e^{ix} = 1 + \theta x$, where $|\theta| \leqslant 1$. The lemma is proved for $k = 1$. We assume that (3.3) is true for a given k, and we show that (3.3) is true when k is replaced by $k+1$. Consider the integral

$$I = \int_0^x \left(e^{it} - \sum_{v=0}^{k-1} \frac{(it)^v}{v!} \right) dt.$$

We have

$$|I| = \left| e^{ix} - \sum_{v=0}^k \frac{(ix)^v}{v!} \right| \leqslant \int_0^{|x|} \frac{t^k}{k!}\, dt = \frac{|x|^{k+1}}{(k+1)!}$$

Therefore, (3.3) is fulfilled when k is replaced by $k + 1$. □

Lemma 3.3 *If the condition of infinite smallness is satisfied, then*

$$\max_k |mX_{nk}| \to 0, \quad \max_k \int_{|x|<\tau} |x|^r\, dF_{nk}(x) \to 0$$

for every positive τ and r.

Proof It is easy to prove that an arbitrary median of the random variable X belongs to any interval I for which $P(X \in I) > \frac{1}{2}$. Let ε be an arbitrary positive number. The condition of infinite smallness implies that $\min_k P(|X_{nk}| < \varepsilon) \to 1$ and, therefore, $\min_k P(|X_{nk}| < \varepsilon) > \frac{1}{2}$ for all sufficiently large n. It follows that $\max_k |mX_{nk}| < \varepsilon$ for all sufficiently large n. Further,

$$\max_k \int_{|x|<\tau} |x|^r\, dF_{nk}(x) \leqslant \varepsilon^r + \tau^r \max_k \int_{|x|\geqslant\varepsilon} dF_{nk}(x).$$

The right-hand side of this inequality can be made arbitrarily small by the choice of sufficiently small ε for all sufficiently large n, in view of (3.2).

3.2 Infinitely divisible distributions as limit laws

The following fundamental result of Khintchine clarifies the role of infinitely divisible distributions in the area of limit theorems of probability theory.

Theorem 3.1 *Let $\{X_{nk}; k = 1, \ldots, k_n; n = 1, 2, \ldots\}$ be a sequence of series of independent random variables satisfying the condition of infinite smallness. Then the set of distribution functions that are limits (in the sense of weak convergence) of the distribution functions of the sums $\sum_{k=1}^{k_n} X_{nk}$ coincides with the set of all infinitely divisible distribution functions.*

In order to prove this theorem we need a chain of lemmas.

From now on, τ will denote an arbitrary fixed positive number, so that $0 < \tau < \infty$. With every d.f. F (with or without subscript) we associate the quantity

$$a = \int_{|x|<\tau} x \, dF(x) \tag{3.4}$$

and the functions

$$\bar{F}(x) = F(x + a), \quad \bar{f}(t) = \int_{-\infty}^{\infty} e^{itx} \, d\bar{F}(x). \tag{3.5}$$

Obviously, $|a| \leq \int_{|x|<\tau} |x| \, dF(x) < \tau$.

Lemma 3.4 *If the sequence $\{X_{nk}\}$ satisfies the condition of infinite smallness, then the sequence $\{\bar{X}_{nk}\}$, where $\bar{X}_{nk} = X_{nk} - a_{nk}$, also satisfies this condition.*

Proof By Lemma 3.3,

$$\max_k |a_{nk}| \leq \max_k \int_{|x|<\tau} |x| \, dF_{nk}(x) \to 0.$$

Hence

$$\max_k P(|\bar{X}_{nk}| \geq \varepsilon) \leq \max_k P(|X_{nk}| + |a_{nk}| \geq \varepsilon) \leq \max_k P(|X_{nk}| \geq \varepsilon/2)$$

for every $\varepsilon > 0$ and sufficiently large n. Applying (3.2), we arrive at the assertion of the lemma. □

Lemma 3.5 *Let the sequence $\{X_{nk}\}$ satisfy the condition of infinite smallness. Then $\log f_{nk}(t)$ is finite in the interval $|t| \leq b$ for every finite $b > 0$ and for all sufficiently large n and, moreover,*

$$\log f_{nk}(t) = f_{nk}(t) - 1 + \theta_{nk}(f_{nk}(t) - 1)^2,$$

where $|\theta_{nk}| \leq 1$. The same assertion is true for $\log \bar{f}_{nk}(t)$.

Proof We have

$$\log(1+z) = z - \tfrac{1}{2}z^2 + o(|z|^2) \qquad (z \to 0).$$

Therefore, $|\log(1+z) - z| \leq |z|^2$ for all sufficiently small $|z|$. Putting $z = z_{nk}(t) = f_{nk}(t) - 1$, we observe that $\max_k |z_{nk}(t)| \to 0$ uniformly in t in an arbitrary finite interval, by Lemma 3.1. The first assertion of the lemma follows. The permissibility of replacing f_{nk} by $\bar{f}_{nk}$ is guaranteed by Lemma 3.4. □

In the following two lemmas, $F(x)$ is an arbitrary distribution function with the characteristic function $f(t)$. We define a, $\bar{F}(x)$, and $\bar{f}(t)$ by the equalities (3.4) and (3.5).

Lemma 3.6 *For every positive $b < \infty$ there exists a positive number $c_1 = c_1(a, b, \tau)$ such that*

$$c_1 \max_{|t| \leq b} |\bar{f}(t) - 1| \leq \int_{-\infty}^{\infty} \frac{x^2}{1+x^2} \, d\bar{F}(x). \qquad (3.6)$$

Proof Lemma 3.2 for $k = 2$ implies that $|e^{ix} - 1 - ix| \leq x^2/2$ for every real x. Using this result, (3.4) and (3.5), we obtain

$$\max_{|t| \leq b} |\bar{f}(t) - 1| = \max_{|t| \leq b} \left| \int_{-\infty}^{\infty} (e^{it(x-a)} - 1) \, dF(x) \right|$$

$$\leq 2 \int_{|x| \geq \tau} dF(x) + b \left| \int_{|x| < \tau} (x-a) \, dF(x) \right|$$

$$+ \tfrac{1}{2} b^2 \int_{|x| < \tau} (x-a)^2 \, dF(x) \leq (2 + |a|b) \int_{|x| \geq \tau} dF(x)$$

$$+ \tfrac{1}{2} b^2 \int_{|x| < \tau} (x-a)^2 \, dF(x).$$

Taking into account the inequality $|a| < \tau$ and the fact that the function $(x-a)^{-2}$ does not increase in the interval $x \geq \tau$, we find that

$$\max_{|t| \leq b} |\bar{f}(t) - 1| \leq \left\{ \frac{2 + |a|b}{(\tau - |a|)^2} + \frac{b^2}{2} \right\} \{1 + (\tau + |a|)^2\} \int_{-\infty}^{\infty} \frac{(x-a)^2}{1 + (x-a)^2} \, dF(x).$$

The integral on the right-hand side coincides with the integral in (3.6). The inequality (3.6) is proved. We may set

$$c_1 = \left\{ \frac{2 + |a|b}{(\tau - |a|)^2} + \frac{b^2}{2} \right\}^{-1} \{1 + (\tau + |a|)^2\}^{-1}. \qquad (3.7)$$

Lemma 3.7 Suppose $0 < b < \infty$. Let m be a median of a random variable with the d.f. $F(x)$. If $|m| < \tau$, there exists a positive number $c_2 = c_2(m, b, \tau)$ such that

$$\int_{-\infty}^{\infty} \frac{x^2}{1+x^2} \, dF(x) \leq c_2 \int_0^b (1 - |f(t)|^2) \, dt. \tag{3.8}$$

If $f(t) \neq 0$ for $0 \leq t \leq b$, we may replace $1 - |f(t)|^2$ by $2|\log |f(t)||$.

Proof Let $F^s(x)$ denote the d.f. of the symmetrized random variable $X^s = X - Y$, where X and Y are independent random variables with the same d.f. $F(x)$. The random variable X^s has the c.f.

$$|f(t)|^2 = \int_{-\infty}^{\infty} \cos tx \, dF^s(x).$$

It follows from inequalities (1.50) (Chapter 1) that

$$\inf_x \left(1 - \frac{\sin bx}{bx}\right) \frac{1+x^2}{x^2} \geq c(b),$$

where $c(b)$ is a positive constant depending only on b. Hence

$$\int_0^b (1 - |f(t)|^2) \, dt = \int_{-\infty}^{\infty} \left(\int_0^b (1 - \cos tx) \, dt\right) F^s(x)$$

$$= b \int_{-\infty}^{\infty} \left(1 - \frac{\sin bx}{bx}\right) dF^s(x) \geq bc(b) \int_{-\infty}^{\infty} \frac{x^2}{1+x^2} \, dF^s(x). \tag{3.9}$$

We put $F_m(x) = P(X - m < x)$, $q_m(t) = P(|X - m| \geq t)$, $q^s(t) = P(|X^s| \geq t)$ for $t \geq 0$. We shall show that

$$q_m(t) \leq 2q^s(t) \quad \text{for every } t \geq 0. \tag{3.10}$$

If $X - m \geq t$ and $Y - m \leq 0$, then $X^s \geq t$. Using the independence of X and Y and the definition of a median, we obtain

$$P(X^s \geq t) \geq P(X - m \geq t, Y - m \leq 0) = P(X - m \geq t) P(Y \leq m)$$

$$\geq \tfrac{1}{2} P(X - m \geq t)$$

and similarly $P(X^s \leq -t) \geq \tfrac{1}{2} P(X - m \leq -t)$ for every $t \geq 0$. The assertion (3.10) follows.

We have $q_m(t) = 1 - F_m(t) + F_m(-t)$ if $-t$ is a point of continuity of F_m.

94 | Limit theorems of probability theory

By integrating by parts we obtain with the help of (3.10)

$$\int_{-\infty}^{\infty} \frac{x^2}{1+x^2} \, dF_m(x) = -\int_0^{\infty} \frac{x^2}{1+x^2} \, dq_m(x) = \int_0^{\infty} q_m(x) \, d\frac{x^2}{1+x^2}$$

$$\leq 2 \int_0^{\infty} q^s(x) \, d\frac{x^2}{1+x^2} = 2 \int_{-\infty}^{\infty} \frac{x^2}{1+x^2} \, dF^s(x). \qquad (3.11)$$

It is clear that

$$\int_{-\infty}^{\infty} \frac{x^2}{1+x^2} \, d\bar{F}(x) = \int_{-\infty}^{\infty} \frac{(x-a)^2}{1+(x-a)^2} \, dF(x)$$

$$\leq \int_{|x|<\tau} (x-a)^2 \, dF(x) + \int_{|x|\geq\tau} dF(x).$$

For every real x, a, and m, we have $(x-a)^2 \leq (x-m)^2 + 2(m-a)(x-a)$. Using this easily verified inequality and (3.4), we get

$$\int_{|x|<\tau} (x-a)^2 \, dF(x) \leq \int_{|x|<\tau} (x-m)^2 \, dF(x) + 2(\tau+|m|) \left| \int_{|x|<\tau} (x-a) \, dF(x) \right|$$

$$\leq \int_{|x|<\tau} (x-m)^2 \, dF(x) + 2|a|(\tau+|m|) \int_{|x|\geq\tau} dF(x).$$

Thus

$$\int_{-\infty}^{\infty} \frac{x^2}{1+x^2} \, d\bar{F}(x) \leq \int_{|x|<\tau} (x-m)^2 \, dF(x) + \{1 + 2\tau(\tau+|m|)\} \int_{|x|\geq\tau} dF(x)$$

by the inequality $|a| < \tau$. Furthermore,

$$\int_{|x|<\tau} (x-m)^2 \, dF(x) \leq \{1 + (\tau+|m|)^2\} \int_{|x|<\tau} \frac{(x-m)^2}{1+(x-m)^2} \, dF(x)$$

$$\leq \{1 + (\tau+|m|)^2\} \int_{-\infty}^{\infty} \frac{x^2}{1+x^2} \, dF_m(x),$$

$$\int_{|x|\geq\tau} dF(x) \leq \frac{1+(\tau+|m|)^2}{(\tau-|m|)^2} \int_{|x|\geq\tau} \frac{(x-m)^2}{1+(x-m)^2} \, dF(x)$$

$$\leq \frac{1+(\tau+|m|)^2}{(\tau-|m|)^2} \int_{-\infty}^{\infty} \frac{x^2}{1+x^2} \, dF_m(x).$$

Therefore,

$$\int_{-\infty}^{\infty} \frac{x^2}{1+x^2} \, d\bar{F}(x) \leq c \int_{-\infty}^{\infty} \frac{x^2}{1+x^2} \, dF_m(x), \qquad (3.12)$$

where

$$c = c(m, \tau) = \{1 + (\tau+|m|)^2\} \left\{ 1 + \frac{1+2\tau(\tau+|m|)}{(\tau-|m|)^2} \right\}. \qquad (3.13)$$

The inequalities (3.9), (3.11) and (3.12) imply (3.8) with

$$c_2 = \frac{2c}{bc(b)}. \tag{3.14}$$

The second assertion of the lemma follows from the elementary inequality $1 + x \leqslant e^x$ for every real x. If we take here $x = |f(t)|^2 - 1$, we obtain

$$1 - |f(t)|^2 \leqslant -\log |f(t)|^2 = 2|\log |f(t)||. \qquad \square$$

Lemma 3.8 *Suppose $0 < b < \infty$. If the sequence $\{X_{nk}\}$ satisfies the condition of infinite smallness, then there exist positive numbers $c_* = c_*(b, \tau)$ and $c^* = c^*(b, \tau)$ such that*

$$c_* \max_{|t| \leqslant b} |\bar{f}_{nk}(t) - 1| \leqslant \int_{-\infty}^{\infty} \frac{x^2}{1 + x^2} d\bar{F}_{nk}(x) \leqslant c^* \int_0^b |\log |f_{nk}(x)|| dt$$

for all k and all sufficiently large n.

Proof By Lemma 3.3 we have

$$|a_{nk}| = \left| \int_{|x| < \tau} x \, dF_{nk}(x) \right| \leqslant \int_{|x| < \tau} |x| \, dF_{nk}(x) \to 0$$

uniformly in k ($1 \leqslant k \leqslant k_n$). It follows that $|a_{nk}| < \tau/2$ for all k and sufficiently large n. Applying Lemma 3.6 to the d.f. $F_{nk}(x)$, we can choose the constant c_* to be the one we obtain from (3.7) for $|a| = \tau/2$.

Furthermore, the condition of infinite smallness and Lemma 3.3 imply that $|mX_{nk}| < \tau/2$ for every positive τ and all k, if n is sufficiently large. Applying Lemma 3.7 to the distribution $F_{nk}(x)$, in (3.13) and (3.14) we replace $|m|$ by $\tau/2$. This replacement gives us an upper bound for c_2 that depends only on b and τ. $\qquad \square$

Lemma 3.9 *Suppose the condition of infinite smallness is satisfied. Let*

$$\prod_k |f_{nk}(t)| \to |f(t)| \qquad \text{for every real } t,$$

where $f(t)$ is a c.f. Then there exists a positive constant c such that

$$\sum_k \int_{-\infty}^{\infty} \frac{x^2}{1 + x^2} d\bar{F}_{nk}(x) \leqslant c \tag{3.15}$$

for all sufficiently large n.

Proof Any c.f. $f(t)$ is continuous and satisfies the condition $f(0) = 1$. Therefore, there exists a positive number b such that $|f(t)| > 0$ for $|t| \leqslant b$. By Lemma 3.5 the functions $\log f_{nk}(t)$ are defined and finite in the interval

96 | Limit theorems of probability theory

$|t| \leq b$ for sufficiently large n. We have

$$\prod_k |f_{nk}(t)|^2 \to |f(t)|^2, \qquad (3.16)$$

where $|f(t)|^2$ is a c.f. By Lemma 1.6 (Chapter 1) the relations (3.16) and $\sum_k \log |f_{nk}(t)| \to \log |f(t)|$ hold uniformly in t in the interval $|t| \leq b$. By Lemma 3.8,

$$\sum_k \int_{-\infty}^{\infty} \frac{x^2}{1+x^2} d\bar{F}_{nk}(x) \leq -c^* \int_0^b \sum_k \log |f_{nk}(t)| \, dt.$$

The right-hand side of this inequality has a finite limit equal to

$$-c^* \int_0^b \log |f(t)| \, dt. \qquad \square$$

Lemma 3.10 *Suppose the condition of infinite smallness is satisfied. Let there exist a positive constant c such that the inequality (3.15) holds for all sufficiently large n. Then*

$$\sum_k \{\log \bar{f}_{nk}(t) - (\bar{f}_{nk}(t) - 1)\} \to 0 \qquad (3.17)$$

for every real t.

Proof Let t be an arbitrary fixed real number. By Lemmas 3.4, 3.5 and 3.1 we have

$$\max_k |\bar{f}_{nk}(t) - 1| \to 0 \qquad (3.18)$$

and

$$\log \bar{f}_{nk}(t) = \bar{f}_{nk}(t) - 1 + \theta_{nk}(\bar{f}_{nk}(t) - 1)^2$$

for all sufficiently large n, where $|\theta_{nk}| \leq 1$. Using Lemma 3.8 and the condition (3.15), we obtain

$$\sum_k |\bar{f}_{nk}(t) - 1| \leq \frac{1}{c_*} \sum_k \int \frac{x^2}{1+x^2} d\bar{F}_{nk}(x) \leq \frac{c}{c_*}$$

for all sufficiently large n. It follows that

$$\left| \sum_k \{\log \bar{f}_{nk}(t) - (\bar{f}_{nk}(t) - 1)\} \right| \leq \sum_k |\bar{f}_{nk}(t) - 1|^2 \leq \frac{c}{c_*} \max_k |\bar{f}_{nk}(t) - 1|.$$

These inequalities and (3.18) imply (3.17). $\square$

Let us complete the proof of Theorem 3.1. First we shall show that an arbitrary infinitely divisible d.f. is a limit of the distribution functions of sums of the form (3.1). Let $F(x)$ be an arbitrary infinitely divisible d.f.

with the c.f. $f(t)$. We have $f(t) = f_n^n(t)$, where $f_n(t)$ is a c.f. for every n. Consider a triangle sequence of series of independent random variables $\{X_{nk}; k = 1, \ldots, n; n = 1, 2, \ldots\}$ such that X_{nk} has the c.f. $f_n(t)$ for all k between 1 and n. The sum $\sum_{k=1}^{n} X_{nk}$ has the c.f. $f_n^n(t) = f(t)$. Thus $f(t) = \lim \prod_{k=1}^{n} f_{nk}(t)$. Therefore, the sequence of the distribution functions of the sums $\sum_{k=1}^{n} X_{nk}$ of independent random variables X_{nk} with the c.f. $f_{nk}(t) = f_n(t)$ converges weakly to $F(x)$, by Theorem 1.10 (Chapter 1). The sequence $\{X_{nk}\}$ satisfies the condition of infinite smallness, since $(f(t))^{1/n} \to 1$ uniformly in t in an arbitrary finite interval, and we can apply Lemma 3.1.

We now consider a sequence of series of independent random variables $\{X_{nk}; k = 1, \ldots, k_n; n = 1, 2, \ldots\}$ satisfying the condition of infinite smallness. Let

$$P\left(\sum_k X_{nk} < x\right) \to F(x)$$

at every point of continuity of the d.f. $F(x)$. We shall prove that F is infinitely divisible.

By Theorem 1.9 (Chapter 1) we have

$$\prod_k f_{nk}(t) \to f(t) = \int_{-\infty}^{\infty} e^{itx} \, dF(x). \tag{3.19}$$

By Lemmas 3.9 and 3.10 the relation (3.17) holds for every real t. According to (3.5) we obtain

$$\bar{f}_{nk}(t) = \int_{-\infty}^{\infty} e^{itx} \, dF_{nk}(x + a_{nk}) = e^{-ita_{nk}} f_{nk}(t).$$

Therefore,

$$\log \bar{f}_{nk}(t) - (\bar{f}_{nk}(t) - 1) = \log f_{nk}(t) - \left\{ita_{nk} + \int_{-\infty}^{\infty} (e^{itx} - 1) \, d\bar{F}_{nk}(x)\right\}$$

$$= \log f_{nk}(t) - ita_{nk} - it \int_{-\infty}^{\infty} \frac{x}{1 + x^2} \, d\bar{F}_{nk}(x)$$

$$- \int_{-\infty}^{\infty} \left(e^{itx} - 1 - \frac{itx}{1 + x^2}\right) d\bar{F}_{nk}(x).$$

We put

$$\gamma_n = \sum_k \left\{a_{nk} + \int_{-\infty}^{\infty} \frac{x}{1 + x^2} \, d\bar{F}_{nk}(x)\right\},$$

$$G_n(x) = \sum_k \int_{-\infty}^{x} \frac{y^2}{1 + y^2} \, d\bar{F}_{nk}(y),$$

$$\psi_n(t) = i\gamma_n t + \int_{-\infty}^{\infty} \left(e^{itx} - 1 - \frac{itx}{1 + x^2}\right) \frac{1 + x^2}{x^2} \, dG_n(x).$$

98 | Limit theorems of probability theory

Thus

$$\sum_k \{\log \bar{f}_{nk}(t) - (\bar{f}_{nk}(t) - 1)\} = \log \prod_k f_{nk}^{(t)} - \psi_n(t). \qquad (3.20)$$

By Lemma 3.9, $G_n(+\infty) \leq c < \infty$ for all sufficiently large n, so that $G_n(x)$ is a bounded non-decreasing function. By Theorem 1.16 (Chapter 1) the function $e^{\psi_n(t)}$ is an infinitely divisible c.f. The relations (3.17), (3.19) and (3.20) imply that

$$e^{\psi_n(t)} \to f(t) \qquad (3.21)$$

for every real t. In accordance with Lemma 1.20 (Chapter 1), the c.f. $f(t)$ is infinitely divisible. □

It may happen that the distribution of the sum (3.1) does not converge, while there exists a sequence of constants $\{b_n\}$ such that the distribution of the sum

$$\sum_{k=1}^{k_n} X_{nk} - b_n \qquad (3.22)$$

does converge to some limiting distribution. The c.f. of the random variable (3.22) is equal to $\exp\{-ib_n t\} \prod_{k=1}^{k_n} f_{nk}(t)$, and differs only by the factor $\exp\{-ib_n t\}$ from the c.f. of the sum (3.1), so that the absolute values of both c.f. coincide. If the function $e^{\psi_n(t)}$ is an infinitely divisible c.f., the function $\exp\{-ib_n t + \psi_n(t)\}$ has the same property. By tracing the proof of Theorem 3.1 it is not hard to arrive at the following proposition.

Theorem 3.2 *Let $\{X_{nk}; k = 1, \ldots, k_n; n = 1, 2, \ldots\}$ be a sequence of series of independent random variables satisfying the condition of infinite smallness, and let $\{b_n; n = 1, 2, \ldots\}$ be a sequence of constants. Then the set of distributions that are limits (in the sense of weak convergence) of distributions of sums (3.22) coincides with the set of all infinitely divisible distributions.*

Together with the condition of infinite smallness, we can consider a more general condition of the existence of a sequence of constants $\{l_{nk}; k = 1, \ldots, k_n; n = 1, 2, \ldots\}$ such that

$$\max_{1 \leq k \leq k_n} P(|X_{nk} - l_{nk}| \geq \varepsilon) \to 0 \qquad (3.23)$$

for every fixed $\varepsilon > 0$. This condition is called the condition of constancy in the limit.

In other words, the random variables X_{nk} are constant in the limit if the differences $X_{nk} - l_{nk}$ satisfy the condition of infinite smallness (3.2) for some constants l_{nk}.

If the random variables X_{nk} satisfy condition (3.23) then $l_{nk} = mX_{nk} + o(1)$

uniformly in k. In fact, (3.23) implies that

$$\min_{1 \leq k \leq k_n} P(|X_{nk} - l_{nk}| < \varepsilon) > \tfrac{1}{2}$$

for every $\varepsilon > 0$ and sufficiently large n. Therefore $|mX_{nk} - l_{nk}| < \varepsilon$ for all k and sufficiently large n by a property of a median (see the beginning of the proof of Lemma 3.3). It follows that if (3.23) is fulfilled for some constants l_{nk}, then it is also fulfilled when the l_{nk} are replaced by mX_{nk}.

Theorems 3.1 and 3.2 remain true if the condition of infinite smallness is replaced by the condition of constancy in the limit.

3.3 Necessary and sufficient conditions for convergence to a given infinitely divisible distribution

The proof of Theorem 3.1 was quite complicated, but its advantage is that some very few additional considerations allow us to find the necessary and sufficient conditions for the convergence of distributions of sums of independent random variables to a given infinitely divisible distribution.

Theorem 3.3 *Let*

$$\{X_{nk}; k = 1, \ldots, k_n; n = 1, 2, \ldots\}$$

be a sequence of series of random variables that are independent within each series and satisfy the condition of infinite smallness. We put $F_{nk}(x) = P(X_{nk} < x)$. Let $F(x)$ be an infinitely divisible distribution function with the characteristic function $f(t)$ having the Lévy–Khintchine representation (1.41) (Chapter 1). In order that the distribution functions of the sums $\sum_{k=1}^{k_n} X_{nk}$ converge weakly to $F(x)$ it is necessary and sufficient that*

$$G_n \rightrightarrows G, \quad \gamma_n \to \gamma \tag{3.24}$$

where

$$G_n(x) = \sum_{k=1}^{k_n} \int_{-\infty}^{x} \frac{y^2}{1+y^2} \, d\bar{F}_{nk}(y), \tag{3.25}$$

$$\gamma_n = \sum_{k=1}^{k_n} \left\{ a_{nk} + \int_{-\infty}^{\infty} \frac{x}{1+x^2} \, d\bar{F}_{nk}(x) \right\}, \tag{3.26}$$

$$a_{nk} = \int_{|x| < \tau} x \, dF_{nk}(x), \quad \bar{F}_{nk}(x) = F_{nk}(x + a_{nk}), \tag{3.27}$$

and τ is an arbitrary positive number.

---

* We recall that the symbol $G_n \rightrightarrows G$ means that G_n converges completely to G (see Section 1.4).

Proof To prove the necessity, we note that the relation $P(\sum_{k=1}^{k_n} X_{nk} < x) \to F(x)$ at every point of continuity of F implies the relation (3.21). Therefore, $\psi_n(t) \to \log f(t)$. By Lemma 1.22 (Chapter 1) we have $G_n \rightrightarrows G$ and $\gamma_n \to \gamma$. The necessity is proved.

Now assume that (3.24) is satisfied. Then $\psi_n \to \psi = (\gamma, G)$ as a consequence of Lemma 1.22 (Chapter 1). It follows from the relation $G_n \rightrightarrows G$ that

$$\sum_k \int_{-\infty}^{\infty} \frac{x^2}{1+x^2} d\bar{F}_{nk}(x) \to G(+\infty) < \infty.$$

By Lemma 3.10 and (3.20) we have

$$\log \prod_k f_{nk}(t) - \psi_n(t) \to 0, \quad \prod_k f_{nk}(t) \to e^{\psi(t)} = f(t)$$

for every t. By Theorem 1.10 (Chapter 1) the distribution functions of the sums $\sum_{k=1}^{k_n} X_{nk}$ converge to the d.f. $F(x)$ with the c.f. $f(t)$. The sufficiency is proved. □

It is easy to generalize Theorem 3.3 in the following way.

Theorem 3.4 *Suppose the conditions of Theorem 3.3 are satisfied, and let $\{b_n\}$ be a sequence of constants. In order that the distributions of the sums $\sum_{k=1}^{k_n} X_{nk} - b_n$ converge weakly to $F(x)$ it is necessary and sufficient that*

$$\sum_{k=1}^{k_n} \int_{-\infty}^{x} \frac{y^2}{1+y^2} dF_{nk}(y + a_{nk}) \rightrightarrows G(x) \tag{3.28}$$

and

$$\sum_{k=1}^{k_n} \left\{ a_{nk} + \int_{-\infty}^{\infty} \frac{x}{1+x^2} dF_{nk}(x + a_{nk}) \right\} - b_n \to \gamma, \tag{3.29}$$

where

$$a_{nk} = \int_{|x|<\tau} x \, dF_{nk}(x) \tag{3.30}$$

and τ is an arbitrary positive constant.

Theorem 3.4 implies a further proposition.

Theorem 3.5 *Let $\{X_{nk}\}$ be a sequence of series of independent random variables satisfying the condition of infinite smallness (3.2). There exists a sequence of constants $\{b_n\}$ such that the distributions of the sums $\sum_k X_{nk} - b_n$ converge weakly to a limiting distribution if and only if the relation (3.28) holds, where $G(x)$ is a non-decreasing bounded function.*

The changes that are introduced into these propositions by the replacement of (3.2) with the condition (3.23) are obvious. In Theorems 3.3, 3.4 and 3.5 it is necessary to replace $F_{nk}(x)$ by $F_{nk}(x + m_{nk})$, where m_{nk} is a median of the random variable X_{nk}. In Theorem 3.3 we have to replace the sum $\sum_k X_{nk}$ by the sum $\sum_k (X_{nk} - m_{nk})$.

It is possible to formulate the necessary and sufficient conditions for the convergence to a given infinitely divisible distribution in terms of the Lévy spectral function $L(x)$ instead of the Lévy–Khintchine spectral function $G(x)$. These formulations can be found in the books of, among others, Gnedenko and Kolmogorov [147] and Petrov [342].

3.4 Limit distributions of class L and stable distributions

Consider a sequence of independent random variables $\{X_n; n = 1, 2, \ldots\}$ and a sequence of positive constants $\{a_n; n = 1, 2, \ldots\}$. Suppose that the condition

$$\max_{1 \leq k \leq n} P(|X_k| \geq \varepsilon a_n) \to 0 \text{ for every fixed } \varepsilon > 0 \tag{3.31}$$

is satisfied. Clearly, this is a particular case of the condition of infinite smallness (3.2) for the sequence of series of random variables $\{X_{nk}; k = 1, \ldots, n; n = 1, 2, \ldots\}$, where $X_{nk} = X_k/a_n$.

We denote by the class L the set of distribution functions that are limits of distributions of sums

$$\frac{1}{a_n} \sum_{k=1}^{n} X_k - b_n, \tag{3.32}$$

where $\{X_n\}$ is a sequence of independent random variables satisfying the condition (3.31), and $\{a_n\}$, $\{b_n\}$ are sequences of constants with $a_n > 0$.

Obviously, the class L is a subset of the set of all infinitely divisible distributions. In order to characterize the distributions of the class L, we need the following proposition.

Lemma 3.11 *Let $X_1, X_2, \ldots$, be a sequence of independent random variables satisfying the condition (3.31) with some sequence of positive constants $a_1, a_2, \ldots$ If the distributions of sums (3.32) converge weakly to a nondegenerate distribution $F(x)$, then $a_n \to \infty$ and $a_{n+1}/a_n \to 1$.*

Proof If $v_k(t)$ denotes the c.f. of the random variable X_k, then the c.f. of the random variable (3.32) has the form

$$f_n(t) = e^{-ib_n t} \prod_{k=1}^{n} v_k(t/a_n). \tag{3.33}$$

We have

$$f_n(t) \to f(t) = \int_{-\infty}^{\infty} e^{itx}\, dF(x). \tag{3.34}$$

Suppose that the condition $a_n \to \infty$ is not satisfied. Then the sequence $\{a_n\}$ contains a bounded sub-sequence, which in turn contains a sub-sequence $\{a_{n_m}\}$ converging to a finite limit a as $m \to \infty$. Let t be an arbitrary fixed number. We put $t_m = a_{n_m} t$. Then $t_m \to at$ as $m \to \infty$. Condition (3.31) and Lemma 3.1 imply that $\max_{1 \leq k \leq n} |v_k(t/a_n) - 1| \to 0$ as $n \to \infty$. Therefore, $|v_k(t)| = |v_k(t_m/a_{n_m})| \to 1$ as $m \to \infty$ for every k. Thus $|v_k(t)| \equiv 1$ and $|f(t)| \equiv 1$, that is, $F(x)$ is a degenerate distribution, contrary to the hypothesis of the lemma. Therefore, $a_n \to \infty$.

Condition (3.31) and Lemma 1.10 (Chapter 1) imply that the distribution of the sum

$$\frac{1}{a_{n+1}} \sum_{k=1}^{n} X_k - b_{n+1}$$

converges to $F(x)$. If $F_n(x)$ denotes the d.f. of the sum (3.32), then

$$P\left(\frac{1}{a_{n+1}} \sum_{k=1}^{n} X_k - b_{n+1} < x\right) = F_n(\alpha_n x + \beta_n),$$

where $\alpha_n = a_{n+1}/a_n$, $\beta_n = b_{n+1} a_{n+1}/a_n - b_n$. It follows from Theorem 1.14 (Chapter 1) that $a_{n+1}/a_n \to 1$. □

Theorem 3.6 *A distribution function $F(x)$ with the characteristic function $f(t)$ belongs to the class L if and only if corresponding to every positive $\alpha < 1$ there exists a c.f. $f_\alpha(t)$ such that*

$$f(t) = f(\alpha t) f_\alpha(t). \tag{3.35}$$

Proof We shall show that (3.35) implies the inequality $f(t) \neq 0$ for all real t. Suppose, to the contrary, that $f(2a) = 0$ and $f(t) \neq 0$ in the interval $0 < t < 2a$. Then $f_\alpha(2a) = 0$ and

$$1 = 1 - |f_\alpha(2a)|^2 \leq 4(1 - |f_\alpha(a)|^2), \tag{3.36}$$

by Lemma 1.3 (Chapter 1). On the other hand, the function $f(t)$ is continuous and, therefore, $f_\alpha(a) = f(a)/f(\alpha a) \to 1$ as $\alpha \to 1$, so that (3.36) cannot be true for α sufficiently close to unity.

Suppose the (3.35) is satisfied. Consider a sequence of independent random variables $\{X_n\}$ with the c.f.

$$v_n(t) = f_{(n-1)/n}(nt) = f(nt)/f((n-1)t)$$

for every n. The c.f. of the sum $\sum_{k=1}^{n} X_k/n$ is equal to $\prod_{k=1}^{n} v_k(t/n) = f(t)$.

Using the continuity of $f(t)$ once more, we conclude that $\max_k |v_k(t/n) - 1| \to 0$ and $\max_k P(|X_k| \geq \varepsilon n) \to 0$ for every $\varepsilon > 0$. Thus $F(x)$ belongs to the class L. The sufficiency is proved.

Let $\{X_n\}$ be a sequence of independent random variables, and $\{a_n\}$ and $\{b_n\}$ be sequences of constants with $a_n > 0$, and such that the distributions of the sums (3.32) converge weakly to $F(x)$ and condition (3.31) is satisfied. We have to prove that for the corresponding c.f. $f(t)$ the representation (3.35) is valid for every positive $\alpha < 1$. We may assume that the distribution $F(x)$ is non-degenerate, since for a degenerate distribution equality (3.35) is obviously true with the degenerate c.f. $f_\alpha(t)$.

Denoting by $v_k(t)$ and $f_n(t)$ respectively the c.f. of X_k and the sum (3.32), we have the relations (3.33) and (3.34). By Theorem 1.15 (Chapter 1), $f(t) \neq 0$ for all t. It follows from Lemma 3.11 that $a_n \to \infty$ and $a_{n+1}/a_n \to 1$. Therefore, for every positive $\alpha < 1$ there exists a sequence of integers $\{m_n\}$ such that $m_n \to \infty$, $n - m_n \to \infty$, and $a_{m_n}/a_n \to \alpha$. Putting

$$f_n^{(1)}(t) = \exp\{-i\alpha b_{n_m} t\} \prod_{k=1}^{m_n} v_k(t/a_n)$$

and

$$f_n^{(2)}(t) = \exp\{-i\alpha(b_n - b_{n_m})t\} \prod_{k=m_n+1}^{n} v_k(t/a_n),$$

we have $f_n(t) = f_n^{(1)}(t) f_n^{(2)}(t)$. Taking into account (3.34) and the properties of the sequence $\{m_n\}$, we conclude that $f_n^{(1)}(t) \to f(\alpha t)$. Hence $f_n^{(2)}(t) \to f_\alpha(t)$, where $f_\alpha(t) = f(t)/f(\alpha t)$. The function $f_\alpha(t)$ is continuous. By Theorem 1.10 (Chapter 1), it is a c.f. □

It is possible to characterize the characteristic functions of the distributions belonging to the class L in terms of the Lévy spectral function $L(x)$ in the Lévy formula for the c.f. of infinitely divisible distributions. The corresponding propositions can be found, for example, in [147] and [342]. It follows from these results that the Lévy spectral function of any distribution of the class L is continuous at every point $x \neq 0$. In turn, this implies that the Poisson distribution with arbitrary parameters (a, b, λ) does not belong to the class L, because its Lévy spectral function has a discontinuity at the point $b \neq 0$. The normal and degenerate distributions belong to the class L.

Under the additional assumption that the independent random variables $X_1, X_2, \ldots$ have identical distributions, the set of all limit distributions for the sums (3.32) is a very narrow subset of the set of all infinitely divisible distributions. This subset coincides with the set of the stable distributions.

A distribution function $F(x)$ and the corresponding c.f. $f(t)$ are called stable if for every positive a_1 and a_2 there exist real numbers $a > 0$ and b such that

$$f(a_1 t) f(a_2 t) = e^{ibt} f(at). \tag{3.37}$$

In other words, a distribution $F(x)$ is stable if and only if for every $a_1 > 0$, $a_2 > 0$, b_1 and b_2 there exist real numbers $a > 0$ and b such that

$$F(a_1 x + b_1) * F(a_2 x + b_2) = F(ax + b).$$

In (3.37) we put $at = u$, $\alpha = a_1/a$, $\alpha_1 = a_2/a$, and we obtain

$$f(u) = \exp\left\{-\frac{i}{a} bu\right\} f(\alpha u) f(\alpha_1 u) = f(\alpha u) f_\alpha(u),$$

where $f_\alpha(u) = \exp\{-(i/a)bu\} f(\alpha_1 u)$ is a c.f. It follows that an arbitrary stable distribution belongs to the class L.

Theorem 3.7 *The set of distributions that are limits of distributions of sums (3.32) of independent identically distributed random variables coincides with the set of stable distributions.*

Proof Let $F(x)$ be a stable d.f. and let $f(t)$ be the corresponding c.f. We consider a sequence of independent random variables X_n with a common d.f. $F(x)$. The sum $S_n = X_1 + \cdots + X_n$ has the c.f. $f^n(t)$. By (3.37) we have $f^n(t) = e^{ib_n t} f(a_n t)$. Therefore, the random variable $S_n/a_n - b_n$ has the c.f. $f(t)$ and the d.f. $F(x)$.

It remains only to show that a distribution function $F(x)$ is stable if it is a limit of distributions of sums (3.32) of independent identically distributed random variables where $a_n > 0$ and b_n are constants. We shall suppose that $F(x)$ is non-degenerate, since degenerate distributions are obviously stable.

Let $g(t)$ be the c.f. of the random variable X. Then

$$e^{-ib_n t} g^n(t/a_n) \to f(t) \tag{3.38}$$

for every real t. We shall show that the random variables X_k/a_n ($k = 1, 2, \ldots, n$) satisfy the condition of infinite smallness, that is, $X_1/a_n \xrightarrow{P} 0$. This is equivalent to the condition

$$g(t/a_n) \to 1 \quad \text{for every real } t. \tag{3.39}$$

Let $\delta > 0$ be such that $f(t) \neq 0$ for $|t| \leq \delta$. It follows from (3.38) that $\exp\{-(i/n)b_n t\} g(t/a_n) \to 1$ for $|t| \leq \delta$. By inequality (1.14) (Chapter 1) we obtain

$$1 - \operatorname{Re} g(2t/a_n) \leq 4(1 - \operatorname{Re} g(t/a_n)).$$

Therefore, $\operatorname{Re} g(t/a_n) \to 1$ in the interval $|t| \leq 2\delta$, and (3.39) follows.

By Lemma 3.11 we have $a_n \to \infty$ and $a_{n+1}/a_n \to 1$. Let c_1 and c_2 be arbitrary positive numbers, and let d_1 and d_2 be arbitrary real numbers. There exists a sequence of integers $\{m_n\}$ such that $a_{m_n}/a_n \to c_1/c_2$. We put

$$\alpha_n = a_n c_1, \quad \beta_n = (a_n b_n + a_{m_n} b_{m_n} + a_n d_1 + a_{m_n} d_2)/\alpha_n.$$

Then
$$\frac{a_n}{\alpha_n}\left(\frac{1}{a_n}\sum_{k=1}^{n} X_k - b_n - d_1\right) + \frac{a_{m_n}}{\alpha_n}\left(\frac{1}{a_{m_n}}\sum_{k=n+1}^{n+m_n} X_k - b_{m_n} - d_2\right) = \frac{1}{\alpha_n}\sum_{k=1}^{n+m_n} X_k - \beta_n. \quad (3.40)$$

The distribution functions of the first and second terms on the left-hand side of (3.40) converge respectively to $F(c_1 x + d_1)$ and $F(c_2 x + d_2)$ by Theorem 1.14 (Chapter 1). It follows that the d.f. of the right-hand side of (3.40) has the limit $F(c_1 x + d_1) * F(c_2 x + d_2)$. On the other hand, by the same theorem the d.f. of the right-hand side of (3.40) can converge only to a function of the form $F(ax + b)$. Thus

$$F(c_1 x + d_1) * F(c_2 x + d_2) = F(ax + b).$$

Therefore, $F(x)$ is a stable d.f. □

Theorem 3.8 *An infinitely divisible c.f. $f(t)$ is stable if and only if the corresponding Lévy spectral function $L(x)$ and the non-negative constant σ^2 in the Lévy formula (equality (1.53), Chapter 1) satisfy one of the two following conditions:*

(A) $L(x) \equiv 0$;
(B) $\sigma^2 = 0$, $L(x) = c_1 |x|^{-\alpha}$ for $x < 0$, $L(x) = -c_2 x^{-\alpha}$ for $x > 0$, where $0 < \alpha < 2$, $c_1 \geq 0$, $c_2 \geq 0$, $c_1 + c_2 > 0$.

Theorem 3.9 *A characteristic function $f(t)$ is stable if and only if it admits the representation*

$$f(t) = \exp\left\{i\gamma t - c|t|^{\alpha}\left(1 + i\beta \frac{t}{|t|}\omega(t, \alpha)\right)\right\}, \quad (3.41)$$

where c, α, β and γ are constants (γ a real constant, $c \geq 0$, $0 < \alpha \leq 2$, $-1 \leq \beta \leq 1$) and

$$\omega(t, \alpha) = \begin{cases} \tan \dfrac{\pi \alpha}{2}, & \text{if } \alpha \neq 1, \\ \dfrac{2}{\pi} \log |t|, & \text{if } \alpha = 1. \end{cases}$$

If $\alpha < 2$, the constant α is the same as in Theorem 3.8.

The proofs of Theorems 3.8 and 3.9 can be found in Gnedenko and Kolmogorov [147] or Ibragimov and Linnik [205]. In what follows, we do not use these theorems.

The value $c = 0$ corresponds to a degenerate distribution, and the value $\alpha = 2$ to a normal distribution. It follows from Theorem 3.9 that if $f(t)$ is the c.f. of a non-degenerate stable distribution $F(x)$, then $|f(t)| = \exp\{-c|t|^\alpha\}$ ($0 < \alpha \leq 2$), and therefore the d.f. $F(x)$ has an everywhere continuous derivative (and, furthermore, derivatives of all orders).

Let $\{X_n\}$ be a sequence of independent random variables with the same d.f. $V(x)$. If there exist sequences of constants $\{a_n\}$ and $\{b_n\}$ such that $a_n > 0$ and the distributions of the sums

$$Z_n = \frac{1}{a_n} \sum_{k=1}^{n} X_k - b_n \qquad (3.42)$$

converge weakly to some d.f. $G(x)$, then we say that $V(x)$ is attracted to $G(x)$. The set of all distribution functions that are attracted to $G(x)$ is called the domain of attraction of the distribution $G(x)$. It follows from Theorem 3.7 that only stable distributions have domains of attraction.

Theorem 3.10 *A distribution function $V(x)$ belongs to the domain of attraction of a normal distribution if and only if*

$$\int_{|x| \geq z} dV(x) = o\left(\frac{1}{z^2} \int_{|x| < z} x^2 \, dV(x)\right) \qquad (z \to +\infty).$$

The constant α in the representation (3.41) of a stable c.f. is called the exponent of this stable distribution.

Theorem 3.11 *A distribution function $V(x)$ belongs to the domain of attraction of a stable distribution with the exponent $\alpha < 2$ if and only if*

$$V(x) = (c_1 + o(1))|x|^{-\alpha} h(|x|) \qquad \text{as } x \to -\infty$$

and

$$1 - V(x) = (c_2 + o(1))x^{-\alpha} h(x) \qquad \text{as } x \to +\infty,$$

where $h(x)$ is a slowly varying function*. Here c_1 and c_2 are the same constants as those defined in Theorem 3.8.

The proofs of Theorems 3.10 and 3.11 can be found in the above-mentioned books [147] and [205].

Some results in the Addenda to Chapter 7 are connected with the concepts of the domain of normal attraction and partial attraction of a stable distribution. We shall present the corresponding definitions.

A d.f. $V(x)$ belongs to the domain of normal attraction of a stable distribution $F(x)$ with the exponent α ($0 < \alpha \leq 2$) if there exist a sequence of constants $\{b_n\}$ and a positive number c such that the distributions of the

---

* A positive function $h(x)$ defined in the interval $x > 0$ is called slowly varying if $h(cx)/h(x) \to 1$ as $x \to +\infty$ for every $c > 0$. Surveys of properties of slowly varying functions and a more general class of regularly varying functions can be found in Feller [133], Seneta [419], and Bingham et al. [41].

sums (3.42), where $X_1, \ldots, X_n$ are independent random variables all having the d.f. $V(x)$, converge weakly to $F(x)$ with $a_n = cn^{1/\alpha}$.

A d.f. $V(x)$ belongs to the domain of normal attraction of a normal distribution if and only if $\int_{-\infty}^{\infty} x^2 \, dV(x) < \infty$. A d.f. $V(x)$ belongs to the domain of normal attraction of a stable distribution with the exponent α ($0 < \alpha < 2$) and the given constants c, c_1, and c_2, if and only if

$$V(x) = (c_1 c^\alpha + o(1))|x|^{-\alpha} \qquad \text{as } x \to -\infty,$$
$$1 - V(x) = (c_2 c^\alpha + o(1))x^{-\alpha} \qquad \text{as } x \to +\infty.$$

Here c_1 and c_2 are the same constants as in Theorems 3.8 and 3.11.

By definition, a d.f. $V(x)$ belongs to the domain of partial attraction of a stable distribution $F(x)$ if there exist sequences of constants $\{a_n\}, \{b_n\}$ ($a_n > 0$) and a sequence of integers $\{n_k\}$ such that $n_k \to \infty$ as $k \to \infty$ and the distributions of the sums Z_{n_k}, where Z_n is defined by (3.42), converge weakly to $F(x)$.

A d.f. $V(x)$ belongs to the domain of partial attraction of a normal distribution if and only if

$$\liminf_{z \to +\infty} \left(\int_{|x| \geq z} dV(x) \bigg/ \frac{1}{z^2} \int_{|x| < z} x^2 \, dV(x) \right) = 0.$$

3.5 Bibliographical notes

The results of this chapter can be found in the classical book by Gnedenko and Kolmogorov [147]. We follow Loève [256] in the proof of a number of the propositions. Theorem 3.1 was proved by Khintchine [221]. Convergence of the distributions of sums of independent random variables to a given infinitely divisible distribution was studied by Gnedenko and Doeblin. The necessary and sufficient conditions for this convergence that were stated in Section 3.3 are due to Gnedenko. Most of the results of Section 3.4 are due to Lévy and Khintchine. A detailed history of the problem can be found in the books by Gnedenko and Kolmogorov [147] and Khintchine [222].

The nearness of the distributions of sums of independent random variables to the class of infinitely devisible laws was investigated by Kolmogorov [233, 235], Prohorov [370], Tsaregradskii [455], LeCam [249], Ibragimov and Presman [206], Arak [9, 10], Zaitsev [475, 476], Zaitsev and Arak [477], and others. A survey of these investigations and new results can be found in the monograph by Arak and Zaitsev [11].

The theory of summation of independent random variables without the condition of infinite smallness was developed by Zolotarev, Kruglov, Rotar and Bergström. Surveys of this development are given by Rotar [401] and Zolotarev [486].

Analytical properties of the distributions of the class L were studied by Zolotarev [481], Wolfe [470], and Yamazato [474]. Analytical properties of the stable distributions take up a significant portion of the books by Ibragimov and Linnik [205], Mijnheer [298], and Zolotarev [485].

Zinger [479, 480] obtained characterizations of the class of the distributions that are limits of the distributions of normed sums of independent random variables which have no more than r different distributions or types of the distributions.

Kruglov [241] obtained a characterization of the class of the distributions that are limits of the distributions of the normed sums $Z_{n_k} = \sum_{j=1}^{n_k} X_j/b_k + a_k$, where $\{X_n\}$ is a sequence of independent identically distributed random variables, $\{b_k\}$ and $\{a_k\}$ are sequences of real numbers, and $\{n_k\}$ is an increasing sequence of integers such that $\lim_{k \to \infty} n_{k+1}/n_k = r$, $1 \leqslant r < \infty$. Analogous results were independently obtained by Mejzler [295]. Moreover, it follows from [295] that the class of limit distributions is the same even if we replace the condition on the sequence $\{n_k\}$ by the weaker condition $\liminf_{k \to \infty} n_k/n_{k+1} > 0$.

A series of papers by Rossberg, Jesiak and Siegel (see, for example, [395–397]) is devoted to finding conditions under which the weak convergence of the distributions of sums of independent random variables on the whole real line is an implication of the convergence of these distributions on some subsets of the real line.

3.6 Addenda

3.6.1 Let $\{X_{nk}\}$ be a sequence of series of random variables that are independent within each series and satisfy the condition of infinite smallness. Let $F_{nk}(x) = P(X_{nk} < x)$, and let $\Pi(x)$ be the Poisson d.f. with the c.f. $\pi(t) = \exp\{\lambda(e^{it} - 1)\}$. In order that the distribution functions of the sums $\sum_k X_{nk}$ converge weakly to $\Pi(x)$ it is necessary and sufficient that for every positive $\varepsilon < 1$ and some positive $\tau < 1$,

$$\sum_k P(|X_{nk}| \geqslant \varepsilon, |X_{nk} - 1| \geqslant \varepsilon) \to 0,$$

$$\sum_k P(|X_{nk} - 1| < \varepsilon) \to \lambda,$$

$$\sum_k \left\{ \int_{|x|<\tau} x^2 \, dF_{nk}(x) - \left(\int_{|x|<\tau} x \, dF_{nk}(x) \right)^2 \right\} \to 0,$$

$$\sum_k \int_{|x|<\tau} x \, dF_{nk}(x) \to 0$$

(see, for instance, Loève [256]).

In Subsections 3.6.2 and 3.6.3 we consider a sequence of independent random variables $\{X_n\}$ having a common d.f. $F(x)$. We shall use the following notation:

$$F^n(x) = P\left(\sum_{k=1}^{n} X_k < x\right),$$

$\mathscr{F}$ is the set of all d.f., $\mathscr{D}$ is the set of all infinitely divisible d.f., $\rho(F, G) = \sup_x |F(x) - G(x)|$ for every d.f. $F(x)$ and $G(x)$,

$$\rho(F^n, \mathscr{D}) = \inf_{D \in \mathscr{D}} \rho(F^n, D), \quad \psi(n) = \sup_{F \in \mathscr{F}} \rho(F^n, \mathscr{D}).$$

3.6.2 For every d.f. $F(x)$ there exists a sequence of infinitely divisible d.f. $\{G_n(x)\}$ such that $\rho(F^n, G_n) \to 0$ (Prohorov [365]).

3.6.3 For every $n \in \mathbb{N}$ and every d.f. F there exists an infinitely divisible d.f. B such that $\rho(F^n, B) \leqslant An^{-2/3}$. For every $n \in \mathbb{N}$ there exists a d.f. $F_n \in \mathscr{F}$ such that $\rho(F_n^n, \mathscr{D}) \geqslant A_0 n^{-2/3}$. Here A and A_0 are absolute positive constants. Thus there exist absolute positive constants A_1 and A_2 such that

$$A_1 n^{-2/3} \leqslant \psi(n) \leqslant A_2 n^{-2/3}$$

for all n (Arak [9, 10]).

3.6.4 Let ε be a positive number. Let the distributions $F_i \in \mathscr{F}$ ($i = 1, \ldots, n$) satisfy the inequalities $L(F_i, E_{\beta_i}) \leqslant \varepsilon$ for some β_i, where $L(F, G)$ is the Lévy distance between the d.f. F and G, and E_a is the degenerate distribution concentrated at the point $a \in \mathbb{R}$. Then there exists an infinitely divisible distribution D such that

$$L(F_1 * \cdots * F_n, D) \leqslant A\varepsilon(|\log \varepsilon| + 1),$$

where A is an absolute positive constant (Zaitsev; see Zaitsev and Arak [477]).

3.6.5 For every positive $\varepsilon \leqslant 1$ there exist $n \in \mathbb{N}$ and a d.f. F such that $L(F, E_0) \leqslant \varepsilon$ and

$$L(F^{*n}, \mathscr{D}) \geqslant A_0 \varepsilon(|\log \varepsilon| + 1),$$

where A_0 is an absolute positive constant (Arak; see Zaitsev and Arak [477]).

The following proposition is a consequence of the results stated in Subsections 3.6.4 and 3.6.5. Putting

$$\Omega(\varepsilon) = \{G \in \mathscr{F} : L(G, E_a) \leqslant \varepsilon \text{ for some } a \in \mathbb{R}\}$$

and

$$\varphi(\varepsilon) = \sup_{n} \sup_{F_i \in \Omega(\varepsilon)} L(F_1 * \cdots * F_n, \mathscr{D}),$$

we have
$$A_1\varepsilon(|\log \varepsilon| + 1) \leq \varphi(\varepsilon) \leq A_2\varepsilon(|\log \varepsilon| + 1),$$
where A_1 and A_2 are absolute positive constants (Arak and Zaitsev [11]).

3.6.6 Let X_n be a sequence of independent random variables, and let $\{a_n\}$ be a sequence of positive numbers. Let the distributions of the normed sums $\sum_{k=1}^{n} X_k/a_n$ converge weakly to some non-degenerate distribution. Then the sequence $\{a_n\}$ either converges to a finite limit or diverges to $+\infty$ almost monotonically in the sense that there exists a sequence of numbers $\{b_n\}$ such that $b_n \geq a_n$, $b_n/a_n \to 1$, and $b_n \uparrow \infty$ (Hsu [197]).

3.6.7 An arbitrary non-degenerate infinitely divisible distribution function, belonging to the class L, is absolutely continuous (Hsu [198]; later this fact was rediscovered by Fisz and Varadarajan [137] and Zolotarev [481]).

A distribution function $F(x)$ is said to be unimodal with mode M if $F(x)$ is convex for $x < M$ and concave for $x > M$. $F(x)$ is said to be unimodal if, for some M, it is unimodal with mode M.

3.6.8 All distribution functions of the class L are unimodal (Yamazato [474]).

A sequence of random variables $\{Y_n\}$ is said to be stochastically compact if every sub-sequence of $\{Y_n\}$ contains another sub-sequence with distributions converging weakly to some non-degenerate distribution.

3.6.9 Let $\{X_n\}$ be a sequence of independent identically distributed random variables. We put
$$F(x) = P(X_1 < x), \quad S_n = \sum_{k=1}^{n} X_k,$$
$$K(x) = x^{-2} \int_{|y|<x} y^2 \, dF(y), \quad H = \left\{F : \limsup_{x \to \infty} P(|X_1| \geq x)/K(x) < \infty\right\}.$$
Then $F \in H$ if and only if there exist sequences of numbers $\{a_n\}$ and $\{b_n\}$ such that $a_n > 0$ and the sequence $\{(S_n - b_n)/a_n\}$ is stochastically compact (Feller [132]). Other criteria of stochastic compactness are found by Maller [272]. Pruitt [372] obtained a characterization of the class of limit distributions for sub-sequences of a stochastically compact sequence of normed sums of independent identically distributed random variables.

3.6.10 If $F(x)$ is a stable d.f. with the exponent $\alpha < 2$, then $\int_{-\infty}^{\infty} |x|^p \, dF(x) < \infty$ for every positive $p < \alpha$ (Gnedenko [144]).

3.6.11 If $V(x)$ is a d.f. that belongs to the domain of attraction of a stable distribution with the exponent $\alpha \leq 2$, then $\int_{-\infty}^{\infty} |x|^p \, dV(x) < \infty$ for every $p < \alpha$. This proposition was proved by Khintchine, Lévy and Feller for $\alpha = 2$ and by Gnedenko [145] for $\alpha < 2$.

3.6.12 Let $V(x)$ be a d.f. that belongs to the domain of normal attraction of a stable distribution with the exponent $\alpha < 2$. Let δ be an arbitrary positive number, and let $\psi(x)$ be an even function that is positive and non-decreasing in the interval $x \geq \delta$. Then the integral $\int_{|x| \geq \delta} [|x|^\alpha / \psi(x)] \, dV(x)$ converges if and only if the integral $\int_\delta^\infty dx / x\psi(x)$ converges. (This proposition fails if we replace the condition that $V(x)$ belongs to the domain of normal attraction of a stable distribution by the weaker condition that $V(x)$ belongs to the domain of attraction of a stable distribution.) In particular, if a random variable X has the d.f. that belongs to the domain of normal attraction of a stable distribution with the exponent $\alpha < 2$, then the moment $E|X|^\alpha (\log(|X| + 2))^{-\beta}$ is finite (infinite) for $\beta > 1$ (respectively, $\beta \leq 1$) (Petrov [337]).

3.6.13 Let $\{X_{nk}\}$ be a sequence of (not necessarily finite) series of random variables that are independent within each series, and let X be a random variable. We put

$$X_n = \sum_k X_{nk}, \quad F_n(x) = P(X_n < x), \quad F(x) = P(X < x),$$

$$F_{nk}(x) = P(X_{nk} < x), \quad \bar{F}_{nk}(x) = F_{nk}(x + m_{nk}),$$

where m_{nk} is a median of X_{nk}. Suppose that $E|X_{nk}|^p < \infty$ for all n and k and some $p > 0$. If $r > 1/p$, then the set of the conditions

$$\int_{-\infty}^{\infty} |F_n(x) - F(x)|^r \, dx \to 0, \quad E|X_n|^p \to E|X|^p$$

is equivalent to the conditions

$$\lim_{T \to \infty} \sup_n \sum_k \int_{|x| \geq T} |x|^p \, d\bar{F}_{nk}(x) = 0$$

and $L(F_n, F) \to 0$, where L is the Lévy metric (Kruglov [243]). Kruglov [244] obtained the necessary and sufficient conditions for the relation

$$\int_{-\infty}^{\infty} \varphi(F_n(x) - F(x)) q(x) \, dx \to 0,$$

where $\varphi(x)$ and $q(x)$ are continuous functions satisfying some conditions of general character.

4

Weak limit theorems: the central limit theorem and the weak law of large numbers

4.1 The central limit theorem for a sequence of series of independent random variables

In Chapter 3 we studied the conditions under which the distributions of sums of independent random variables will converge to a limit distribution. We were interested in the weak convergence of distributions; the corresponding limit theorems are called the weak limit theorems, contrary to the strong limit theorems making use of the convergence almost surely.

The necessary and sufficient conditions were found for the weak convergence of distributions of sums of independent random variables to a given infinitely divisible distribution. These conditions are quite complicated but are very helpful when we are looking for simpler conditions of convergence to a prescribed infinitely divisible distribution such as a normal, a degenerate, or a Poisson distribution.

First we consider the case when the limit distribution is normal. We denote the parameters of this distribution by (a, σ). For this case we have in the Lévy–Khintchine formula (equality (1.41), Chapter 1) $\gamma = a$ and

$$G(x) = \begin{cases} 0 & \text{if } x \leq 0, \\ \sigma^2 & \text{if } x > 0. \end{cases} \tag{4.1}$$

Therefore, Theorem 3.3 (Chapter 3) implies the following proposition. Let $\{X_{nk}\}$ be a sequence of series of independent random variables satisfying the condition of infinite smallness, and let $F_{nk}(x)$ be the distribution function of the random variable X_{nk}. We define $G_n(x)$, γ_n, a_{nk} and $\bar{F}_{nk}(x)$ by the equalities (3.25), (3.26), and (3.27) (Chapter 3) respectively. In order that the distributions of the sums $\sum_k X_{nk}$ converge weakly to the normal distribution with parameters (a, σ), it is necessary and sufficient that

$$G_n \Rightarrow G \tag{4.2}$$

and

$$\gamma_n \to a, \tag{4.3}$$

where G is defined by (4.1).

With the help of this result we shall prove the following much more pleasing proposition.

Theorem 4.1 *Let $\{X_{nk}; k = 1, \ldots, k_n; n = 1, 2, \ldots\}$ be a sequence of series of random variables that are independent within each series, and let $F_{nk}(x)$ be the distribution function of X_{nk}. The condition of infinite smallness (3.2) of Chapter 3 will be satisfied and the distribution of the sums $\sum_k X_{nk}$ will converge weakly to the normal (a, σ) distribution if and only if for every fixed $\varepsilon > 0$ the following conditions are satisfied:*

$$\sum_k P(|X_{nk}| \geq \varepsilon) \to 0, \qquad (4.4)$$

$$\sum_k \left\{ \int_{|x|<\varepsilon} x^2 \, dF_{nk}(x) - \left(\int_{|x|<\varepsilon} x \, dF_{nk}(x) \right)^2 \right\} \to \sigma^2, \qquad (4.5)$$

$$\sum_k \int_{|x|<\varepsilon} x \, dF_{nk}(x) \to a. \qquad (4.6)$$

Proof First we shall prove the sufficiency. Suppose that the conditions (4.4)–(4.6) are satisfied. Then

$$\max_k P(|X_{nk}| \geq \varepsilon) \leq \sum_k P(|X_{nk}| \geq \varepsilon) \to 0$$

for every $\varepsilon > 0$ by (4.4), that is, the condition of infinite smallness is satisfied. Taking into account the above-mentioned corollary to Theorem 3.3 (Chapter 3), we conclude that the sufficiency will be proved if we prove the relations (4.2) and (4.3).

We now prove (4.2). First of all we shall prove that

$$\sum_k \int_{|x|<\varepsilon} x^2 \, d\bar{F}_{nk}(x) \to \sigma^2 \qquad \text{for every } \varepsilon > 0. \qquad (4.7)$$

Let us consider the sums

$$T_1 = \sum_k \left\{ \int_{|x|<\varepsilon} x^2 \, d\bar{F}_{nk}(x) - \int_{|x|<\varepsilon} (x - a_{nk})^2 \, dF_{nk}(x) \right\} \qquad (4.8)$$

and

$$T_2 = \sum_k \left\{ \int_{|x|<\varepsilon} (x - a_{nk})^2 \, dF_{nk}(x) - \int_{|x|<\varepsilon} x^2 \, dF_{nk}(x) + \left(\int_{|x|<\varepsilon} x \, dF_{nk}(x) \right)^2 \right\} \qquad (4.9)$$

We shall show that

$$T_1 \to 0, \quad T_2 \to 0 \quad (n \to \infty). \qquad (4.10)$$

114 | Limit theorems of probability theory

It follows from the equality (3.27) (Chapter 3) that

$$T_1 = \sum_k \left\{ \int_{|x-a_{nk}|<\varepsilon} (x-a_{nk})^2 \, dF_{nk}(x) - \int_{|x|<\varepsilon} (x-a_{nk})^2 \, dF_{nk}(x) \right\}.$$

The set $\{x \in \mathbb{R}: |x-a_{nk}| < \varepsilon\}$ is the union of the disjunct sets $\{x: |x-a_{nk}| < \varepsilon, |x| < \varepsilon\}$ and $\{x: |x-a_{nk}| < \varepsilon, |x| \geq \varepsilon\}$. A similar representation is true for the set $\{x: |x| < \varepsilon\}$. Every integral is equal to the sum of the integrals over these disjunct sets, therefore, we obtain, after some simplifications,

$$T_1 = \sum_k \left\{ \int_{\substack{|x-a_{nk}|<\varepsilon \\ |x| \geq \varepsilon}} (x-a_{nk})^2 \, dF_{nk}(x) - \int_{\substack{|x|<\varepsilon \\ |x-a_{nk}| \geq \varepsilon}} (x-a_{nk})^2 \, dF_{nk}(x) \right\}.$$

By (3.27) and Lemma 1.1 (Chapter 1) we have

$$\max_k |a_{nk}| \leq \max_k \int_{|x|<\tau} |x| \, dF_{nk}(x) \to 0,$$

since the condition of infinite smallness is satisfied. Hence

$$\{x: |x-a_{nk}| < \varepsilon, |x| \geq \varepsilon\} \subset \{x: \varepsilon \leq |x| < 2\varepsilon\}$$

and

$$\{x: |x| < \varepsilon, |x-a_{nk}| \geq \varepsilon\} \subset \{x: \varepsilon/2 \leq |x| < \varepsilon\}$$

for all k and sufficiently large n. It follows that

$$|T_1| \leq \sum_k \int_{\varepsilon/2 \leq |x| < 2\varepsilon} (x-a_{nk})^2 \, dF_{nk}(x) \leq 9\varepsilon^2 \sum_k P(|X_{nk}| \geq \varepsilon/2) \to 0$$

for every $\varepsilon > 0$ by (4.4). Furthermore,

$$T_2 = \sum_k \left\{ -2a_{nk} \int_{|x|<\varepsilon} x \, dF_{nk}(x) + a_{nk}^2 \int_{|x|<\varepsilon} dF_{nk}(x) + \left(\int_{|x|<\varepsilon} x \, dF_{nk}(x) \right)^2 \right\}$$

$$= \sum_k \left\{ \left(a_{nk} - \int_{|x|<\varepsilon} x \, dF_{nk}(x) \right)^2 - a_{nk}^2 + a_{nk}^2 \int_{|x|<\varepsilon} dF_{nk}(x) \right\}.$$

If $\varepsilon < \tau$, where τ is the number from the definition of a_{nk}, then

$$|T_2| \leq \left| \sum_k \left\{ \left(\int_{\varepsilon \leq |x| < \tau} x \, dF_{nk}(x) \right)^2 - a_{nk}^2 \int_{|x| \geq \varepsilon} dF_{nk}(x) \right\} \right|$$

$$\leq \tau^2 \sum_k \int_{|x| \geq \varepsilon} dF_{nk}(x) + \max_k a_{nk}^2 \sum_k P(|X_{nk}| \geq \varepsilon) \to 0,$$

by (4.4). The case when $\varepsilon \geq \tau$ can be treated in a similar way. Relations (4.10) are proved. Note that we obtained (4.10) as a consequence of condition (4.4). In the proof of (4.10) we have not used conditions (4.5) and (4.6).

Obviously, (4.5) and (4.10) imply (4.7). We write

$$I_n(\varepsilon) = \sum_k \int_{|x|<\varepsilon} \frac{x^2}{1+x^2} \, d\bar{F}_{nk}(x). \qquad (4.11)$$

It can be proved that

$$I_n(\varepsilon) \to \sigma^2 \qquad \text{for every } \varepsilon > 0. \qquad (4.12)$$

In fact, if $0 < \underline{\varepsilon} < \varepsilon$, then

$$(1+\underline{\varepsilon}^2)^{-1} \sum_k \int_{|x|<\underline{\varepsilon}} x^2 \, d\bar{F}_{nk}(x) \leq I_n(\varepsilon) \leq \sum_k \int_{|x|<\varepsilon} x^2 \, d\bar{F}_{nk}(x).$$

Applying (4.7), we have

$$(1+\underline{\varepsilon}^2)^{-1} \sigma^2 \leq \liminf I_n(\varepsilon) \leq \limsup I_n(\varepsilon) \leq \sigma^2.$$

Passing to the limit as $\underline{\varepsilon} \downarrow 0$, we find that $\lim I_n(\varepsilon) = \sigma^2$.

The next step of the proof is to show that

$$\sum_k \int_{|x| \geq \varepsilon} \frac{x^2}{1+x^2} \, d\bar{F}_{nk}(x) \to 0 \qquad \text{for every } \varepsilon > 0. \qquad (4.13)$$

The sum in (4.13) does not exceed the quantity

$$\sum_k \int_{|x| \geq \varepsilon} d\bar{F}_{nk}(x) = \sum_k P(|X_{nk} - a_{nk}|) \geq \varepsilon).$$

We have $\max_k |a_{nk}| < \varepsilon/2$ for every $\varepsilon > 0$ and all sufficiently large n, therefore,

$$\sum_k P(|X_{nk} - a_{nk}| \geq \varepsilon) \leq \sum_k P(|X_{nk}| \geq \varepsilon/2) \to 0.$$

The assertion (4.13) is proved.

Taking account of (4.11), (4.12), (4.13), and (4.1), we arrive at the relation (4.2). To complete the proof of the sufficiency, we need to prove (4.3). This relation is an obvious consequence of (4.6),

$$\sum_k \int_{-\infty}^{\infty} \frac{x}{1+x^2} \, d\bar{F}_{nk}(x) \to 0 \qquad (4.14)$$

and the definition of γ_n given by the equality (3.26) (Chapter 3). Thus, the sufficiency will be proved if we prove (4.14).

The left-hand side of (4.14) is equal to the sum $L_1 - L_2 + L_3$, where

$$L_1 = \sum_k \int_{|x|<\tau} x \, d\bar{F}_{nk}(x), \quad L_2 = \sum_k \int_{|x|<\tau} \frac{x^3}{1+x^2} \, d\bar{F}_{nk}(x),$$

$$L_3 = \sum_k \int_{|x| \geq \tau} \frac{x}{1+x^2} \, d\bar{F}_{nk}(x).$$

We shall show that $L_1 \to 0$, $L_2 \to 0$, and $L_3 \to 0$. The relation (4.14) follows from these relations.

By (4.2) and Theorem 1.8 (Chapter 1) we have

$$L_2 = \int_{|x|<\tau} x \, dG_n(x) \to \int_{|x|<\tau} x \, dG(x) = 0.$$

Here we have made use of the definitions of the functions $G(x)$ and $G_n(x)$ given by (4.1) and (3.25) (Chapter 3). Furthermore,

$$L_3 = \int_{|x| \geq \tau} \frac{1}{x} dG_n(x) \to \int_{|x| \geq \tau} \frac{1}{x} dG(x)$$

by the same reasoning. It is clear that

$$L_1 = \sum_k \int_{|x|<\tau} x \, dF_{nk}(x + a_{nk}) = \sum_k \int_{|x-a_{nk}|<\tau} (x - a_{nk}) \, dF_{nk}(x)$$

$$= \sum_k \left\{ \int_{\substack{|x-a_{nk}|<\tau \\ |x|<\tau}} + \int_{\substack{|x-a_{nk}|<\tau \\ |x| \geq \tau}} \right\} (x - a_{nk}) \, dF_{nk}(x).$$

Hence

$$L_1 = \sum_k \left\{ \int_{|x|<\tau} - \int_{\substack{|x| \leq \epsilon \\ |x-a_{nk}| \geq \tau}} + \int_{\substack{|x| \geq \tau \\ |x-a_{nk}|<\tau}} \right\} (x - a_{nk}) \, dF_{nk}(x).$$

Taking account of (4.4) and the relation $\max_k |a_{nk}| \to 0$, we obtain

$$|L_1| \leq \left| \sum_k \int_{|x|<\tau} (x - a_{nk}) \, dF_{nk}(x) \right|$$

$$+ \sum_k \int_{\substack{|x|<\tau \\ |x-a_{nk}| \geq \tau}} |x - a_{nk}| \, dF_{nk}(x) + \sum_k \int_{\substack{|x| \geq \tau \\ |x-a_{nk}|<\tau}}$$

$$\leq \left| \sum_k a_{nk} \int_{|x| \geq \tau} dF_{nk}(x) \right| + 2\tau \sum_k P(|X_{nk}| \geq \tau/2) + \tau \sum_k P(|X_{nk}| \geq \tau) \to 0.$$

The sufficiency is proved.

We now turn to the proof of the necessity. We have to show that the condition of infinite smallness and and the relation

$$P\left(\sum_k X_{nk} < x\right) \to \frac{1}{\sigma\sqrt{2\pi}} \int_{-\infty}^x \exp\left\{-\frac{(t-a)^2}{2\sigma^2}\right\} dt \quad (4.15)$$

imply the relations (4.4)–(4.6). By the above-mentioned corollary to Theorem 3.3 (Chapter 3), the condition of infinite smallness and (4.15) imply (4.2) and (4.3). In accordance with the definitions of the functions $G_n(x)$ and $G(x)$,

we have

$$\sum_k \int_{|x|\geq \varepsilon} \frac{x^2}{1+x^2} d\bar{F}_{nk}(x) = \sum_k \left\{ \int_{-\infty}^{\infty} - \int_{-\infty}^{\varepsilon} + \int_{-\infty}^{-\varepsilon} \right\} \frac{x^2}{1+x^2} d\bar{F}_{nk}(x)$$

$$\to \sigma^2 - \sigma^2 = 0.$$

The function $x^2/(1+x^2)$ does not decrease in the interval $x > 0$, therefore,

$$\int_{|x|\geq \varepsilon} \frac{x^2}{1+x^2} d\bar{F}_{nk}(x) \geq \frac{\varepsilon^2}{1+\varepsilon^2} \int_{|x|\geq \varepsilon} d\bar{F}_{nk}(x)$$

and

$$\sum_k \int_{|x|\geq \varepsilon} d\bar{F}_{nk}(x) \to 0 \quad \text{for every } \varepsilon > 0. \tag{4.16}$$

It is easy to see that $\{x: |x| \geq \varepsilon\} \subset \{x: |x - a_{nk}| \geq \varepsilon/2\}$ for every $\varepsilon > 0$ and k and for all sufficiently large n. Consequently,

$$\sum_k P(|X_{nk}| \geq \varepsilon) \leq \sum_k P(|X_{nk} - a_{nk}| \geq \varepsilon/2) = \sum_k \int_{|x|\geq \varepsilon/2} d\bar{F}_{nk}(x) \to 0$$

by (4.16). The relation (4.4) is proved.

In the proof of the sufficiency we remarked that (4.4) implies (4.10). Therefore, (4.5) will be proved if we prove (4.7). It follows from (4.2) that

$$I_n(\varepsilon) \to \sigma^2 \quad \text{for every } \varepsilon > 0, \tag{4.17}$$

where $I_n(\varepsilon)$ is defined by (4.11). Putting

$$J_n(\varepsilon) = \sum_k \int_{|x|<\varepsilon} x^2 d\bar{F}_{nk}(x),$$

consider the difference $J_n(\varepsilon) - I_n(\varepsilon)$. We have

$$0 \leq J_n(\varepsilon) - I_n(\varepsilon) = \sum_k \int_{|x|<\varepsilon} \frac{x^4}{1+x^2} dF_{nk}(x)$$

$$= \sum_k \left\{ \int_{|x|<\varepsilon'} + \int_{\varepsilon'\leq |x|<\varepsilon} \right\} \frac{x^4}{1+x^2} d\bar{F}_{nk}(x),$$

where $0 < \varepsilon' < \varepsilon$. Accordingly,

$$0 \leq J_n(\varepsilon) - I_n(\varepsilon)$$

$$\leq \varepsilon'^2 \sum_k \int_{|x|<\varepsilon'} \frac{x^2}{1+x^2} d\bar{F}_{nk}(x) + \varepsilon^4 \sum_k \int_{|x|\geq \varepsilon'} d\bar{F}_{nk}(x).$$

Passing to the limit as $n \to \infty$ and using (4.16), (4.17), and (4.11), we obtain

$$0 \leq \liminf \{J_n(\varepsilon) - I_n(\varepsilon)\} \leq \limsup \{J_n(\varepsilon) - I_n(\varepsilon)\} \leq \varepsilon^2 \sigma^2.$$

Making use of the limit passage as $\varepsilon \downarrow 0$, we find that

$$0 \leq \liminf \{J_n(\varepsilon) - I_n(\varepsilon)\} \leq \limsup \{J_n(\varepsilon) - I_n(\varepsilon)\} \leq 0.$$

Hence there exists $\lim \{J_n(\varepsilon) - I_n(\varepsilon)\}$, and this limit is equal to zero. Therefore,

$$\lim J_n(\varepsilon) = \lim I_n(\varepsilon) = \sigma^2 \quad \text{for every } \varepsilon > 0,$$

by (4.17). We have proved the assertion (4.7) which implies (4.5).

To complete the proof we need only show that (4.6) is fulfilled. By (4.3) it is sufficient to prove the relation

$$\sum_k \int_{-\infty}^{\infty} \frac{x}{1+x^2} \, d\bar{F}_{nk}(x) \to 0.$$

In the course of the proof of sufficiency it was stated that this relation is a consequence of (4.2) and (4.4). □

Theorem 4.1 is a general form of the central limit theorem for sums of independent variables. In accordance with the established terminology, we shall call a central limit theorem any assertion to the effect that under some conditions the distribution functions of sums of unboundedly increasing numbers of random variables converge to a normal distribution function.

We shall now give another useful formulation of Theorem 4.1.

Theorem 4.2 *Let $\{X_{nk}\}$ be a sequence of series of random variables that are independent within each series, and let $F_{nk}(x)$ be the distribution function of X_{nk}. The condition of infinite smallness will be satisfied and the distributions of the sums $\sum_k X_{nk}$ will converge weakly to the normal (a, σ) distribution if and only if for every fixed $\varepsilon > 0$ and some $\tau > 0$ the following conditions hold:*

$$\sum_k P(|X_{nk}| \geq \varepsilon) \to 0,$$

$$\sum_k \left\{ \int_{|x|<\tau} x^2 \, dF_{nk}(x) - \left(\int_{|x|<\tau} x \, dF_{nk}(x) \right)^2 \right\} \to \sigma^2,$$

$$\sum_k \int_{|x|<\tau} x \, dF_{nk}(x) \to a.$$

Proof Obviously, we need only show that if the last three relations hold for every $\varepsilon > 0$ and some $\tau > 0$, conditions (4.5) and (4.6) will be satisfied for every $\varepsilon > 0$.

Suppose $0 < \varepsilon < \tau$. Putting

$$R_n(\tau) = \sum_k \left\{ \int_{|x|<\tau} x^2 \, dF_{nk}(x) - \left(\int_{|x|<\tau} x \, dF_{nk}(x) \right)^2 \right\},$$

we find that

$$|R_n(\tau) - R_n(\varepsilon)| \leq \sum_k \int_{\varepsilon \leq |x| < \tau} x^2 \, dF_{nk}(x) + 2\tau \sum_k \left| \int_{\varepsilon \leq |x| < \tau} x \, dF_{nk}(x) \right|$$

$$\leq 3\tau^2 \sum_k \int_{\varepsilon \leq |x| < \tau} dF_{nk}(x) \to 0,$$

since $\sum_k P(|X_{nk}| \geq \varepsilon) \to 0$. Interchanging ε and τ, we obtain the same result in the case when $0 < \tau < \varepsilon$. Similarly,

$$\left| \sum_k \int_{|x|<\tau} x \, dF_{nk}(x) - \sum_k \int_{|x|<\varepsilon} x \, dF_{nk}(x) \right| \leq \sum_k \int_{\min(\tau,\varepsilon) \leq |x| < \max(\tau,\varepsilon)} |x| \, dF_{nk}(x) \to 0.$$

The relations (4.5) and (4.6) follow for every $\varepsilon > 0$. □

Theorem 4.3 *Let $\{X_{nk}\}$ be a sequence of series of random variables that are independent within each series. The condition of infinite smallness is satisfied and the distributions of the sums $\sum_k X_{nk} - b_n$ converge weakly to the standard normal distribution for some sequence of constants $\{b_n\}$ if and only if the following conditions hold:*

$$\sum_k P(|X_{nk}| \geq \varepsilon) \to 0 \quad \text{for every } \varepsilon > 0$$

and

$$\sum_k \left\{ \int_{|x|<\tau} x^2 \, dF_{nk}(x) - \left(\int_{|x|<\tau} x \, dF_{nk}(x) \right)^2 \right\} \to 1$$

for some $\tau > 0$. If the latter conditions are satisfied, we can set

$$b_n = \sum_k \int_{|x|<H} x \, dF_{nk}(x) + o(1),$$

where H is an arbitrary positive number.

Theorem 4.3 follows from the previous considerations. It remains valid if we replace the words 'for some $\tau > 0$' by 'for every $\tau > 0$'. We note that Theorem 4.3 will be used only in the proof of Theorem 4.6.

Let us show that the condition $\sum_k P(|X_{nk}| \geq \varepsilon) \to 0$ from Theorems 4.1 to 4.3 is equivalent to the condition

$$P\left(\max_k |X_{nk}| \geq \varepsilon \right) \to 0.$$

We write $p_{nk} = P(|X_{nk}| \geq \varepsilon)$. Clearly,

$$P\left(\max_k |X_{nk}| \geq \varepsilon\right) = 1 - P\left(\max_k |X_{nk}| < \varepsilon\right)$$

$$= 1 - \prod_k P(|X_{nk}| < \varepsilon) = 1 - \prod_k (1 - p_{nk}).$$

Our assertion follows immediately from the inequalities

$$1 - \exp\left\{-\sum_k p_{nk}\right\} \leq 1 - \prod_k (1 - p_{nk}) \leq \sum_k p_{nk}.$$

4.2 Classical forms of the central limit theorem

From the general theorems that we have proved, it is easy to deduce the classical results of Lindeberg, Bernstein, Feller, Lévy, and Lyapunov on the central limit theorem for sums of independent random variables. We write

$$\Phi(x) = \frac{1}{\sqrt{2\pi}} \int_{-\infty}^{x} e^{-t^2/2} \, dt.$$

Theorem 4.4 (Feller's theorem) *Let $\{X_n; n = 1, 2, \ldots\}$ be a sequence of independent random variables, $V_n(x)$ be the distribution function of X_n, and $\{a_n\}$ be a sequence of positive constants. In order that*

$$\max_{1 \leq k \leq n} P(|X_k| \geq \varepsilon a_n) \to 0 \quad \text{for every fixed } \varepsilon > 0 \qquad (4.18)$$

and

$$\sup_x \left| P\left(a_n^{-1} \sum_{k=1}^n X_k < x\right) - \Phi(x) \right| \to 0, \qquad (4.19)$$

it is necessary and sufficient that

$$\sum_{k=1}^n \int_{|x| \geq \varepsilon a_n} dV_k(x) \to 0 \quad \text{for every fixed } \varepsilon > 0, \qquad (4.20)$$

$$a_n^{-2} \sum_{k=1}^n \left\{ \int_{|x| < a_n} x^2 \, dV_k(x) - \left(\int_{|x| < a_n} x \, dV_k(x) \right)^2 \right\} \to 1 \qquad (4.21)$$

and

$$a_n^{-1} \sum_{k=1}^n \int_{|x| < a_n} x \, dV_k(x) \to 0. \qquad (4.22)$$

This result follows easily from Theorems 4.1 and 4.2. We apply Theorems 4.1 and 4.2 to the triangle sequence of series of random variables

$\{X_{nk}; k = 1, \ldots, n; n = 1, 2, \ldots\}$, where $X_{nk} = X_k/a_n$ for $k = 1, \ldots, n$. If $F_{nk}(x)$ is the d.f. of X_{nk}, then $F_{nk}(x) = V_k(xa_n)$ and the left-hand sides of (4.4), (4.5), and (4.6) coincide with the left-hand sides of (4.20), (4.21), and (4.22) respectively. Relation (4.19) is equivalent to the weak convergence of the distributions of the sums $\sum_{k=1}^{n} X_k/a_n$ to the standard normal distribution by Theorem 1.11 (Chapter 1), since the standard normal distribution function $\Phi(x)$ is continuous. □

In accordance with Theorem 4.1 the conditions (4.21) and (4.22) in Theorem 4.4 can be replaced by the conditions

$$a_n^{-2} \sum_{k=1}^{n} \left\{ \int_{|x|<\varepsilon a_n} x^2 \, dV_k(x) - \left(\int_{|x|<\varepsilon a_n} x \, dV_k(x) \right)^2 \right\} \to 1 \quad (4.23)$$

and

$$a_n^{-1} \sum_{k=1}^{n} \int_{|x|<\varepsilon a_n} x \, dV_k(x) \to 0 \quad (4.24)$$

for every fixed $\varepsilon > 0$.

The following theorem contains simpler sufficient conditions for the normal convergence (4.19). It is an easy consequence of Theorem 4.4.

Theorem 4.5 *Let $\{X_n\}$ be a sequence of independent random variables, and $\{a_n\}$ a sequence of positive constants. If*

$$a_n^{-2} \sum_{k=1}^{n} \int_{|x|<a_n} x^2 \, dV_k(x) \to 1,$$

$$a_n^{-1} \sum_{k=1}^{n} \left| \int_{|x|<a_n} x \, dV_k(x) \right| \to 0,$$

and condition (4.20) is satisfied, the relation (4.19) holds. Here $V_k(x)$ is the d.f. of X_k.

Theorem 4.6 (the Bernstein–Feller theorem) *Let $\{X_n\}$ be a sequence of independent random variables, and $V_n(x)$ be the d.f. of X_n. In order that there exist sequences of constants $\{a_n > 0\}$ and $\{b_n\}$ such that*

$$\sup_x \left| P\left(a_n^{-1} \sum_{k=1}^{n} X_k - b_n < x \right) - \Phi(x) \right| \to 0 \quad (4.25)$$

and that the condition (4.18) shall hold, it is necessary and sufficient that there exist a sequence of constants $\{c_n\}$ such that $c_n \to +\infty$ and

$$\sum_{k=1}^{n} \int_{|x|\geq c_n} dV_k(x) \to 0 \quad (4.26)$$

and

$$c_n^{-2} \sum_{k=1}^{n} \left\{ \int_{|x|<c_n} x^2 \, dV_k(x) - \left(\int_{|x|<c_n} x \, dV_k(x) \right)^2 \right\} \to \infty \qquad (4.27)$$

In the case of the existence of a sequence of constants $\{c_n\}$ with the indicated properties, we can put in (4.25)

$$a_n^2 = \sum_{k=1}^{n} \left\{ \int_{|x|<c_n} x^2 \, dV_k(x) - \left(\int_{|x|<c_n} x \, dV_k(x) \right)^2 \right\}, \qquad (4.28)$$

$$b_n = a_n^{-1} \sum_{k=1}^{n} \int_{|x|<c_n} x \, dV_k(x). \qquad (4.29)$$

To prove Theorem 4.6 we need the following proposition.

Lemma 4.1 Let $x_n(\varepsilon) \to x$ for every fixed $\varepsilon > 0$, where $|x| < \infty$. Then there exists a sequence of positive numbers $\{\varepsilon_n\}$ such that $\varepsilon_n \to 0$ and $x_n(\varepsilon_n) \to x$.

Proof The substitution $y_n(\varepsilon) = |x_n(\varepsilon) - x|$ reduces the proof of the lemma to the case when $x = 0$ and $x_n(\varepsilon) \geq 0$. We put $z_n = \max_{k \geq n} x_k(\varepsilon)$, so that $z_n(\varepsilon) \downarrow 0$. Clearly, there exists an increasing sequence of integers $\{n_m\}$ such that $z_n(2^{-m}) \leq 2^{-m}$ for $n \geq n_m$ ($m = 1, 2, \ldots$). Putting $\varepsilon_n = 2^{1-m}$ if $n_{m-1} \leq n < n_m$, we have

$$\max_{n_{m-1} \leq n < n_m} z_n(\varepsilon_n) \leq z_{n_{m-1}}(\varepsilon_n) \leq z_{n_{m-1}}(2^{1-m}) \leq 2^{1-m} \to 0$$

as $m \to \infty$. The lemma is proved. □

We begin with the proof of necessity. We shall assume that there exist sequences $\{a_n\}$ and $\{b_n\}$ such that $a_n > 0$ and (4.25) and (4.18) are satisfied. Lemma 3.11 (Chapter 3) implies that $a_n \to \infty$ and $a_{n+1}/a_n \to 1$. By Theorem 4.4 and the accompanying note, conditions (4.20) and (4.23) hold. In accordance with Lemma 4.1 there exists a sequence of positive numbers $\{\varepsilon_n\}$ such that $\varepsilon_n \to 0$, $\varepsilon_n a_n \to \infty$,

$$\sum_{k=1}^{n} P(|X_k| \geq \varepsilon_n a_n) \to 0,$$

and

$$a_n^{-2} \sum_{k=1}^{n} \left\{ \int_{|x|<\varepsilon_n a_n} x^2 \, dV_k(x) - \left(\int_{|x|<\varepsilon_n a_n} x \, dV_k(x) \right)^2 \right\} \to 1.$$

We put $c_n = \varepsilon_n a_n$, so that $c_n \to \infty$, and we obtain the relations (4.26) and (4.27).

Let us now turn to the proof of sufficiency. Suppose there exists a sequence of constants $\{c_n\}$ such that the conditions (4.26), (4.27), and $c_n \to +\infty$ are satisfied. We define a_n by (4.28). The relation (4.27) implies that $c_n = o(a_n)$. Therefore,

$$\sum_{k=1}^{n} \int_{|x| \geq c_n} dV_k(x) \geq \sum_{k=1}^{n} \int_{|x| \geq \varepsilon a_n} dV_k(x)$$

for every fixed $\varepsilon > 0$ and all sufficiently large n. It follows from (4.26) that

$$\sum_{k=1}^{n} \int_{|x| \geq \varepsilon a_n} dV_k(x) \to 0$$

for every fixed $\varepsilon > 0$. The following relations are obvious:

$$a_n^{-2} \sum_{k=1}^{n} \left\{ \int_{c_n \leq |x| \leq \varepsilon a_n} x^2 \, dV_k(x) - \left(\int_{c_n \leq |x| < \varepsilon a_n} x \, dV_k(x) \right)^2 \right\}$$

$$\leq 2\varepsilon^2 \sum_{k=1}^{n} \int_{|x| \geq c_n} dV_k(x) \to 0,$$

$$a_n^{-2} \sum_{k=1}^{n} \left| \int_{|x| < c_n} x \, dV_k(x) \right| \cdot \left| \int_{c_n \leq |x| < \varepsilon a_n} x \, dV_k(x) \right|$$

$$\leq a_n^{-2} \varepsilon a_n c_n \sum_{k=1}^{n} \int_{|x| \geq c_n} dV_k(x) \to 0.$$

Hence

$$a_n^{-2} \sum_{k=1}^{n} \left\{ \int_{|x| < \varepsilon a_n} x^2 \, dV_k(x) - \left(\int_{|x| < \varepsilon a_n} x \, dV_k(x) \right)^2 \right\} \to 1,$$

by (4.28). Applying Theorem 4.3 to the random variables $X_{nk} = X_k/a_n$ ($k = 1, \ldots, n; n = 1, 2, \ldots$), we arrive at the desired assertions. □

So far we have made no assumptions about the existence of moments of the random variables under considerations. The rest of this section contains several classical results with some moment conditions.

Theorem 4.7 (the Lindeberg–Feller theorem) *Let $\{X_n\}$ be a sequence of independent random variables, at least one of which has a non-degenerate distribution. Let X_n have the finite variance σ_n^2 for all n. We put*

$$V_n(x) = P(X_n < x), \quad m_n = EX_n,$$

$$B_n = \sum_{k=1}^{n} \sigma_k^2, \quad F_n(x) = P\left(B_n^{-1/2} \sum_{k=1}^{n} (X_k - m_k) < x\right).$$

In order that

$$B_n^{-1} \max_{1 \leq k \leq n} \sigma_k^2 \to 0 \qquad (4.30)$$

and
$$\sup_n |F_n(x) - \Phi(x)| \to 0, \qquad (4.31)$$

it is necessary and sufficient that the following condition (the Lindeberg condition) be satisfied:

$$B_n^{-1} \sum_{k=1}^n \int_{|x - m_k| \geq \varepsilon B_n^{1/2}} (x - m_k)^2 \, dV_k(x) \to 0 \qquad (4.32)$$

for every fixed $\varepsilon > 0$.

Proof Without loss of generality we may assume that $m_n = 0$ for all n. If the theorem is proved for this special case, then we can prove it for the general case by introducing the random variables $Y_n = X_n - m_n$ ($n = 1, 2, \ldots$), which have zero means and the same variances as X_n.

We shall suppose that conditions (4.30) and (4.31) are satisfied. By Chebyshev's inequality,

$$\max_{1 \leq k \leq n} P(|X_k| \geq \varepsilon B_n^{1/2}) \leq \max_{1 \leq k \leq n} \sigma_k^2 \varepsilon^{-2} B_n^{-1}.$$

Therefore, (4.30) implies (4.18) for $a_n = B_n^{1/2}$. Clearly, for this choice of a_n the relations (4.31) and (4.19) coincide. By Theorem 4.4 and the accompanying note we have

$$B_n^{-1} \sum_{k=1}^n \left\{ \int_{|x| < \varepsilon B_n^{1/2}} x^2 \, dV_k(x) - \left(\int_{|x| < \varepsilon B_n^{1/2}} x \, dV_k(x) \right)^2 \right\} \to 1.$$

Recalling that $B_n = \sum_{k=1}^n \int_{-\infty}^\infty x^2 \, dV_k(x)$, we arrive at the relation

$$B_n^{-1} \sum_{k=1}^n \left\{ \int_{|x| \geq \varepsilon B_n^{1/2}} x^2 \, dV_k(x) + \left(\int_{|x| < \varepsilon B_n^{1/2}} x \, dV_k(x) \right)^2 \right\} \to 0$$

for every fixed $\varepsilon > 0$. The Lindeberg condition follows. The necessity is proved.

Now suppose that the Lindeberg condition holds. Put

$$\Lambda_n(\varepsilon) = B_n^{-1} \sum_{k=1}^n \int_{|x| \geq \varepsilon B_n^{1/2}} x^2 \, dV_k(x).$$

We shall call $\Lambda_n(\varepsilon)$ the Lindeberg ratio. If $1 \leq k \leq n$, then

$$\sigma_k^2 = \int_{|x| < \varepsilon B_n^{1/2}} x^2 \, dV_k(x) + \int_{|x| \geq \varepsilon B_n^{1/2}} x^2 \, dV_k(x)$$

$$\leq \varepsilon^2 B_n + \sum_{k=1}^n \int_{|x| \geq \varepsilon B_n^{1/2}} x^2 \, dV_k(x).$$

Hence
$$B_n^{-1} \max_{1 \leq k \leq n} \sigma_k^2 \leq \varepsilon^2 + \Lambda_n(\varepsilon).$$

The right-hand side of this inequality can be made arbitrarily small for all sufficiently large n if we choose ε sufficiently small, since the Lindeberg condition is satisfied. Thus (4.30) is proved.

Furthermore,
$$\sum_{k=1}^{n} \int_{|x| \geq \varepsilon B_n^{1/2}} dV_k(x) \leq \varepsilon^{-2} B_n^{-1} \sum_{k=1}^{n} \int_{|x| \geq \varepsilon B_n^{1/2}} x^2 \, dV_k(x) = \varepsilon^{-2} \Lambda_n(\varepsilon) \to 0$$

for every $\varepsilon > 0$, and
$$B_n^{-1} \sum_{k=1}^{n} \int_{|x| < B_n^{1/2}} x^2 \, dV_k(x) = B_n^{-1} \left(B_n - \sum_{k=1}^{n} \int_{|x| \geq B_n^{1/2}} x^2 \, dV_k(x) \right)$$
$$= 1 - \Lambda_n(1) \to 1.$$

Making use of the assumption $EX_n = 0$ for all n, we obtain
$$B_n^{-1/2} \sum_{k=1}^{n} \left| \int_{|x| < B_n^{1/2}} x \, dV_k(x) \right| = B_n^{-1/2} \sum_{k=1}^{n} \left| \int_{|x| \geq B_n^{1/2}} x \, dV_k(x) \right|$$
$$\leq B_n^{-1} \sum_{k=1}^{n} \int_{|x| \geq B_n^{1/2}} x^2 \, dV_k(x) = \Lambda_n(1) \to 0.$$

Consequently, the conditions of Theorem 4.5 are satisfied for $a_n = B_n^{1/2}$, and the relation (4.31) follows. □

In Theorem 4.7 the assumption was set of the existence of at least one non-degenerate distribution among the distributions of the random variables $X_1, X_2, \ldots$. This is equivalent to the following condition: Var $X_n > 0$ for at least one value n. It follows that $B_n > 0$ for all sufficiently large n. The above assumption excludes the trivial case when all the summands have degenerate distributions.

It is easy to show that (4.30) implies the relation $B_n \to \infty$. In fact, if this relation does not hold, then $B_n \leq C$ for all n and some positive constant C, since $\{B_n\}$ is a non-decreasing sequence. Let k_0 be an integer such that $\sigma_{k_0}^2 > 0$. We have $\max_{1 \leq k \leq n} \sigma_k^2 \geq \sigma_{k_0}^2$ for sufficiently large n. Hence
$$\max_{1 \leq k \leq n} \sigma_k^2 / B_n \geq c_0$$

for sufficiently large n, where $c_0 = \sigma_{k_0}^2 / C > 0$. This condradicts (4.30). Thus $B_n \to \infty$.

By Theorem 4.7 it follows that if the Lindeberg condition is satisfied, then $B_n \to \infty$.

In the particular case when the random variables $X_1, X_2, \ldots$ have a common distribution with finite variance, the Lindeberg condition is obviously satisfied. The following proposition is a consequence of Theorem 4.7.

Theorem 4.8 (the Lévy theorem) *If $\{X_n\}$ is a sequence of independent identically distributed random variables having the finite positive variance σ^2, then*

$$P\left(\sigma^{-1}n^{-1/2}\left(\sum_{k=1}^{n} X_k - nm\right) < x\right) \to \Phi(x)$$

uniformly in x. Here $m = EX_n$.

Let us now return to the general case of not necessarily identical distributions.

Theorem 4.9 (the Lyapunov theorem) *Let $\{X_n\}$ be a sequence of independent random variables, at least one of which has a non-degenerate distribution. Let $E|X_n|^{2+\delta} < \infty$ for some $\delta > 0$ ($n = 1, 2, \ldots$). We put*

$$m_n = EX_n, \quad \sigma_n^2 = \operatorname{Var} X_n, \quad B_n = \sum_{k=1}^{n} \sigma_k^2,$$

$$F_n(x) = P\left(B_n^{-1/2} \sum_{k=1}^{n} (X_k - m_k) < x\right).$$

If

$$B_n^{-1-\delta/2} \sum_{k=1}^{n} E|X_k - m_k|^{2+\delta} \to 0, \tag{4.33}$$

then

$$\sup_x |F_n(x) - \Phi(x)| \to 0.$$

Proof The theorem will be proved if we show that (4.33) implies the Lindeberg condition. This follows from the inequalities

$$B_n^{-1} \sum_{k=1}^{n} \int_{|x-m_k| \geq \varepsilon B_n^{1/2}} (x - m_k)^2 \, dV_k(x) \leq B_n^{-1}(\varepsilon B_n^{1/2})^{-\delta}$$

$$\times \sum_{k=1}^{n} \int_{|x-m_k| \geq \varepsilon B_n^{1/2}} |x - m_k|^{2+\delta} \, dV_k(x)$$

$$\leq \varepsilon^{-\delta} B_n^{-1-\delta/2} \sum_{k=1}^{n} E|X_k - m_k|^{2+\delta}. \quad \square$$

4.3 The weak law of large numbers for a sequence of series of independent random variables

We consider a sequence of random variables $X, X_1, X_2, \ldots$ defined on a common probability space $(\Omega, \mathscr{A}, P)$. We say that $\{X_n\}$ converges in probability to X if $P(|X_n - X| \geq \varepsilon) \to 0$ for every fixed $\varepsilon > 0$. For this kind of convergence we shall use the notation $X_n \xrightarrow{P} X$.

By definition, the sequence $\{X_n\}$ is stable if there exists a sequence of constants $\{b_n\}$ such that $X_n - b_n \xrightarrow{P} 0$. If this condition holds, then $P(|X_n - b_n| < \varepsilon) > 1/2$ for every $\varepsilon > 0$ and for all sufficiently large n. From the definition of a median mX of the random variable X, it follows that $|mX_n - b_n| < \varepsilon$ for all sufficiently large n. Thus, for a stable sequence $\{X_n\}$ we have

$$P(|X_n - mX_n| \geq \varepsilon) \leq P(|X_n - b_n| + |b_n - mX_n| \geq \varepsilon) \leq P(|X_n - b_n| \geq \varepsilon/2)$$

for every $\varepsilon > 0$ and for all sufficiently large n, and the latter probability tends to zero, so that $X_n - mX_n \xrightarrow{P} 0$. Furthermore, $b_n - mX_n \to 0$.

Let $[X_{nk}; k = 1, \ldots, k_n; n = 1, 2, \ldots]$ be a sequence of series of random variables that are independent within each series. By definition, the sequence of series $\{X_{nk}\}$ obeys the weak law of large numbers if the sequence $\{\sum_{k=1}^{k_n} X_{nk}; n = 1, 2, \ldots\}$ is stable.

Lemma 4.2 *Let $Y_1, Y_2, \ldots$ be a sequence of random variables defined on a common probability space $(\Omega, \mathscr{A}, P)$. We put*

$$D(x) = \begin{cases} 0 & \text{if } x \leq 0, \\ 1 & \text{if } x > 0. \end{cases}$$

Then the relation $Y_n \xrightarrow{P} 0$ is equivalent to the weak convergence of the distributions of Y_n to the degenerate distribution $D(x)$.

The lemma follows immediately from the definition of the convergence in probability.

Making use of this lemma, we observe that a sequence of series of random variables $\{X_{nk}\}$ obeys the weak law of large numbers if and only if there exists a sequence of constants $\{b_n\}$ such that the distribution of the sum $\sum_k X_{nk} - b_n$ converges weakly to the degenerate distribution $D(x)$. Chapter 3 contains the necessary and sufficient conditions for the weak convergence of the distributions of sums of independent random variables to an arbitrary infinitely divisible distribution. A degenerate distribution is infinitely divisible, therefore, we can derive the conditions under which the sequence $\{X_{nk}\}$ obeys the weak law of large numbers, using these results.

Lemma 4.3 *Let the sequence $\{X_{nk}\}$ obey the weak law of large numbers. Then $\{X_{nk}\}$ satisfies the condition of constancy in the limit.*

Proof If $f_{nk}(t)$ is the c.f. of X_{nk}, then $e^{-ibnt} \prod_k f_{nk}(t) \to 1$ for every real t, since the c.f. of the degenerate distribution $D(x)$ is identically equal to unity.

It follows that $\prod_k |f_{nk}(t)| \to 1$ and $\min_k |f_{nk}(t)| \to 1$. But $|f_{nk}(t)|^2$ is the c.f. of the difference of two independent identically distributed random variables X_{nk} and Y_{nk}, hence

$$\max_k P(|X_{nk} - Y_{nk}| \geq \varepsilon) \to 0 \tag{4.34}$$

for every $\varepsilon > 0$. Let $\{a_{nk}\}$ be a number sequence such that $P(X_{nk} < a_{nk}) \leq \frac{1}{2} \leq P(X_{nk} \leq a_{nk})$. Then for every $\varepsilon > 0$ we have

$$P(X_{nk} - Y_{nk} \geq \varepsilon) \geq P(X_{nk} - a_{nk} \geq \varepsilon, Y_{nk} - a_{nk} \leq 0) \geq \tfrac{1}{2}(1 - F_{nk}(a_{nk} + \varepsilon)),$$

where $F_{nk}(x)$ is the d.f. of X_{nk}. Similarly,

$$P(X_{nk} - Y_{nk} \leq -\varepsilon) \geq \tfrac{1}{2} F_{nk}(a_{nk} - \varepsilon + 0).$$

Therefore,

$$P(|X_{nk} - a_{nk}| \geq \varepsilon) \leq 2 P(|X_{nk} - Y_{nk}| \geq \varepsilon) \to 0$$

uniformly in k, by (4.34). Thus the random variables X_{nk} satisfy the condition of constancy in the limit (condition (3.23), Chapter 3). □

As we remarked at the end of Section 3.2, if the X_{nk} are constant in the limit, then the random variables $X_{nk} - \mathrm{m}X_{nk}$ satisfy the condition of infinite smallness. This remark, together with Lemma 4.3, yields the following proposition which corresponds to Theorems 4.2 and 4.3 for the special case $a = 0$ and $\sigma = 0$.

Lemma 4.4 *Let $F_{nk}(x)$ be the d.f. of X_{nk}. We write $m_{nk} = \mathrm{m}X_{nk}$. The sequence $\{X_{nk}\}$ obeys the weak law of large numbers if and only if*

$$\sum_k \int_{|x| \geq \varepsilon} dF_{nk}(x + m_{nk}) \to 0 \tag{4.35}$$

for every fixed $\varepsilon > 0$ and

$$\sum_k \left\{ \int_{|x| < 1} x^2 \, dF_{nk}(x + m_{nk}) - \left(\int_{|x| < 1} x \, dF_{nk}(x + m_{nk}) \right)^2 \right\} \to 0. \tag{4.36}$$

Lemma 4.5 *Let (4.35) be satisfied. Then the conditions (4.36) and*

$$\sum_k \int_{|x| < 1} x^2 \, dF_{nk}(x + m_{nk}) \to 0 \tag{4.37}$$

are equivalent.

Proof Suppose that (4.36) holds. We denote by I that one of the intervals $(-1, 0)$ and $(0, 1)$ on which $\int_I x \, dF_{nk}(x + m_{nk})$ attains its largest absolute value. (I may depend on n and k.) We have

$$\sum_k \left(\int_{|x|<1} x \, dF_{nk}(x + m_{nk}) \right)^2 \leq \sum_k \left(\int_I x \, dF_{nk}(x + m_{nk}) \right)^2$$

$$\leq \sum_k \int_I x^2 \, dF_{nk}(x + m_{nk}) \int_I dF_{nk}(x + m_{nk})$$

$$\leq \sum_k \int_{|x|<1} x^2 \, dF_{nk}(x + m_{nk}) \int_I dF_{nk}(x + m_{nk}).$$

Taking into account the definition of a median, we obtain $\int_I dF_{nk}(x + m_{nk}) \leq \frac{1}{2}$. Hence

$$\sum_k \left(\int_{|x|<1} x \, dF_{nk}(x + m_{nk}) \right)^2 \leq \frac{1}{2} \sum_k \int_{|x|<1} x^2 \, dF_{nk}(x + m_{nk})$$

$$\leq \sum_k \left\{ \int_{|x|<1} x^2 \, dF_{nk}(x + m_{nk}) - \left(\int_{|x|<1} x \, dF_{nk}(x + m_{nk}) \right)^2 \right\}.$$

Thus (4.36) implies (4.37).

The converse is also true, since

$$0 \leq \int_{|x|<1} x^2 \, dF_{nk}(x + m_{nk}) - \left(\int_{|x|<1} x \, dF_{nk}(x + m_{nk}) \right)^2$$

$$\leq \int_{|x|<1} x^2 \, dF_{nk}(x + m_{nk}). \qquad \square$$

Theorem 4.10 *The sequence $\{X_{nk}\}$ obeys the weak law of large numbers if and only if it satisfies the conditions (4.37) and*

$$\sum_k \int_{|x|\geq 1} dF_{nk}(x + m_{nk}) \to 0. \tag{4.38}$$

Proof Clearly by Lemmas 4.4 and 4.5 we need only show that (4.37) and (4.38) imply (4.35) for every $\varepsilon > 0$. If $\varepsilon \geq 1$, this implication is obvious. If $0 < \varepsilon < 1$, then

$$\sum_k \int_{|x|\geq \varepsilon} dF_{nk}(x + m_{nk}) = \sum_k \left\{ \int_{\varepsilon \leq |x|<1} + \int_{|x|\geq 1} \right\}$$

$$\leq \varepsilon^{-2} \sum_k \int_{\varepsilon \leq |x|<1} x^2 \, dF_{nk}(x + m_{nk})$$

$$+ \sum_k \int_{|x|\geq 1} dF_{nk}(x + m_{nk}) \to 0,$$

by (4.37) and (4.38). $\square$

It is possible to show that if the conditions (4.37) and (4.38) are satisfied, then $\sum_k X_{nk} - b_n \xrightarrow{P} 0$ with the constants b_n satisfying the equality (which all admissible constants satisfy)

$$b_n = \sum_k \left(m_{nk} + \int_{|x|<\tau} x \, dF_{nk}(x + m_{nk}) \right) + o(1),$$

where τ is an arbitrary positive number.

Lemma 4.6 *Let X be a random variable and b a positive constant. We put*

$$Z = \begin{cases} X & \text{if } |X| < b, \\ 0 & \text{if } |X| \geq b. \end{cases}$$

Then

$$\frac{1}{2}\left(\frac{EZ^2}{b^2} + P(|X| \geq b) \right) \leq E \frac{X^2}{X^2 + b^2} \leq \frac{EZ^2}{b^2} + P(|X| \geq b).$$

Proof If $F(x)$ is the d.f. of X, then

$$E \frac{X^2}{X^2 + b^2} = \int_{|x|<b} \frac{x^2}{x^2 + b^2} dF(x) + \int_{|x|\geq b} \frac{x^2}{x^2 + b^2} dF(x)$$

$$\leq b^{-2} \int_{|x|<b} x^2 \, dF(x) + \int_{|x|\geq b} dF(x),$$

$$\int_{-\infty}^{\infty} \frac{x^2}{b^2 + x^2} dF(x) \geq 2^{-1} \int_{|x|\geq b} dF(x) + (2b^2)^{-1} \int_{|x|<b} x^2 \, dF(x). \quad \square$$

Applying Lemma 4.6, we can give another formulation of Theorem 4.10.

Theorem 4.11 *The sequence $\{X_{nk}\}$ obeys the weak law of large numbers if and only if*

$$\sum_k \int_{-\infty}^{\infty} \frac{x^2}{1 + x^2} dF_{nk}(x + m_{nk}) \to 0. \tag{4.39}$$

In fact, (4.37) and (4.38) are equivalent to (4.39) by Lemma 4.6.

Let us note an immediate consequence of Theorems 4.1 and 4.2 for the special case when $a = 0$ and $\sigma = 0$.

Theorem 4.12 *We shall have*

$$\sum_k X_{nk} \xrightarrow{P} 0, \tag{4.40}$$

and the sequence $\{X_{nk}\}$ will obey the condition of infinite smallness if and *only if*

$$\sum_k P(|X_{nk}| \geq \varepsilon) \to 0, \qquad (4.41)$$

$$\sum_k \left\{ \int_{|x|<\tau} x^2 \, dF_{nk}(x) - \left(\int_{|x|<\tau} x \, dF_{nk}(x) \right)^2 \right\} \to 0, \qquad (4.42)$$

$$\sum_k \int_{|x|<\tau} x \, dF_{nk}(x) \to 0 \qquad (4.43)$$

for every $\varepsilon > 0$ and some $\tau > 0$. We can replace the words 'for every $\varepsilon > 0$ and some $\tau > 0$' by the words 'for every $\varepsilon > 0$ and every $\tau > 0$'.

4.4 Classical forms of the weak law of large numbers

We now consider a sequence of independent random variables $\{X_n; n = 1, 2, \ldots\}$. Let $V_n(x)$ be the d.f. of X_n, and let $\{a_n\}$ be a sequence of positive numbers such that $a_n \uparrow \infty$. The following theorem was found by Kolmogorov for the special case $a_n = n$ and by Feller for arbitrary sequences $\{a_n\}$.

Theorem 4.13 *The relation*

$$a_n^{-1} \sum_{k=1}^n X_k \xrightarrow{P} 0 \qquad (4.44)$$

holds if and only if

$$\sum_{k=1}^n P(|X_k| \geq a_n) \to 0, \qquad (4.45)$$

$$a_n^{-2} \sum_{k=1}^n \left\{ \int_{|x|<a_n} x^2 \, dV_k(x) - \left(\int_{|x|<a_n} x \, dV_k(x) \right)^2 \right\} \to 0, \qquad (4.46)$$

$$a_n^{-1} \sum_{k=1}^n \int_{|x|<a_n} x \, dV_k(x) \to 0. \qquad (4.47)$$

Proof We introduce the triangle sequence of series of random variables $[X_{nk}; k = 1, \ldots, n; n = 1, 2, \ldots]$, putting $X_{nk} = X_k/a_n$. Then (4.40) and (4.44) coincide, and we can use Theorem 4.12. However, we prefer to give a straightforward proof of sufficiency, avoiding Theorem 4.12. We shall apply this theorem only for the proof of necessity.

We first show that the conditions (4.45)–(4.47) imply (4.44).
Consider the truncated random variables

$$\bar{X}_{nk} = \begin{cases} X_{nk} & \text{if } |X_{nk}| < 1, \\ 0 & \text{if } |X_{nk}| \geq 1. \end{cases}$$

Since $F_{nk}(x) = P(X_{nk} < x) = V_k(a_n x)$, we have

$$E\bar{X}_{nk} = \int_{|x|<1} x\, dF_{nk}(x) = a_n^{-1}\int_{|x|<a_n} x\, dV_k(x)$$

and

$$E\bar{X}_{nk}^2 = \int_{|x|<1} x^2\, dF_{nk}(x) = a_n^{-2}\int_{|x|<a_n} x^2\, dV_k(x).$$

The left-hand side of (4.46) coincides with $\sum_{k=1}^{n} \text{Var } \bar{X}_{nk}$. By Chebyshev's inequality and (4.46) we have

$$P\left(\left|\sum_k \bar{X}_{nk} - \sum_k E\bar{X}_{nk}\right| \geq \delta\right) \leq \delta^{-2}\text{Var}\left(\sum_k \bar{X}_{nk}\right) = \delta^{-2}\sum_k \text{Var } \bar{X}_{nk} \to 0 \quad (4.48)$$

for every $\delta > 0$. Hence $\sum_k \bar{X}_{nk} - \sum_k E\bar{X}_{nk} \xrightarrow{P} 0$. Using (4.47), we find that $\sum_k \bar{X}_{nk} \xrightarrow{P} 0$. If $\delta > 0$, then

$$P\left(\left|\sum_k X_{nk} - \sum_k \bar{X}_{nk}\right| \geq \delta\right) \leq P\left(\sum_k X_{nk} \neq \sum_k \bar{X}_{nk}\right) \leq P\left(\bigcup_k [X_{nk} \neq \bar{X}_{nk}]\right)$$

$$\leq \sum_k P(X_{nk} \neq \bar{X}_{nk}) = \sum_k P(|X_{nk}| \geq 1)$$

$$= \sum_k P(|X_k| \geq a_n) \to 0,$$

by (4.45). Therefore, $\sum_k X_{nk} - \sum_k \bar{X}_{nk} \xrightarrow{P} 0$. Together with $\sum_k \bar{X}_{nk} \xrightarrow{P} 0$, this implies the relation $\sum_k X_{nk} \xrightarrow{P} 0$, which is equivalent to (4.44). The sufficiency is proved.

To prove the necessity we need only show that (4.44) implies the condition of infinite smallness for the sequence of series $\{X_{nk}\}$ where $X_{nk} = X_k/a_n$. After that, we apply Theorem 4.12 to complete the proof.

Suppose (4.44) holds. Put $S_n = \sum_{k=1}^{n} X_k$. We have

$$\frac{X_n}{a_n} = \frac{S_n}{a_n} - \frac{S_{n-1}}{a_{n-1}} \cdot \frac{a_{n-1}}{a_n} \xrightarrow{P} 0,$$

since the sequence of positive numbers $\{a_n\}$ does not decrease. Let ε and δ be arbitrary positive numbers. If $N \leq k \leq n$ and N is sufficiently large, then $P(|X_k| \geq \varepsilon a_n) \leq P(|X_k| \geq \varepsilon a_k) < \delta$ by the same reasoning. If, however, $1 \leq k < N$, then $X_k/a_n \to 0$ by the hypothesis $a_n \uparrow \infty$ and, consequently, $\max_{1 \leq k < N} P(|X_k| \geq \varepsilon a_n) < \delta$ for all sufficiently large n. Thus $\max_{1 \leq k \leq n} P(|X_k| \geq \varepsilon a_n) < \delta$ for all sufficiently large n, that is, the condition of infinite smallness is satisfied. □

Note that we actually proved a somewhat more general result than Theorem 4.13. The conditions (4.45)–(4.47) and $a_n \uparrow \infty$ imply (4.44) even if we replace the independence condition by the weaker condition of the pairwise independence of the random variables $X_1, X_2, \ldots$. In fact, in the proof of the sufficiency we made use of the independence condition only in (4.48) writing the equality $\operatorname{Var}(\sum_k \bar{X}_{nk}) = \sum_k \operatorname{Var} \bar{X}_{nk}$. But this equality holds if $\bar{X}_{n1}, \bar{X}_{n2}, \ldots$ are pairwise independent random variables. The hypothesis about the pairwise independence of the random variables $X_1, X_2, \ldots$ implies that $\bar{X}_{n1}, \ldots, \bar{X}_{nn}$ are pairwise independent variables being the functions of $X_1, \ldots, X_n$ correspondingly.

In the special case when the random variables under consideration are identically distributed and $a_n = n$, Theorem 4.13 admits a much simpler formulation.

Theorem 4.14 *Let $\{X_n\}$ be a sequence of independent random variables with a common d.f. $V(x)$. The relation*

$$n^{-1} \sum_{k=1}^{n} X_k \xrightarrow{P} 0 \qquad (4.49)$$

holds if and only if

$$nP(|X_1| \geq n) \to 0 \qquad (4.50)$$

and

$$\int_{|x|<n} x \, dV(x) \to 0. \qquad (4.51)$$

Proof Taking account of Theorem 4.13, we need only show that (4.50) and (4.51) imply (4.49). First we shall show that (4.50) implies the relation

$$n^{-1} \int_{|x|<n} x^2 \, dV(x) \to 0. \qquad (4.52)$$

In fact,

$$\int_{|x|<n} x^2 \, dV(x) = \sum_{m=1}^{n} \int_{m-1 \leq |x| < m} x^2 \, dV(x) \leq \sum_{m=1}^{n} m^2 P(m-1 \leq |X_1| < m)$$

$$\leq 2 \sum_{m=1}^{n} \sum_{l=1}^{m} lP(m-1 \leq |X_1| < m)$$

$$= 2 \sum_{l=1}^{n} lP(l-1 \leq |X_1| < n) \leq 2 \sum_{l=1}^{n} lP(|X_1| \geq l-1).$$

We note that if $\{x_n\}$ is a number sequence such that $x_n \to 0$, then $n^{-1} \sum_{k=1}^{n} x_k \to 0$. Accordingly, the inequalities just obtained, taken together with (4.50) imply (4.52). Clearly, (4.50)–(4.52) imply (4.45)–(4.47) for $a_n = n$ and $V_n(x) \equiv V(x)$. There remains only to refer to Theorem 4.13. □

Theorem 4.15 *Let $\{X_n\}$ be a sequence of pairwise independent identically distributed random variables such that the mean value EX_1 exists. Then $n^{-1}\sum_{k=1}^{n} X_k \xrightarrow{P} EX_1$.*

Proof Without loss of generality we may assume that $EX_1 = 0$. Denoting by $V(x)$ the d.f. of X_1, we observe that the condition (4.51) is satisfied and, moreover,

$$P(|X_1| \geq n) = \int_{|x| \geq n} dV(x) \leq n^{-1} \int_{|x| \geq n} |x|\, dV(x).$$

Therefore, (4.50) is also satisfied. By Theorem 4.14 and the note accompanying Theorem 4.13, the relation (4.49) holds. □

This theorem was found by Khintchine under the stronger condition that $\{X_n\}$ is a sequence of independent random variables. The following theorem of Markov is free from any assumptions about the independence.

Theorem 4.16 *Let $\{X_n\}$ be a sequence of random variables such that the mean values EX_n exist for all n, and let*

$$n^{-2} \operatorname{Var}\left(\sum_{k=1}^{n} X_k\right) \to 0. \tag{4.53}$$

Then

$$n^{-1} \sum_{k=1}^{n} X_k - n^{-1} \sum_{k=1}^{n} EX_k \xrightarrow{P} 0. \tag{4.54}$$

Proof By Chebyshev's inequality,

$$P\left(\left|n^{-1}\sum_{k=1}^{n} X_k - n^{-1}\sum_{k=1}^{n} EX_k\right| \geq \delta\right) \leq \delta^{-2} n^{-2} \operatorname{Var}\left(\sum_{k=1}^{n} X_k\right)$$

for every $\delta > 0$. Hence (4.53) implies (4.54). □

This theorem is a generalization of a previous result of Chebyshev which can be stated as follows. If $\{X_n\}$ is a sequence of pairwise independent random variables with uniformly bounded variances (that is, $\operatorname{Var} X_n \leq C$ for all n and some constant C), then the relation (4.54) holds.

Chebyshev's theorem is a consequence of Markov's theorem, since the conditions of the first theorem obviously imply (4.53).

4.5 Bibliographical notes

Theorems of this chapter are well-known limit theorems of probability theory. They can be found in the books by Gnedenko and Kolmogorov [147], Feller [133], Loève [256], Lévy [252] and Khintchine [222].

The suffiency in Theorems 4.6 and 4.7 is due to Bernstein and Lindeberg respectively; the necessity in both theorems is due to Feller.

It is quite reasonable to accompany every limit theorem by estimates of the speed of convergence. There is a vast literature devoted to the rates of convergence in the central limit theorem. We shall present a sample from this literature in the next chapter. The next section of this chapter contains, among others, some results concerning the rates of convergence in the weak law of large numbers.

4.6 Addenda

In Subsections 4.6.1–4.6.3 we consider a sequence of series of random variables $\{X_{nk}\}$ that are independent within each series and satisfy the condition of infinite smallness. Moreover, we suppose that $EX_{nk} = 0$, $\sigma_{nk}^2 = \text{Var } X_{nk} < \infty$ for all n and k, and $\sum_k \sigma_{nk}^2 = 1$. We put $F_n(x) = P(\sum_k X_{nk} < x)$.

4.6.1 The weak convergence of $F_n(x)$ to $\Phi(x)$ is equivalent to the relation $\sum_k X_{nk}^2 \xrightarrow{P} 1$ (Raikov [376]).

4.6.2 Let $\Pi(x)$ be the Poisson d.f. with the c.f. $\pi(t) = \exp(e^{it} - 1)$. The weak convergence of $F_n(x)$ to $\Pi(x)$ is equivalent to the relation $\sum_k (X_{nk}^2 - X_{nk}) \xrightarrow{P} 1$ (Alda [2]).

4.6.3 Let $F_{nk}(x) = P(X_{nk} < x)$ and $\Phi_{nk}(x) = \Phi(x/\sigma_{nk})$. The weak convergence of $F_n(x)$ to $\Phi(x)$ is equivalent to the relation

$$\sum_k \int_{|x| \geq \varepsilon} |x| |F_{nk}(x) - \Phi_{nk}(x)| \, dx \to 0$$

for every fixed $\varepsilon > 0$ (Rotar [399]).

4.6.4 Let $\{X_n\}$ be a sequence of independent random variables with zero means and finite variances. Let $p > 2$. We put

$$B_n = \sum_{k=1}^n \text{Var } X_k, \quad Z_n = B_n^{-1/2} \sum_{k=1}^n X_k, \quad F_n(x) = P(Z_n < x).$$

The condition

$$B_n^{-p/2} \sum_{k=1}^n E|X_k|^p \to 0$$

is sufficient for the relations

$$\sup_x |F_n(x) - \Phi(x)| \to 0 \quad \text{and} \quad E|Z_n|^p \to \int_{-\infty}^{\infty} |x|^p \, d\Phi(x);$$

it is necessary for these relations if the additional condition

$$\max_{1 \leq k \leq n} \operatorname{Var} X_k / B_n \to 0$$

is satisfied (Bernstein [29]; see also [31], p. 358).

4.6.5 Let $\{X_n\}$ be a sequence of independent random variables, and let $\{a_n\}$ be a sequence of positive constants. We put $V_n(x) = P(X_n < x)$. Suppose that the condition (4.23) and

$$a_n^{-p} \sum_{k=1}^{n} \int_{|x| \geq \varepsilon a_n} |x|^p \, dV_k(x) \to 0$$

are satisfied for every fixed $\varepsilon > 0$ and some non-negative integer p. Then the distribution of the random variable

$$Z_n = a_n^{-1} \sum_{k=1}^{n} \left(X_k - \int_{|x| < a_n} x \, dV_k(x) \right)$$

converges weakly to the normal distribution $\Phi(x)$ and, if $p > 0$, the moments of Z_n up to the order p converge to the corresponding moments of $\Phi(x)$ (Zaremba [478]).

4.6.6 Let $\{X_n\}$ be a sequence of independent identically distributed random variables and $F_n(x) = P(1/a_n \sum_{k=1}^{n} X_k - b_n < x)$. If there exist sequences of real numbers $\{a_n\}$ and $\{b_n\}$ such that $a_n > 0$ and $F_n(x) \to \Phi(x)$ for all $x \leq 0$, then $F_n(x) \to \Phi(x)$ for all real x (Rossberg and Siegel [398]).

4.6.7 Let $\{X_{nk}\}$ be a sequence of series of random variables that are independent within each series. Suppose that $EX_{nk} = 0$, $\operatorname{Var} X_{nk} < \infty$ for all n and k, $\sum_k \operatorname{Var} X_{nk} = 1$. Let $x_1 \in \mathbb{R}$, $x_2 \in \mathbb{R}$, $x_1 \neq x_2$, $F_n(x) = P(\sum_k X_{nk} - b_n < x)$. In order that the sequence $\{F_n(x)\}$ converge weakly to a normal distribution and the condition of infinite smallness be satisfied, it is necessary and sufficient that the condition (4.4) be satisfied for every $\varepsilon > 0$ and the sequences $\{F_n(x_1)\}$ and $\{F_n(x_2)\}$ converge to limits different from 0 and 1 (Jésiak and Rossberg [208]).

4.6.8 Let $\{X_n\}$ be a sequence of independent identically distributed random variables, $F_n(x) = P(1/a_n \sum_{k=1}^{n} X_k - b_n < x)$, $a_n > 0$. We put $\bar{F}(x) = \limsup F_n(x)$, $\underline{F}(x) = \liminf F_n(x)$. If

$$\lim_{x \to -\infty} \frac{\underline{F}(x)}{\Phi(x)} = \lim_{x \to -\infty} \frac{\bar{F}(x)}{\Phi(x)} = 1,$$

then $F_n(x) \to \Phi(x)$ for every $x \in \mathbb{R}$ (Riedel [385]).

In Subsections 4.6.9–4.6.18 we consider a sequence of independent identically distributed random variables $\{X_n\}$. We put $S_n = \sum_{k=1}^n X_k$.

4.6.9 Let $t \geq 0$. The relation

$$P(|S_n| \geq n\varepsilon) = o(n^{-t}) \quad \text{for every } \varepsilon > 0 \quad (4.55)$$

holds if and only if $P(|X_1| \geq n) = o(n^{-t-1})$ and $\int_{|x|<n} x\, dV(x) = o(1)$. Here $V(x) = P(X_1 < x)$ (Baum and Katz [16]). It follows that the conditions $EX_1 = 0$ and $E|X_1|^r < \infty$ for some $r \geq 1$ are sufficient for the relation (4.55) with $t = r - 1$.

4.6.10 The following conditions are equivalent:
(A) there exist positive constants $\rho < 1$, ε and C such that $P(S_n \geq n\varepsilon) \leq C\rho^n$ for all sufficiently large n;
(B) there exists a positive number T such that $Ee^{tX_1} < \infty$ in the interval $0 \leq t < T$ (Petrov and Shirokova [357]).

4.6.11 If $t \geq 1$, then the condition

$$\sum_{n=1}^\infty n^{t-2} P(|S_n/n - b| \geq \varepsilon) < \infty \quad \text{for every } \varepsilon > 0 \quad (4.56)$$

is equivalent to the set of the conditions $E|X_1|^t < \infty$ and $EX_1 = b$ (Katz [211]). In the case when $t = 2$ this result was obtained by Erdös [113]; earlier, Hsu and Robbins [200] proved that the condition (4.56) for $t = 2$ follows from the conditions $EX_1 = b$ and $EX_1^2 < \infty$. For $t = 1$ this result was obtained by Spitzer [432].

4.6.12 If $EX_1 = 0$ and $EX_1^2 = \sigma^2 < \infty$, then

$$\lim_{\varepsilon \to 0+} \varepsilon^2 \sum_{n=1}^\infty P(|S_n| \geq n\varepsilon) = \sigma^2$$

(Heyde [188]).

4.6.13 If $0 < t < 1$, the condition

$$\sum_{n=1}^\infty n^{-1} P(|S_n| \geq n^{1/t}\varepsilon) < \infty \quad \text{for every } \varepsilon > 0$$

is equivalent to the condition $E|X_1|^t < \infty$. If $1 \leq t < 2$, the condition

$$\sum_{n=1}^\infty n^{-1} P(|S_n - nb| \geq n^{1/t}\varepsilon) < \infty \quad \text{for every } \varepsilon > 0$$

is equivalent to the conditions $E|X_1|^t < \infty$ and $EX_1 = b$ (Baum and Katz [16]).

4.6.14 If $\alpha t > 1$ and $\alpha > \frac{1}{2}$, the following conditions are equivalent:
(A) $E|X_1|^t < \infty$ and (in the case when $t \geq 1$) $EX_1 = 0$;
(B) $\sum_{n=1}^{\infty} n^{\alpha t - 2} P(|S_n| \geq n^{\alpha}\varepsilon) < \infty$ for every $\varepsilon > 0$;
(C) $\sum_{n=1}^{\infty} n^{\alpha t - 2} P\left(\sup_{k \geq n} k^{-\alpha}|S_k| \geq \varepsilon\right) < \infty$ for every $\varepsilon > 0$.

If $0 < t < 2$, the following conditions are equivalent:
(D) $E|X_1|^t \log(1 + |X_1|) < \infty$ and (in the case when $t \geq 1$) $EX_1 = 0$;
(E) $\sum_{n=1}^{\infty} n^{-1} \log n\, P(|S_n| \geq n^{1/t}\varepsilon) < \infty$ for every $\varepsilon > 0$;
(F) $\sum_{n=1}^{\infty} n^{-1} P\left(\sup_{k \geq n} k^{-1/t}|S_k| \geq \varepsilon\right) < \infty$ for every $\varepsilon > 0$

(Baum and Katz [16]).

4.6.15 Let $t > 1$ and $EX_1 = 0$. The following conditions are equivalent:
(1) $\int_0^{\infty} x^t \, dV(x) < \infty$, where $V(x)$ is the d.f. of X_1;

(2) $\sum_{n=1}^{\infty} n^{t-2} P(S_n \geq n\varepsilon) < \infty$ for every $\varepsilon > 0$;

(3) $\sum_{n=1}^{\infty} n^{t-2} P\left(\sup_{k \geq n} S_k/k \geq \varepsilon\right) < \infty$ for every $\varepsilon > 0$

(Amosova [4]). Other results concerning the rates of convergence in the one-sided laws of large numbers can be found in the papers by Chow and Lai [70] and Amosova [5].

4.6.16 Suppose $EX_1 = 0$. If $t \geq 2$ and $\alpha > \frac{1}{2}$, then

$$\sum_{n=1}^{\infty} n^{\alpha t - 2} P\left(\max_{k \leq n} |S_k| \geq n^{\alpha}\right) \leq c_1\{E|X_1|^t + (EX_1^2)^{(\alpha t - 1)/(2\alpha - 1)}\},$$

$$1 + \sum_{n=1}^{\infty} n^{\alpha t - 2} P(|S_n| \geq n^{\alpha}) \geq c_2\{E|X_1|^t + (EX_1^2)^{(\alpha t - 1)/(2\alpha - 1)}\},$$

where c_1 and c_2 are positive constants depending only on t and α (Chow and Lai [70, 71]).

4.6.17 Let $r \geq 0$ and $0 < t < 2$. Let $L(x)$ be a non-negative non-decreasing continuous slowly varying function.
(A) The relation

$$n^r L(n) P(|S_n| \geq n^{1/t}\varepsilon) \to 0 \quad \text{for every } \varepsilon > 0 \quad (4.57)$$

holds if and only if
$$n^{r+1}L(n)P(|X_1| \geq n^{1/t}) \to 0$$
and
$$n^{1-1/t}\int_{|x|<n^{1/t}} x\,dV(x) \to 0, \qquad (4.58)$$

where $V(x)$ is the d.f. of X_1. If $r > 0$, the relation (4.57) holds if and only if

$$n^r L(n) P\left(\sup_{k \geq n} k^{-1/t}|S_k| \geq \varepsilon\right) \to 0 \qquad \text{for every } \varepsilon > 0.$$

(B) The relation
$$\sum_{n=1}^{\infty} n^{r-1}L(n)P(|S_n| \geq n^{1/t}\varepsilon) < \infty \qquad \text{for every } \varepsilon > 0 \qquad (4.59)$$

holds if and only if the conditions (4.58) and

$$\sum_{n=1}^{\infty} n^r L(n) P(|X_1| \geq n^{1/t}) < \infty$$

are satisfied. If $r > 0$, the relation (4.59) is equivalent to the condition

$$\sum_{n=1}^{\infty} n^{r-1}L(n)P\left(\sup_{k \geq n} k^{-1/t}|S_k| \geq \varepsilon\right) < \infty \qquad \text{for every } \varepsilon > 0$$

(Heyde and Rohatgi [191]).

4.6.18 For any constants a and b, define
$$\Delta(a,b) = E\min\{1, ((S_n - a)/b)^2\}.$$
Suppose that X_1 has a non-degenerate distribution and $(S_n - a_n)/b_n \xrightarrow{P} 0$ for some sequences $\{a_n\}$ and $\{b_n\}$ such that $b_n > 0$, $b_n \to \infty$. Then $\inf_a \Delta(a, b_n) \geq c\delta_n$, where
$$\delta_n = nb_n^{-2} EX_1^2 I(|X_1| < b_n) + nP(|X_1| \geq b_n).$$
Define $S_n' = \sum_{k=1}^n X_k I(|X_k| < b_n)$. When a_n is equal to any one of mS_n, mS_n' or $n\, EX_1 I(|X_1| < b_n)$, we have $\Delta(a_n, b_n) \leq C\delta_n$. In particular, with this choice of a_n, $\Delta(a_n, b_n) \asymp \delta_n$. If $S_n/b_n \xrightarrow{P} 0$, then
$$\Delta(0, b_n) \asymp \delta_n + (nb_n^{-1} EX_1 I(|X_1| < b_n))^2$$
(Hall [161]).

In Subsections 4.6.19–4.6.24 we consider a sequence of independent (not necessarily identically distributed) random variables $\{X_n\}$. We put $S_n = \sum_{k=1}^n X_k$.

4.6.19 If
$$\sum_{n=1}^{\infty} n^{-1} P(|S_n| \geq n\varepsilon) < \infty \quad \text{for every } \varepsilon > 0,$$
then $E \log(1 + |X_n|) < \infty$ for all n. If $t > 1$ and
$$\sum_{n=1}^{\infty} n^{t-2} P(|S_n| \geq n\varepsilon) < \infty \quad \text{for every } \varepsilon > 0,$$
then $E|X_n|^{t-1} < \infty$ (Baum and Katz [16]).

4.6.20 If $t \geq 0$, the conditions
$$\sum_{n=1}^{\infty} n^t P(|S_n| \geq n\varepsilon) < \infty \quad \text{for every } \varepsilon > 0$$
and
$$\sum_{n=1}^{\infty} n^t P\left(\sup_{k \geq n} |S_k|/k \geq \varepsilon\right) < \infty \quad \text{for every } \varepsilon > 0$$
are equivalent (Baum and Katz [16]).

4.6.21 Suppose that $EX_n = 0$ for all n and $\sum_{n=1}^{\infty} n^{-r-1} E|X_n|^{2r} < \infty$ for some $r > 1$. Then
$$\sum_{n=1}^{\infty} n^{-1} P(|S_n| \geq n\varepsilon) < \infty \quad \text{for every } \varepsilon > 0$$
(Baum and Katz [16]).

4.6.22 Let there exist positive constants $\rho < 1$, ε and C such that $P(S_n \geq n\varepsilon) \leq C\rho^n$ for all sufficiently large n. If
$$S_n/n \xrightarrow{P} 0, \tag{4.60}$$
then $E\,e^{tX_j} < \infty$ for $j \in \mathbb{N}$ and $0 \leq t < (1/2\varepsilon) \log(1/\rho)$. In this proposition one cannot omit the condition (4.60) (Petrov and Shirokova [357]).

4.6.23 Let $a_n \uparrow \infty$ and $a_{n+1}/a_n = O(1)$. Let $u(x)$ be a positive function such that $u(x)x^{-\gamma} \downarrow 0$ as $x \uparrow \infty$, where γ is a non-negative constant and, moreover,
$$\frac{1}{u(x)} \int_x^{\infty} \frac{u(y)}{y} dy = O(1) \quad \text{as } x \to +\infty.$$
Then the following conditions are equivalent:
$$P(|S_n| \geq \varepsilon a_n) = O(u(a_n)) \quad \text{for every } \varepsilon > 0$$
and
$$P\left(\sup_{k \geq n} |S_k|/a_k \geq \varepsilon\right) = O(u(a_n)) \quad \text{for every } \varepsilon > 0$$
(Rozovsky [408]).

4.6.24 Suppose that $EX_n = 0$, $\operatorname{Var} X_n < \infty$ for all n. We put

$$B_n = \sum_{k=1}^{n} \operatorname{Var} X_k \quad \text{and} \quad Z_n = B_n^{-1/2} \sum_{k=1}^{n} X_k.$$

Furthermore, suppose that $\liminf B_n/n > 0$ and there exists a random variable X with zero mean and finite variance such that

$$n^{-1} \sum_{k=1}^{n} P(X_k \geqslant x) \leqslant b P(X \geqslant cx)$$

for all sufficiently large n and x and some positive constants b and c. Then

$$\lim_{\varepsilon \to 0+} \varepsilon^2 \sum_{n=1}^{\infty} P(Z_n \geqslant \varepsilon n^{1/2}) = \tfrac{1}{2}$$

(Sirazdinov and Gafurov [427]). This is a generalization of some results of Heyde (see Subsection 4.6.12), Chen [65], and Petrov [347].

5

Rates of convergence in the central limit theorem

5.1 Estimating the difference of distribution functions by the nearness of characteristic functions

Let $F(x)$ and $G(x)$ be arbitrary distribution functions, and let $f(t)$ and $g(t)$ be the corresponding characteristic functions. If $f(t)$ and $g(t)$ are identically equal, then $F(x)$ and $G(x)$ are identically equal. If, however, we know only that $f(t)$ and $g(t)$ are close to each other in some sense, then we may expect a certain nearness of $F(x)$ and $G(x)$. There are many results which elucidate this point. The following theorem of Esseen is one of the most useful of these.

Theorem 5.1 *Let $F(x)$ and $G(x)$ be distribution functions with the characteristic functions $f(t)$ and $g(t)$. Suppose that $G(x)$ has a bounded derivative on the real line, so that $\sup_x G'(x) \leq K$. Then for every $T > 0$ and every $b > 1/(2\pi)$ we have*

$$\sup_x |F(x) - G(x)| \leq b \int_{-T}^{T} \left| \frac{f(t) - g(t)}{t} \right| dt + c(b) \frac{K}{T}, \quad (5.1)$$

where $c(b)$ is a positive constant depending only on b.

This theorem, which is of interest in itself, will play a fundamental role in obtaining estimates of the rates of convergence in the central limit theorem. We shall apply Theorem 5.1 to the special case when $G(x)$ is the standard normal distribution function $\Phi(x)$, and $F(x)$ is the distribution function of a normalized sum of random variables. If we are able to obtain appropriate estimates of the difference of the corresponding characteristic functions on an interval $(-T, T)$ with sufficiently large T, then Theorem 5.1 allows us to estimate the nearness of the distribution functions $F(x)$ and $\Phi(x)$ in a non-trivial way.

We shall formulate and prove a generalization of Theorem 5.1 where $F(x)$ and $G(x)$ are not necessarily distribution functions. This generalization is also due to Esseen.

Theorem 5.2 *Let $F(x)$ be a bounded non-decreasing function, and $G(x)$ a differentiable function of bounded variation on the real line. Let $F(-\infty) =$*

$G(-\infty)$. Put

$$f(t) = \int_{-\infty}^{\infty} e^{itx}\, dF(x), \quad g(t) = \int_{-\infty}^{\infty} e^{itx}\, dG(x).$$

Suppose $\sup_x |G'(x)| \leqslant K$. Then for every $T > 0$ and every $b > 1/(2\pi)$, the inequality (5.1) holds, where $c(b)$ is a positive constant depending only on b.

We note that if $F(x)$ and $G(x)$ are distribution functions, then their Fourier–Stieltjes transforms become characteristic functions, and Theorem 5.2 reduces to Theorem 5.1. In what follows, we shall apply Theorem 5.2 instead of Theorem 5.1 if we wish to approximate the d.f. of a normalized sum of independent random variables by some function $G(x)$ which is not necessarily a d.f.

Proof Suppose $T > 0$ and $a > 0$. The function

$$p(x) = \frac{T}{\pi} \cdot \frac{1 - \cos(Tx - a)}{(Tx - a)^2}$$

is the density of a distribution with the c.f.

$$h(t) = \begin{cases} (1 - |t|/T)\, e^{ita/T} & \text{if } |t| < T, \\ 0 & \text{if } |t| \geqslant T. \end{cases}$$

In fact, if a random variable X has the density $p_0(x)$ given by equality (1.26) (Chapter 1), then the random variable $Y = (X + a)/T$ has the density $Tp_0(Tx - a) = p(x)$ and, accordingly, the c.f. $h(t)$.

It is clear that

$$p(x) \leqslant T(2\pi)^{-1} \sin^2((Tx - a)/2)/((Tx - \alpha)/2)^2 \leqslant T(2\pi)^{-1} \quad (5.2)$$

for all x, a and T.

We put

$$\gamma = \gamma(a) = \int_0^{2a/T} p(x)\, dx \quad (5.3)$$

so that

$$\gamma = 2\pi^{-1} \int_0^{a/2} \frac{\sin^2 u}{u^2}\, du. \quad (5.4)$$

The function $F(x)$ does not decrease, hence

$$F(x) = \gamma^{-1} \int_x^{x+2a/T} F(x) p(u-x)\, du \leq \gamma^{-1} \int_x^{x+2a/T} F(u) p(u-x)\, du$$

$$= G(x) + \gamma^{-1} \int_x^{x+2a/T} (G(u) - G(x)) p(u-x)\, du$$

$$+ \gamma^{-1} \int_x^{x+2a/T} (F(u) - G(u)) p(u-x)\, du$$

$$\leq G(x) + T(2\pi\gamma)^{-1} \int_0^{2a/T} |G(x+y) - G(x)|\, dy$$

$$+ \gamma^{-1} \int_x^{x+2a/T} (F(u) - G(u)) p(u-x)\, du$$

$$\leq G(x) + Ka^2(\pi\gamma T)^{-1} + \gamma^{-1} \int_x^{x+2a/T} (F(u) - G(u)) p(u-x)\, du,$$

by (5.2) and the condition of boundedness of the derivative of $G(x)$. Put

$$F_1(x) = \int_{-\infty}^{\infty} F(x-z) p(z)\, dz, \quad F_2(x) = \int_{-\infty}^{\infty} F(x+z) p(z)\, dz.$$

We define the functions $G_1(x)$ and $G_2(x)$ by similar equalities, replacing F by G. We have

$$F_1(x) = \int_{-\infty}^{\infty} F(u) p(x-u)\, du, \quad F_2(x) = \int_{-\infty}^{\infty} F(u) p(u-x)\, du \quad (5.5)$$

and

$$\int_{-\infty}^{\infty} e^{itx}\, dF_k(x) = f(t) h_k(t) \quad (k = 1, 2), \tag{5.6}$$

where $h_1(t) = h(t)$ and $h_2(t) = h(-t)$. We note that (5.6) is clear in the special case when F and G are d.fs. In fact, in this case $F_1(x)$ is the d.f. of the random variable $X + Y$, where X and Y are independent random variables with the d.f. $F(x)$ and the density $p(x)$, respectively, and $F_2(x)$ is the d.f. of $X - Y$. The general case reduces to this special case, since any bounded non-decreasing function $F(x)$ admits the representation $F(x) = c_0 F_0(x) + c$, where $F_0(x)$ is a d.f., $c_0 \geq 0$, and c, c_0 are constants. On the other hand, any function of bounded variation $G(x)$ can be represented as the difference of two bounded non-decreasing functions.

Equalities that are similar to (5.6) hold for the functions $G_1(x)$ and $G_2(x)$. Using the inversion formula (Theorem 1.4, Chapter 1) and the fact that

Rates of convergence in the central limit theorem | 145

$h(t) = 0$ for $|t| > T$, we obtain

$$F_k(x) - F_k(y) = (2\pi)^{-1} \int_{-T}^{T} \frac{e^{itx} - e^{ity}}{-it} f(t) h_k(t)\, dt$$

and

$$G_k(x) - G_k(y) = (2\pi)^{-1} \int_{-T}^{T} \frac{e^{itx} - e^{ity}}{-it} g(t) h_k(t)\, dt$$

for every x and y ($k = 1, 2$), since the functions $F_k(x)$ and $G_k(x)$ are continuous.

We shall suppose that

$$\int_{-T}^{T} \left| \frac{f(t) - g(t)}{t} \right| dt < \infty,$$

because in the contrary case the inequality (5.1) is obvious. Then by the Riemann–Lebesgue lemma we have

$$\lim_{y \to -\infty} \int_{-T}^{T} \frac{f(t) - g(t)}{t} h_k(t)\, e^{-ity}\, dt = 0.$$

The hypothesis $F(-\infty) = G(-\infty)$ implies that $F_k(-\infty) = G_k(-\infty)$ for $k = 1$ and $k = 2$. Passing to the limit as $y \to -\infty$, we obtain

$$F_k(x) - G_k(x) = (2\pi)^{-1} \int_{-T}^{T} \frac{f(t) - g(t)}{-it} h_k(t)\, e^{-itx}\, dt \qquad (k = 1, 2). \quad (5.7)$$

Therefore,

$$\left| \int_{-\infty}^{\infty} (F(u) - G(u)) p(x - u)\, du \right| \leq (2\pi)^{-1} \int_{-T}^{T} \left| \frac{f(t) - g(t)}{t} \right| dt \quad (5.8)$$

and

$$\left| \int_{-\infty}^{\infty} (F(u) - G(u)) p(u - x)\, du \right| \leq (2\pi)^{-1} \int_{-T}^{T} \left| \frac{f(t) - g(t)}{t} \right| dt \quad (5.9)$$

for all x by (5.5), since $|h(t)| \leq 1$ for all t.

We put $\Delta = \sup_x |F(x) - G(x)|$. It is easy to see that

$$\left| \int_x^{x + 2a/T} (F(u) - G(u)) p(u - x)\, du \right|$$

$$\leq \left| \int_{-\infty}^{\infty} (F(u) - G(u)) p(u - x)\, du \right| + \Delta \int_{-\infty}^{x} p(u - x)\, du + \Delta \int_{x + 2a/T}^{\infty} p(u - x)\, du$$

$$\leq (2\pi)^{-1} \int_{-T}^{T} \left| \frac{f(t) - g(t)}{t} \right| dt + \Delta \left(1 - \int_0^{2a/T} p(u)\, du \right).$$

It follows from the previous estimates and (5.3) that

$$F(x) - G(x) \leq (2\pi\gamma)^{-1} \int_{-T}^{T} \left|\frac{f(t) - g(t)}{t}\right| dt + Ka^2(\pi\gamma T)^{-1} + \Delta(1 - \gamma)/\gamma. \tag{5.10}$$

We shall now estimate the difference $F(x) - G(x)$ from below. In a similar way we find that

$$F(x) \geq \gamma^{-1} \int_{x-2a/T}^{x} F(u)p(x-u) du$$

$$= G(x) + \gamma^{-1} \int_{x-2a/T}^{x} (G(u) - G(x))p(x-u) du$$

$$+ \gamma^{-1} \int_{x-2a/T}^{x} (F(u) - G(u))p(x-u) du$$

$$\geq G(x) - Ka^2(\pi\gamma T)^{-1} - (2\pi\gamma)^{-1} \int_{-T}^{T} \left|\frac{f(t)-g(t)}{t}\right| dt - \Delta(1-\gamma)/\gamma.$$

for every real x. Together with (5.10) this implies that

$$\Delta \leq (2\pi\gamma)^{-1} \int_{-T}^{T} \left|\frac{f(t) - g(t)}{t}\right| dt + Ka^2(\pi\gamma T)^{-1} + \Delta(1 - \gamma)/\gamma.$$

By (5.4), $0 < \gamma < 1$ for every $a > 0$ and $\lim_{a \to +\infty} \gamma(a) = 1$. Therefore, by choosing a large enough we can guarantee that $\gamma > \frac{1}{2}$. Then we obtain

$$\Delta \leq (2\pi(2\gamma - 1))^{-1} \int_{-T}^{T} \left|\frac{f(t) - g(t)}{t}\right| dt + Ka^2(\pi T(2\gamma - 1))^{-1}.$$

Suppose $b > (2\pi)^{-1}$ is given. We define γ by the equation $2\pi(2\gamma - 1)b = 1$, so that $0 < \gamma < \frac{1}{2}$. Then in the inequality for Δ we can choose for a the solution of the equation $2\gamma(a) - 1 = 1/(2\pi b)$, or

$$\int_0^{a/2} \frac{\sin^2 u}{u^2} du = \frac{\pi}{4} + \frac{1}{8b}. \tag{5.11}$$

The assertion of the theorem follows. □

By following the lines of the proof of Theorem 5.2, it is not hard to generalize this result. Small changes in the proof allow us to obtain a proposition without the assumption about the existence of the derivative of function $G(x)$.

Theorem 5.3 *Let $F(x)$ be a bounded non-decreasing function, and $G(x)$ be a function of bounded variation on the real line. Let $F(-\infty) = G(-\infty)$. Put*

$$f(t) = \int_{-\infty}^{\infty} e^{itx}\, dF(x), \quad g(t) = \int_{-\infty}^{\infty} e^{itx}\, dG(x).$$

Then for every $T > 0$ and every $b > 1/(2\pi)$ we have

$$\sup_x |F(x) - G(x)| \leq b \int_{-T}^{T} \left|\frac{f(t) - g(t)}{t}\right| \left(1 - \frac{|t|}{T}\right) dt$$

$$+ bT \sup_x \int_{|y| \leq c(b)/T} |G(x+y) - G(x)|\, dy, \qquad (5.12)$$

where $c(b)$ is a positive constant depending only on b. In the inequality (5.12) we can put $c(b)$ equal to the root of the equation

$$\int_0^{c(b)/4} \frac{\sin^2 u}{u^2}\, du = \frac{\pi}{4} + \frac{1}{8b}.$$

For a proof of Theorem 5.3 we refer to [342] and [351]. Note that (5.7) and the inequality $|h_k(t)| \leq 1 - |t|/T$ for $|t| < T$ allow us to replace (5.8) and (5.9) by stronger estimates such as

$$\left|\int_{-\infty}^{\infty} (F(u) - G(u)) p(x-u)\, du\right| \leq (2\pi)^{-1} \int_{-T}^{T} \left|\frac{f(t) - g(t)}{t}\right| \left(1 - \frac{|t|}{T}\right) dt.$$

If the conditions of Theorem 5.3 are satisfied and if the function $G(x)$ satisfies the Lipschitz condition $|G(x) - G(y)| \leq C|x - y|^\alpha$ for all x and y, where C and α are positive constants, the second term on the right-hand side of (5.12) can be replaced by $2bC(c(b))^{1+\alpha} T^{-\alpha}/(1 + \alpha)$.

5.2 Esseen's inequality

We shall be interested in estimates of the deviation of the distribution of a sum of independent random variables from the normal law. We exclude the trivial case when all the terms of the sum have a degenerate distribution.

Lemma 5.1 Let $X_1, \ldots, X_n$ be independent random variables, $EX_j = 0$, $E|X_j|^3 < \infty$ $(j = 1, \ldots, n)$. Put

$$\sigma_j^2 = EX_j^2, \quad B_n = \sum_{j=1}^n \sigma_j^2, \quad L_n = B_n^{-3/2} \sum_{j=1}^n E|X_j|^3.$$

Let $f_n(t)$ be the c.f. of the random variable $B_n^{-1/2} \sum_{j=1}^n X_j$. Then

$$|f_n(t) - e^{-t^2/2}| \leq 16 L_n |t|^3\, e^{-t^2/3} \qquad (5.13)$$

for $|t| \leq 1/(4L_n)$.

Proof First we shall prove (5.13) under the additional assumption that $|t| \geq L_n^{-1/3}/2$. In this case, $8L_n|t|^3 \geq 1$, and we need only show that

$$|f_n(t)|^2 \leq e^{-2t^2/3}, \tag{5.14}$$

since (5.14) implies the inequalities

$$|f_n(t) - e^{-t^2/2}| \leq |f_n(t)| + e^{-t^2/2} \leq 2 e^{-t^2/3}.$$

Put $v_j(t) = E e^{itX_j}$ ($j = 1, \ldots, n$) and consider the symmetrized random variable $X_j^s = X_j - Y_j$, where Y_j does not depend on X_j and has the same distribution. It has the c.f. $|v_j(t)|^2$ and the variance $2\sigma_j^2$. Furthermore,

$$E|X_j^s|^3 \leq 4(E|X_j|^3 + E|Y_j|^3) = 8 E|X_j|^3,$$

since $|a + b|^3 \leq 4(|a|^3 + |b|^3)$ for any a and b, and

$$|v_j(t)|^2 = \int_{-\infty}^{\infty} e^{itx} \, dV_j^s(x) = \int_{-\infty}^{\infty} \left(1 + itx - \frac{t^2 x^2}{2} + \frac{1}{6} \theta t^3 x^3\right) dV_j^s(x)$$

by Lemma 3.2 (Chapter 3), where $V_j^s(x)$ is the d.f. of X_j^s and $|\theta| \leq 1$. Taking into account the condition $EX_j = 0$ and the inequality

$$\left|\int_{-\infty}^{\infty} \theta t^3 x^3 \, dV_j^s(x)\right| \leq |t|^3 \int_{-\infty}^{\infty} |x|^3 \, dV_j^s(x) \leq 8|t|^3 E|X_j|^3,$$

we obtain

$$|v_j(t)|^2 = 1 - \sigma_j^2 t^2 + \tfrac{4}{3}\theta|t|^3 E|X_j|^3,$$

where $|\theta| \leq 1$. Together with the inequality $1 + x \leq e^x$ for every real x, this implies that

$$|v_j(t)|^2 \leq 1 - \sigma_j^2 t^2 + \tfrac{4}{3}|t|^3 E|X_j|^3 \leq \exp\{-\sigma_j^2 t^2 + \tfrac{4}{3}|t|^3 E|X_j|^3\}.$$

Therefore, in the interval $|t| \leq 1/(4L_n)$ we have

$$|f_n(t)|^2 = \prod_{j=1}^{n} |v_j(B_n^{-1/2}t)|^2 \leq \exp\{-t^2 + \tfrac{4}{3}|t|^3 L_n\} \leq e^{-2t^2/3},$$

and (5.14) is proved.

Now suppose that $|t| \leq 1/(4L_n)$ and $|t| < 1/(2L_n^{1/3})$. If $k = 1, \ldots, n$, then

$$\sigma_j |t| B_n^{-1/2} \leq (E|X_j|^3)^{1/3}|t| B_n^{-1/2} < L_n^{1/3}|t| < \tfrac{1}{2},$$

$$v_j(B_n^{-1/2} t) = 1 - r_j,$$

where

$$r_j = \sigma_j^2 t^2/(2B_n) + \theta_j|t|^3 E|X_j|^3 B_n^{-3/2}/6, \qquad |\theta_j| \leq 1,$$

so that $|r_j| < 1/6$ and

$$|r_j|^2 \leq 2(\sigma_j^2 t^2/(2B_n))^2 + 2(|t|^3 E|X_j|^3 B_n^{-3/2}/6)^2 \leq |t|^3 E|X_j|^3 B_n^{-3/2}/3.$$

Using the well-known Taylor expansion of the function $\log(1+z)$, we can easily find that $\log(1+z) = z + \theta z$, where $|\theta| \leq 1$ if $|z| < \frac{1}{2}$. Hence

$$\log v_j(B_n^{-1/2}t) = -\sigma_j^2 t^2/(2B_n) + \theta_j |t|^3 \, \mathrm{E}|X_j|^3 B_n^{-3/2}/2$$

and

$$\log f_n(t) = \sum_{j=1}^{n} \log v_j(B_n^{-1/2}t) = -t^2/2 + \theta |t|^3 L_n/2, \qquad |\theta| \leq 1.$$

Furthermore, $|e^z - 1| \leq |z| \, e^{|z|}$ for every complex value z. The condition $L_n|t|^3 < 1/8$ implies that $\exp\{L_n|t|^3/2\} < 2$. Therefore,

$$|f_n(t) - e^{-t^2/2}| \leq e^{-t^2/2} |\exp\{\theta L_n|t|^3/2\} - 1|$$
$$\leq \tfrac{1}{2} L_n |t|^3 \exp\{-t^2/2 + L_n|t|^3/2\} \leq L_n |t|^3 \, e^{-t^2/2}.$$

This inequality is even stronger than (5.13). □

We recall that by agreement the symbols $A, A_1, A_2, \ldots$ denote absolute positive constants and that $\Phi(x)$ denotes the normal $(0,1)$ distribution function.

Theorem 5.4 Let $X_1, \ldots, X_n$ be independent random variables such that $\mathrm{E}X_j = 0$, $\mathrm{E}|X_j|^3 < \infty$ $(j = 1, \ldots, n)$. Put

$$\sigma_j^2 = \mathrm{Var}\, X_j, \quad B_n = \sum_{j=1}^{n} \sigma_j^2, \quad F_n(x) = P\!\left(B_n^{-1/2} \sum_{j=1}^{n} X_j < x\right),$$

$$L_n = B_n^{-3/2} \sum_{j=1}^{n} \mathrm{E}|X_j|^3.$$

Then

$$\sup_x |F_n(x) - \Phi(x)| \leq A L_n. \tag{5.15}$$

Proof We shall use Theorem 5.1 with $F(x) \equiv F_n(x)$ and $G(x) \equiv \Phi(x)$. Setting $b = 1/\pi$ and $T = 1/(4L_n)$, and taking account of the inequality $\Phi'(x) = (2\pi)^{-1/2} e^{-x^2/2} \leq (2\pi)^{-1/2}$ for all x, we find by (5.1) that

$$\sup_x |F_n(x) - \Phi(x)| \leq \pi^{-1} \int_{|t| \leq 1/(4L_n)} \left|\frac{f_n(t) - e^{-t^2/2}}{t}\right| dt + A_1 L_n.$$

Here $f_n(t)$ is the c.f. that corresponds to $F_n(x)$. Applying Lemma 5.1, we obtain (5.15). □

We shall call (5.15) Esseen's inequality.

In the special case of identical distributions, Theorem 5.4 reduces to the following.

Theorem 5.5 *Let $X_1, \ldots, X_n$ be independent identically distributed random variables. Let*

$$EX_1 = 0, \quad \text{Var } X_1 = \sigma^2 > 0, \quad E|X_1|^3 < \infty. \quad \rho = E|X_1|^3/\sigma^3.$$

Then

$$\sup_x \left| P\left(\sigma^{-1} n^{-1/2} \sum_{j=1}^n X_j < x\right) - \Phi(x) \right| \leq A\rho n^{-1/2}. \tag{5.16}$$

We shall call (5.16) the Berry–Esseen inequality.

The assumption about zero means of the random variables $X_1, X_2, \ldots, X_n$ does not diminish the generality of Theorems 5.4 and 5.5. If this condition is not satisfied, then we can consider the random variables $Y_j = X_j - EX_j$ ($j = 1, \ldots, n$) having zero means so that Theorems 5.4 and 5.5 are applicable.

We cannot improve the order of the estimates (5.15) and (5.16) without imposing additional conditions on the distributions of our random variables. This will follow from a theorem to be proved in Section 5.7, on asymptotic expansions of the d.f. $F_n(x)$, but it can also be proved by elementary methods.

Let us consider a sequence of independent identically distributed random variables $\{X_n\}$ with two values -1 and 1 and the corresponding probabilites $\frac{1}{2}$ and $\frac{1}{2}$ (the symmetric Bernoulli scheme). Clearly, $EX_1 = 0$, $\text{Var } X_1 = 1$, $E|X_1|^3 = 1$, and $L_n = n^{-1/2}$. Suppose that n is even. We can interpet $P\left(\sum_{j=1}^n X_j = 0\right)$ as the probability of having $n/2$ successes in n independent trials with a probability of success in every trial that is equal to $\frac{1}{2}$. Therefore,

$$P\left(\sum_{j=1}^n X_j = 0\right) = \binom{n}{n/2}\left(\frac{1}{2}\right)^n \sim 2(2\pi n)^{-1/2},$$

by Stirling's formula. Thus the function $F_n(x) = P(n^{-1/2} \sum_{j=1}^n X_j < x)$ has a jump of magnitude $2(2\pi n)^{-1/2}(1 + o(1))$ at the point $x = 0$. It follows that in the neighbourhood of this point the function $F_n(x)$ cannot be approximated by a continuous function to an accuracy exceeding $(2\pi n)^{-1/2}(1 + o(1))$.

It also follows that the absolute constant A in the Berry–Esseen inequality (5.16) is not less than $(2\pi)^{-1/2}$. This is also true for the constant A in (5.15).

5.3 Generalizations of Esseen's inequality

Esseen's inequality (5.15) was proved under the assumption about the existence of moments of the third order. It is desirable to obtain some non-trivial estimates of the difference of the distribution function of a normalized sum of independent random variables and the standard normal distribution function under weaker moment assumptions.

Let G be the set of functions, defined for all real x, that satisfy the following conditons: (a) $g(x)$ is non-negative, even, and non-decreasing in the interval $x > 0$; (b) $x/g(x)$ is non-decreasing in the interval $x > 0$.

Theorem 5.6 *Let $X_1, \ldots, X_n$ be independent random variables such that $EX_j = 0$, $EX_j^2 g(X_j) < \infty$ for $j = 1, \ldots, n$ and for some $g \in G$. We put*

$$\sigma_j^2 = \operatorname{Var} X_j, \quad B_n = \sum_{j=1}^n \sigma_j^2, \quad F_n(x) = P\left(B_n^{-1/2} \sum_{j=1}^n X_j < x\right).$$

Then

$$\sup_x |F_n(x) - \Phi(x)| \leq \frac{A}{B_n g(B_n^{1/2})} \sum_{j=1}^n EX_j^2 g(X_j). \quad (5.17)$$

Proof We introduce the truncated random variables

$$\bar{X}_j = \begin{cases} X_j & \text{if } |X_j| < B_n^{1/2}, \\ 0 & \text{if } |X_j| \geq B_n^{1/2} \end{cases} \quad (j = 1, \ldots, n).$$

Being bounded, these random variables have moments of all orders. We write

$$\bar{m}_j = E\bar{X}_j, \quad \bar{\sigma}_j^2 = \operatorname{Var} \bar{X}_j,$$

$$\bar{B}_n = \sum_{j=1}^n \bar{\sigma}_j^2, \quad V_j(x) = P(X_j < x).$$

Since $EX_j = 0$, we have

$$0 \leq \sigma_j^2 - \bar{\sigma}_j^2 = \int_{|x| \geq B_n^{1/2}} x^2 \, dV_j(x) + \left(\int_{|x| < B_n^{1/2}} x \, dV_j(x)\right)^2$$

$$\leq 2 \int_{|x| \geq B_n^{1/2}} x^2 \, dV_j(x) \leq \frac{2}{g(B_n^{1/2})} \int_{|x| \geq B_n^{1/2}} x^2 g(x) \, dV_j(x)$$

$$\leq \frac{2}{g(B_n^{1/2})} EX_j^2 g(X_j) \quad (5.18)$$

for $j = 1, \ldots, n$. If $\bar{B}_n \leq B_n/4$, then we obtain $B_n - \bar{B}_n \geq 3B_n/4$ and

$$1 \leq \frac{8}{3 B_n g(B_n^{1/2})} \sum_{j=1}^n EX_j^2 g(X_j),$$

so that (5.17) holds with $A = 8/3$. Therefore, we shall suppose that $\bar{B}_n > B_n/4$. We put

$$Z_n = B_n^{-1/2} \sum_{j=1}^n X_j, \quad Y_n = B_n^{-1/2} \sum_{j=1}^n \bar{X}_j, \quad \bar{Z}_n = \bar{B}_n^{-1/2} \sum_{j=1}^n (\bar{X}_j - \bar{m}_j).$$

The event $Z_n < x$ implies the union of the events

$$(Y_n < x) \cup (|X_1| \geq B_n^{1/2}) \cup \cdots \cup (|X_n| \geq B_n^{1/2}),$$

hence

$$P(Z_n < x) \leq P(Y_n < x) + \sum_{j=1}^{n} P(|X_j| \geq B_n^{1/2}).$$

The event $Y_n < x$ implies

$$(Z_n < x) \cup (|X_1| \geq B_n^{1/2}) \cup \cdots \cup (|X_n| \geq B_n^{1/2}),$$

hence

$$P(Y_n < x) \leq P(Z_n < x) + \sum_{j=1}^{n} P(|X_j| \geq B_n^{1/2}).$$

Here x is an arbitrary real number. Accordingly,

$$\sup_x |F_n(x) - P(Y_n < x)| \leq \sum_{j=1}^{n} P(|X_j| \geq B_n^{1/2}).$$

It is clear that

$$|F_n(x) - \Phi(x)| \leq |F_n(x) - P(Y_n < x)|$$
$$+ |P(pY_n + q < px + q) - \Phi(px + q)|$$
$$+ |\Phi(px + q) - \Phi(x)|$$

for arbitrary numbers x, $p > 0$ and q. We put here $p = (B_n/\bar{B}_n)^{1/2}$ and $q = -\sum_{j=1}^{n} \bar{m}_j/\bar{B}_n^{1/2}$. Then $pY_n + q = \bar{Z}_n$ and

$$|F_n(x) - \Phi(x)| \leq T_1 + T_2 + T_3 \tag{5.19}$$

for every x, where

$$T_1 = \sum_{j=1}^{n} P(|X_j| \geq B_n^{1/2}),$$
$$T_2 = \sup_x \left| P(\bar{Z}_n < px + q) - \Phi\left((B_n/\bar{B}_n)^{1/2} x - \sum_{j=1}^{n} \bar{m}_j \Big/ \bar{B}_n^{1/2}\right) \right|,$$
$$T_3 = \sup_x \left| \Phi\left((B_n/\bar{B}_n)^{1/2} x - \sum_{j=1}^{n} \bar{m}_j \Big/ \bar{B}_n^{1/2}\right) - \Phi(x) \right|.$$

The random variable $\bar{Z}_n$ is a normalized sum of independent random variables having finite absolute moments of any order. Consequently, Theorem 5.4 is applicable, and we have

$$T_2 \leq A_1 \bar{B}_n^{-3/2} \sum_{j=1}^{n} E|\bar{X}_j - \bar{m}_j|^3.$$

Furthermore,

$$E|\bar{X}_j - \bar{m}_j|^3 \leqslant 4(E|\bar{X}_j|^3 + |\bar{m}_j|^3) \leqslant 8\, E|\bar{X}_j|^3$$

$$= 8 \int_{|x|<B_n^{1/2}} \frac{|x|}{g(x)} x^2 g(x)\, dV_j(x) \leqslant \frac{8 B_n^{1/2}}{g(B_n^{1/2})} EX_j^2 g(X_j),$$

since the function $x/g(x)$ does not decrease in the interval $x > 0$. Hence, and from the inequality $\bar{B}_n > B_n/4$, we conclude that

$$T_2 \leqslant \frac{A_2}{B_n g(B_n^{1/2})} \sum_{j=1}^{n} EX_j^2 g(X_j). \tag{5.20}$$

By Chebyshev's inequality,

$$T_1 \leqslant \frac{1}{B_n g(B_n^{1/2})} \sum_{j=1}^{n} EX_j^2 g(X_j). \tag{5.21}$$

It remains to prove that

$$T_3 \leqslant \frac{A_3}{B_n g(B_n^{1/2})} \sum_{j=1}^{n} EX_j^2 g(X_j), \tag{5.22}$$

since the inequalities (5.19)–(5.22) imply (5.17).

It is easy to prove the following elementary proposition.

Lemma 5.2

$$\sup_x |\Phi(px) - \Phi(x)| \leqslant \begin{cases} (p-1)/(2\pi\, e)^{1/2} & \text{if } p \geqslant 1 \\ (p^{-1} - 1)/(2\pi\, e)^{1/2} & \text{if } 0 < p < 1, \end{cases}$$

$$\sup_x |\Phi(x+q) - \Phi(x)| \leqslant |q|/(2\pi)^{1/2} \qquad \textit{for every } q.$$

Lemma 5.2 implies that

$$T_3 \leqslant (2\pi)^{-1/2} \left((B_n/\bar{B}_n)^{1/2} - 1 + \left| \sum_{j=1}^{n} \bar{m}_j \right| \bigg/ \bar{B}_n^{1/2} \right).$$

Furthermore,

$$|\bar{m}_j| = \left| \int_{|x|<B_n^{1/2}} x\, dV_j(x) \right| \leqslant \int_{|x| \geqslant B_n^{1/2}} |x|\, dV_j(x) \leqslant \frac{1}{B_n^{1/2} g(B_n^{1/2})} EX_j^2 g(X_j),$$

since the function $g(x)$ does not decrease in the interval $x > 0$, and

$$(B_n/\bar{B}_n)^{1/2} - 1 = (B_n - \bar{B}_n)/(B_n^{1/2} + \bar{B}_n^{1/2})\bar{B}_n^{1/2} \leq \frac{A_4}{B_n g(B_n^{1/2})} \sum_{j=1}^{n} EX_j^2 g(X_j)$$

by (5.18) and the inequality $\bar{B}_n > B_n/4$. Thus (5.22) holds. □

If $0 < \delta \leq 1$, the function $g(x) = |x|^\delta$ belongs to G. Therefore Theorem 5.6 implies Theorem 5.4, which corresponds to $\delta = 1$ and the following generalization of it.

Theorem 5.7 Let $X_1, \ldots, X_n$ be independent random variables such that $EX_j = 0$ and $E|X_j|^{2+\delta} < \infty$ ($j = 1, \ldots, n$) for some positive $\delta \leq 1$. Then

$$\sup_x |F_n(x) - \Phi(x)| \leq AB_n^{-1-\delta/2} \sum_{j=1}^{n} E|X_j|^{2+\delta}. \tag{5.23}$$

Theorem 5.8 Let $X_1, \ldots, X_n$ be independent random variables with zero means, finite variances and the d.f. $V_1(x), \ldots, V_n(x)$. We put

$$\Lambda_n = B_n^{-1} \sum_{j=1}^{n} \int_{|x| \geq B_n^{1/2}} x^2 \, dV_j(x), \quad l_n = B_n^{-3/2} \sum_{j=1}^{n} \int_{|x| < B_n^{1/2}} |x|^3 \, dV_j(x)$$

Then

$$\sup_x |F_n(x) - \Phi(x)| \leq A(\Lambda_n + l_n). \tag{5.24}$$

Proof The function

$$g(x) = |x| \quad \text{if } |x| < B_n^{1/2}, \quad g(x) = B_n^{1/2} \quad \text{if } |x| \geq B_n^{1/2}$$

belongs to G. We can apply Theorem 5.6 and (5.24) follows from (5.17). □

In turn, Theorem 5.8 implies that if the conditions of Theorem 5.8 are satisfied and if

$$\Lambda_n(\varepsilon) = B_n^{-1} \sum_{j=1}^{n} \int_{|x| \geq \varepsilon B_n^{1/2}} x^2 \, dV_j(x), \quad l_n(\varepsilon) = B_n^{-3/2} \sum_{j=1}^{n} \int_{|x| < \varepsilon B_n^{1/2}} |x|^3 \, dV_j(x)$$

then

$$\sup_x |F_n(x) - \Phi(x)| \leq A(\Lambda_n(\varepsilon) + l_n(\varepsilon)) \tag{5.25}$$

for every $\varepsilon > 0$.

In fact, $\Lambda_n(1) = \Lambda_n$, $l_n(1) = l_n$; if $0 < \varepsilon < 1$, then $\Lambda_n \leq \Lambda_n(\varepsilon)$ and $l_n \leq l_n(\varepsilon) + \Lambda_n(\varepsilon)$. If $\varepsilon > 1$, then $l_n \leq l_n(\varepsilon)$ and $\Lambda_n \leq l_n(\varepsilon) + \Lambda_n(\varepsilon)$. Hence, $\Lambda_n + l_n \leq 2\Lambda_n(\varepsilon) + 2l_n(\varepsilon)$ for every $\varepsilon > 0$ and (5.24) implies (5.25).

We note that $\Lambda_n(\varepsilon)$ is the Lindeberg ratio and that $l_n(\varepsilon) \leq \varepsilon$. Therefore,

(5.25) implies the inequality

$$\sup_x |F_n(x) - \Phi(x)| \leq A(\Lambda_n(\varepsilon) + \varepsilon) \qquad (5.26)$$

for every $\varepsilon > 0$. In turn, the Lindeberg theorem follows: if $\Lambda_n \to 0$ for every fixed $\varepsilon > 0$, then $F_n(x) \to \Phi(x)$.

Theorem 5.7 is also a consequence of Theorem 5.8, since

$$\Lambda_n + l_n \leq B_n^{-1-\delta/2} \sum_{j=1}^n E|X_j|^{2+\delta}.$$

Theorem 5.7 provides an estimate of the rate of convergence in Lyapunov's theorem (Theorem 4.9, Chapter 4).

In Theorems 5.6–5.8 we did not assume the existence of third-order moments, but the variances were assumed to exist. It is possible to obtain a generalization of Esseen's inequality without any assumptions about the existence of moments.

Let $X_1, \ldots, X_n$ be random variables with the distribution functions $V_1(x), \ldots, V_n(x)$. (We do not assume that the random variables $X_1, \ldots, X_n$ are independent.) Let $t_1, \ldots, t_n$ be positive numbers. We introduce the truncated random variables

$$\bar{X}_j = \begin{cases} X_j & \text{if } |X_j| < t_j \\ 0 & \text{if } |X_j| \geq t_j, \end{cases}$$

where $j = 1, \ldots, n$. We put

$$M_n = \sum_{j=1}^n E\bar{X}_j = \sum_{j=1}^n \int_{|x|<t_j} x \, dV_j(x), \quad N_n = \operatorname{Var} \sum_{j=1}^n \bar{X}_j,$$

$$\Delta_n = \sup_x \left| P\left(N_n^{-1/2} \sum_{j=1}^n (\bar{X}_j - E\bar{X}_j) < x \right) - \Phi(x) \right|,$$

$$\Gamma_n = \sum_{j=1}^n P(|X_j| \geq t_j).$$

Theorem 5.9 *For all numbers $a > 0$ and b we have*

$$\sup_x \left| P\left(\frac{1}{a} \sum_{j=1}^n X_j - b < x \right) - \Phi(x) \right|$$

$$\leq \Delta_n + \Gamma_n + \frac{|ab - M_n|}{\sqrt{2\pi N_n}} + \frac{1}{2\sqrt{2\pi e}} \left| 1 - \frac{N_n}{a^2} \right| \max\left(1, \frac{a^2}{N_n}\right).$$

Proof The event $\sum_{j=1}^{n} X_j < x$ implies the event

$$\left(\sum_{j=1}^{n} \bar{X}_j < x\right) \cup (|X_1| \geq t_1) \cup \cdots \cup (|X_n| \geq t_n),$$

and the event $\sum_{j=1}^{n} \bar{X}_j < x$ implies the event

$$\left(\sum_{j=1}^{n} X_j < x\right) \cup (|X_1| \geq t_1) \cup \cdots \cup (|X_n| \geq t_n).$$

Hence

$$\left| P\left(\sum_{j=1}^{n} X_j < x\right) - P\left(\sum_{j=1}^{n} \bar{X}_j < x\right) \right| \leq \sum_{j=1}^{n} P(|X_j| \geq t_j) = \Gamma_n.$$

for every x.

Setting

$$Y_n = N_n^{-1/2} \sum_{j=1}^{n} (\bar{X}_j - E\bar{X}_j), \quad Z_n = \frac{1}{a} \sum_{j=1}^{n} X_j - b,$$

we have

$$|P(Z_n < x) - \Phi(x)| \leq \left| P(Z_n < x) - P\left(\frac{1}{a} \sum_{j=1}^{n} \bar{X}_j - b < x\right) \right|$$

$$+ \left| P\left(\frac{1}{a} \sum_{j=1}^{n} \bar{X}_j - b < x\right) - \Phi(px + q) \right|$$

$$+ |\Phi(px + q) - \Phi(px)| + |\Phi(px) - \Phi(x)|$$

for every real x, p and q. We put $p = aN_n^{-1/2}$, $q = (ab - M_n)N_n^{-1/2}$. Then

$$\left| P\left(\frac{1}{a} \sum_{j=1}^{n} \bar{X}_j - b < x\right) - \Phi(px + q) \right| = |P(Y_n < px + q) - \Phi(px + q)| \leq \Delta_n$$

for every x. Accordingly,

$$\sup_x |P(Z_n < x) - \Phi(x)| \leq \Delta_n + \Gamma_n + T_1 + T_2,$$

where

$$T_1 = \sup_x |\Phi(x + q) - \Phi(x)|, \quad T_2 = \sup_x |\Phi(px) - \Phi(x)|.$$

By Lemma 5.2 we obtain

$$T_1 \leq \frac{|ab - M_n|}{\sqrt{2\pi N_n}},$$

$$T_2 \leq \frac{1}{\sqrt{2\pi e}}\left(\frac{a}{\sqrt{N_n}} - 1\right) \quad \text{if } a \geq \sqrt{N_n},$$

$$T_2 \leq \frac{1}{\sqrt{2\pi e}}\left(\frac{\sqrt{N_n}}{a} - 1\right) \quad \text{if } a < \sqrt{N_n}.$$

Obviously,
$$\frac{a}{\sqrt{N_n}} - 1 = \frac{a^2 - N_n}{\sqrt{N_n}(a + \sqrt{N_n})} \leq \frac{1}{2}\left(\frac{a^2}{N_n} - 1\right)$$
for $a \geq \sqrt{N_n}$. If, however, $a < \sqrt{N_n}$,
$$\frac{\sqrt{N_n}}{a} - 1 = \frac{N_n - a^2}{a(\sqrt{N_n} + a)} \leq \frac{1}{2}\left(\frac{N_n}{a^2} - 1\right).$$
Thus
$$T_2 \leq \frac{1}{2\sqrt{2\pi e}}\left|1 - \frac{N_n}{a^2}\right| \max\left(1, \frac{a^2}{N_n}\right). \quad \square$$

In Theorem 5.9 there are no assumptions about the independence or any type of dependence of the random variables $X_1, \ldots, X_n$ and about the existence of any moments of these random variables. The condition $0 < t_j < \infty$ ($j = 1, \ldots n$) implies the existence of the moments of an arbitrary order of $\bar{X}_1, \ldots, \bar{X}_n$. Under the additional assumption about the independence of the random variables $X_1, \ldots, X_n$, the truncated variables $\bar{X}_1, \ldots, \bar{X}_n$ are also independent, and we can apply, for example, Theorems 5.4 and 5.6 in order to obtain estimates of Δ_n. By Theorem 5.4 we have

$$\Delta_n \leq A N_n^{-3/2} \sum_{j=1}^{n} E|\bar{X}_j - E\bar{X}_j|^3 \tag{5.27}$$

if the random variables $X_1, \ldots, X_n$ are independent. It is not hard to show that the inequality which follows from Theorem 5.9 with the replacement of Δ_n by the right-hand side of (5.27) is a generalization of the estimates (5.15), (5.17), and (5.23). Some other estimates also follow from it.

5.4 Upper and lower estimates having the same order

We consider a sequence of independent random variables $\{X_n\}$ with a common distribution function $V(x)$, the characteristic function $v(t)$, and finite variance $\sigma^2 > 0$. Without loss of generality we may assume that $EX_1 = 0$. We denote
$$F_n(x) = P\left(\sigma^{-1} n^{-1/2} \sum_{j=1}^{n} X_j < x\right),$$
$$\psi_n = \sigma^{-2} \int_{|x| \geq \sigma n^{1/2}} x^2 \, dV(x) + \sigma^{-3} n^{-1/2} \left|\int_{|x| < \sigma n^{1/2}} x^3 \, dV(x)\right|$$
$$+ \sigma^{-4} n^{-1} \int_{|x| < \sigma n^{1/2}} x^4 \, dV(x).$$

Theorem 5.10 *If* $E|X_1|^3 = \infty$ *and* $\lim \sup_{|t| \to \infty} |v(t)| < 1$, *then*

$$\sup_x |F_n(x) - \Phi(x)| \asymp \psi_n. \qquad (5.28)$$

Theorem 5.11 *If X_1 has a lattice distribution, then*

$$\sup_x |F_n(x) - \Phi(x)| \asymp \psi_n + n^{-1/2}. \qquad (5.29)$$

We shall prove these theorems which are due to Osipov. We put

$$\Delta_n = \sup_x |F_n(x) - \Phi(x)|, \quad f_n(t) = v^n\left(\frac{t}{\sigma\sqrt{n}}\right),$$

$$T_n = \min(n^{1/4}, \psi_n^{-1/4}).$$

Lemma 5.3 *The following relations hold:*

$$n\left(v\left(\frac{t}{\sigma\sqrt{n}}\right) - 1 + \frac{t^2}{2n}\right) = (t^2 + t^4)O(\psi_n) \qquad (5.30)$$

uniformly in $t \in \mathbb{R}$ and

$$f_n(t) - e^{-t^2/2} = e^{-t^2/2}\left[n\left(v\left(\frac{t}{\sigma\sqrt{n}}\right) - 1 + \frac{t^2}{2n}\right) + (t^4 + t^8)O(\psi_n^2 + n^{-1})\right] \qquad (5.31)$$

uniformly in t in the interval $|t| < T_n$.

Proof Using the conditions $EX_1 = 0$, $\text{Var } X_1 = \sigma^2$ and Lemma 3.2 (Chapter 3), we obtain

$$n\left(v\left(\frac{t}{\sigma\sqrt{n}}\right) - 1 + \frac{t^2}{2n}\right) = n\int_{-\infty}^{\infty}\left(e^{itx/(\sigma\sqrt{n})} - 1 - \frac{itx}{\sigma\sqrt{n}} + \frac{t^2x^2}{2\sigma^2 n}\right)dV(x)$$

$$= \theta_1 \int_{|x| \geq \sigma n^{1/2}} \frac{t^2 x^2}{\sigma^2} dV(x)$$

$$+ \int_{|x| < \sigma n^{1/2}} \left(\frac{i^3 t^3 x^3}{6\sigma^3 \sqrt{n}} + \theta_2 \frac{t^4 x^4}{24\sigma^4 n}\right) dV(x)$$

$$= (t^2 + t^4)O(\psi_n),$$

where $|\theta_j| \leq 1$. The relation (5.30) is proved. Furthermore,

$$\left|v\left(\frac{t}{\sigma\sqrt{n}}\right) - 1\right| = \left|E\left(e^{(it/\sigma\sqrt{n})X_1} - 1 - \frac{it}{\sigma\sqrt{n}}X_1\right)\right| \leq \frac{t^2}{2n} < \frac{1}{2}$$

for $|t| < n^{1/2}.$ Hence

$$\log f_n(t) = n \log v\left(\frac{t}{\sigma\sqrt{n}}\right) = n\left(v\left(\frac{t}{\sigma\sqrt{n}}\right) - 1\right) + \theta_3 n\left(v\left(\frac{t}{\sigma\sqrt{n}}\right) - 1\right)^2$$

$$= n\left(v\left(\frac{t}{\sigma\sqrt{n}}\right) - 1\right) + t^4 O(n^{-1})$$

uniformly in t in the interval $|t| < n^{1/2}$. Taking account of (5.30), we obtain

$$f_n(t) - e^{-t^2/2} = e^{-t^2/2}\left[\exp\left\{n\left(v\left(\frac{t}{\sigma\sqrt{n}}\right) - 1 + \frac{t^2}{2n}\right) + t^4 O(n^{-1})\right\} - 1\right]$$

$$= e^{-t^2/2}\left[n\left(v\left(\frac{t}{\sigma\sqrt{n}}\right) - 1 + \frac{t^2}{2n}\right) + (t^4 + t^8) O(\psi_n^2 + n^{-1})\right]$$

uniformly in t in the interval $|t| < T_n$. □

In order to obtain an upper estimate for Δ_n under the conditions of Theorem 5.11, we apply Theorem 5.1, setting $T = \varepsilon \sigma n^{1/2}$, where $\varepsilon > 0$ and $F(x) \equiv F_n(x)$, $G(x) \equiv \Phi(x)$. By Lemma 5.3 we have

$$\int_{-T}^{T} |f_n(t) - e^{-t^2/2}| \frac{dt}{|t|} = O(\psi_n + n^{-1}). \tag{5.32}$$

The relation $v(t) = 1 - (\sigma^2 t^2/2) + o(t^2)$ as $t \to 0$ implies that

$$|v(t)| \leq 1 - \tfrac{1}{4}\sigma^2 t^2 \leq e^{-\sigma^2 t^2/4}$$

for $|t| \leq \varepsilon$ and sufficiently small $\varepsilon > 0$. Accordingly,

$$|f_n(t)| = \left|v\left(\frac{t}{\sigma\sqrt{n}}\right)\right|^n \leq e^{-t^2/4}$$

for $|t| < \varepsilon \sigma n^{1/2}$ and

$$\int_{T_n}^{\varepsilon \sigma n^{1/2}} |f_n(t)| \frac{dt}{|t|} = O(\psi_n + n^{-1}). \tag{5.33}$$

It is clear that

$$\int_{|x| \geq T_n} e^{-t^2/2} \frac{dt}{t} = O(\psi_n + n^{-1}). \tag{5.34}$$

These estimates and Theorem 5.1 imply that $\Delta_n = O(\psi_n + n^{-1/2})$.

In order to obtain an upper estimate for Δ_n under the conditions of Theorem 5.10 we shall apply Theorem 5.1, putting $T = n$ and choosing $F(x)$ and $G(x)$ as before. It follows from the condition $\limsup_{|t| \to \infty} |v(t)| < 1$ and a consequence of Lemma 1.4 (Chapter 1) that

$$\int_{\varepsilon\sigma n^{1/2}}^{n} |f_n(t)| \frac{dt}{t} = \int_{\varepsilon}^{n^{1/2}/\sigma} |v(t)|^n \frac{dt}{t} \leqslant e^{-cn} \int_{\varepsilon}^{n^{1/2}/\sigma} \frac{dt}{t} = O(n^{-1}).$$

Relations (5.32)–(5.34) and Theorem 5.1 imply that $\Delta_n = O(\psi_n + n^{-1})$. Since $E|X_1|^3 = \infty$, we have $\lim n\psi_n = \infty$. Hence the upper estimate of Theorem 5.10 follows.

It remains to prove the lower estimates in Theorems 5.10 and 5.11. Let $A(x)$ and $a(t)$ be bounded functions such that

$$\int_{-\infty}^{\infty} |A(x)|\, dx < \infty, \quad a(t) = \int_{-\infty}^{\infty} e^{itx} A(x)\, dx.$$

Making use of the equality

$$f_n(t) - e^{-t^2/2} = -it \int_{-\infty}^{\infty} e^{itx}(F_n(x) - \Phi(x))\, dx,$$

we conclude that the functions

$$B(x) = F_n(x) - \Phi(x), \quad b(t) = -\frac{1}{it}(f_n(t) - e^{-t^2/2})$$

form a pair of Fourier transforms. The functions $A(x)$, $a(t)$, $B(x)$, and $b(t)$ belong to the space $L_2(\mathbb{R})$. By Parseval's formula we obtain

$$\int_{-\infty}^{\infty} (F_n(x) - \Phi(x))\overline{A(x)}\, dx = -(2\pi)^{-1} \int_{-\infty}^{\infty} (f_n(t) - e^{-t^2/2}) \frac{\overline{a(t)}}{it}\, dt,$$

Hence

$$\Delta_n \geqslant \left(2\pi \int_{-\infty}^{\infty} |A(x)|\, dx\right)^{-1} \left|\int_{-\infty}^{\infty} (f_n(t) - e^{-t^2/2}) \frac{\overline{a(t)}}{it}\, dt\right|. \qquad (5.35)$$

Suppose that $|a(t)| \leqslant (|t| + t^2) e^{-t^2/2}$ for every real t. We consider the real line as the union of the areas $|t| < T_n$ and $|t| \geqslant T_n$. By Lemma 5.3 we have

$$\int_{-\infty}^{\infty} (f_n(t) - e^{-t^2/2}) \frac{\overline{a(t)}}{t}\, dt$$

$$= n \int_{|t|<T_n} \frac{\overline{a(t)}}{it} e^{-t^2/2} \left(v\left(\frac{t}{\sigma\sqrt{n}}\right) - 1 + \frac{t^2}{2n}\right) dt + O(\psi_n^2 + n^{-1})$$

$$= n \int_{-\infty}^{\infty} \frac{\overline{a(t)}}{it} e^{-t^2/2} \left(v\left(\frac{t}{\sigma\sqrt{n}}\right) - 1 + \frac{t^2}{2n}\right) dt + O(\psi_n^2 + n^{-1})$$

$$= I(a(t)) + O(\psi_n^2 + n^{-1}), \qquad (5.36)$$

where

$$I(a(t)) = n \int_{-\infty}^{\infty} \int_{-\infty}^{\infty} \frac{\overline{a(t)}}{it} e^{-t^2/2} \left(e^{itx/\sigma\sqrt{n}} - 1 + \frac{t^2 x^2}{2\sigma^2 n}\right) dt\, dV(x).$$

We set $a_1(t) = t\,e^{-t^2/2}$. Then

$$I(a_1(t)) = \frac{n}{i}\int_{-\infty}^{\infty}\int_{-\infty}^{\infty} e^{-t^2}\left(e^{itx/\sigma\sqrt{n}} - 1 + \frac{t^2 x^2}{2\sigma^2 n}\right) dt\,dV(x)$$

$$= n\frac{\sqrt{\pi}}{i}\int_{-\infty}^{\infty}\left(e^{-x^2/4\sigma^2 n} - 1 + \frac{x^2}{4\sigma^2 n}\right) dV(x).$$

It is easy to show that $e^{-x} - 1 + x \geqslant \frac{1}{2}e^{-1/4}x^2$ for $0 \leqslant x \leqslant \frac{1}{4}$ and $e^{-x} - 1 + x \geqslant \frac{1}{8}e^{-1/4}x$ for $x > \frac{1}{4}$. We have

$$|I(a_1(t))| \geqslant \frac{\sqrt{\pi}}{32} e^{-1/4}(\Lambda_n + L_{n,4}),$$

where

$$\Lambda_n = \sigma^{-2}\int_{|x| \geqslant \sigma n^{1/2}} x^2\,dV(x), \quad L_{n,k} = \sigma^{-k} n^{-(k-2)/2}\int_{|x| < \sigma n^{1/2}} x^k\,dV(x).$$

It follows from (5.35), (5.36), and the equalities

$$A_1(x) = -(2\pi)^{-1}\int_{-\infty}^{\infty} e^{-itx} a_1(t)\,dt = xi^{-1}(2\pi)^{-1/2} e^{-x^2/2},$$

$$\int_{-\infty}^{\infty}|A_1(x)|\,dx = 2(2\pi)^{-1/2}\int_{0}^{\infty} x e^{-x^2/2}\,dx = 2(2\pi)^{-1/2}$$

that

$$\Delta_n \geqslant \frac{e^{-1/4}}{64\sqrt{2}}(\Lambda_n + L_{n,4}) - |\beta_n|, \quad \beta_n = O(\psi_n^2 + n^{-1}). \qquad (5.37)$$

Furthermore, set $a_2(t) = t^2 e^{-t^2/2}$. We have

$$I(a_2(t)) = \frac{n}{i}\int_{-\infty}^{\infty}\int_{-\infty}^{\infty} t e^{-t^2}\left(e^{itx/\sigma\sqrt{n}} - 1 + \frac{t^2 x^2}{2\sigma^2 n}\right) dt\,dV(x)$$

$$= n\pi^{1/2}\int_{-\infty}^{\infty}\frac{x}{2\sigma\sqrt{n}} e^{-x^2/4\sigma^2 n}\,dV(x)$$

$$= n\pi^{1/2}\left\{\int_{|x| \geqslant \sigma n^{1/2}}\frac{x}{2\sigma\sqrt{n}} e^{-x^2/4\sigma^2 n}\,dV(x)\right.$$

$$\left. + \int_{|x| < \sigma n^{1/2}}\left(\frac{x}{2\sigma\sqrt{n}} - \frac{x^3}{8\sigma^3 n^{3/2}} + \frac{\theta x^5}{64\sigma^5 n^{5/2}}\right) dV(x)\right\}.$$

These relations and the inequalities

$$\sigma^{-1}n^{1/2}\left|\int_{|x|<\sigma n^{1/2}} x\,dV(x)\right| = \sigma^{-1}n^{1/2}\left|\int_{|x|\geq \sigma n^{1/2}} x\,dV(x)\right| \leq \Lambda_n,$$

$$\sigma^{-5}n^{-3/2}\int_{|x|<\sigma n^{1/2}} |x|^5\,dV(x) \leq \sigma^{-4}n^{-1}\int_{|x|<\sigma n^{1/2}} x^4\,dV(x) = L_{n,4}$$

imply that

$$|I(a_2(t))| \geq \pi^{1/2}(\tfrac{1}{8}|L_{n,3}| - \Lambda_n - L_{n,4}).$$

Hence, and from the relations (5.35), (5.36), and

$$A_2(x) = (2\pi)^{-1}\int_{-\infty}^{\infty} e^{-itx}\,a_2(t)\,dt = (2\pi)^{-1/2}(1-x^2)\,e^{-x^2/2},$$

$$\int_{-\infty}^{\infty} |A_2(x)|\,dx = 4(2\pi\,e)^{-1/2},$$

we find that

$$\Delta_n \geq \frac{1}{4}\sqrt{\frac{e}{2}}\,(\tfrac{1}{8}|L_{n,3}| - \Lambda_n - L_{n,4}) - |\beta_n|,$$

where $\beta_n = O(\psi_n^2 + n^{-1})$. Obviously,

$$\psi_n = \Lambda_n + |L_{n,3}| + L_{n,4}. \tag{5.38}$$

Taking into account (5.37), we obtain

$$\Delta \geq c\psi_n - |\gamma_n|, \tag{5.39}$$

where c is a positive constant and $\gamma_n = O(n^{-1})$.

In the proof of (5.35)–(5.37) and (5.39) we did not use the specific conditions of Theorems 5.10 and 5.11; we made use of the conditions listed before the formulations of these theorems in the beginning of Section 5.4.

Now suppose that $E|X_1|^3 = \infty$. This condition of Theorem 5.10 implies that $\lim n\psi_n = \infty$. Hence and from the inequality (5.39) we conclude that $\Delta_n \geq c_1\psi_n$ for all sufficiently large n, where c_1 is a positive constant. The lower estimate in Theorem 5.10 is proved.

Let us turn to Theorem 5.11. Suppose that the random variable X_1 has a lattice distribution with the possible values of the form. $a + kh$ ($k = 0, \pm 1, \pm 2,\ldots$), where $h > 0$ and a are fixed numbers. If

$$a_3(t) = \exp\{-(t + 2\pi\sigma n^{1/2}h^{-1})^2/2\},$$

then

$$A_3(x) = (2\pi)^{-1}\int_{-\infty}^{\infty} e^{-itx}\,a_3(t)\,dt = (2\pi)^{-1/2}\exp\{2\pi ix\sigma n^{1/2}h^{-1} - x^2/2\}$$

and
$$\int_{-\infty}^{\infty} |A_3(x)|\,dx = 1.$$

Furthermore,
$$f_n(t - 2\pi\sigma n^{1/2}h^{-1}) = \exp\{-2\pi i a n h^{-1}\} f_n(t).$$

Therefore, we have by (5.35) and Lemma 5.3,

$$\Delta_n \geq (2\pi)^{-1} \left| \int_{-\infty}^{\infty} (f_n(t) - e^{-t^2/2}) \exp\left\{-(t + 2\pi\sigma n^{1/2}h^{-1})^2/2\right\} \frac{dt}{t} \right|$$

$$= (2\pi)^{-1} \left| \int_{|t + 2\pi\sigma n^{1/2}h^{-1}| < T_n} f_n(t) \exp\{-(t + 2\pi\sigma n^{1/2}h^{-1})^2/2\} \frac{dt}{t} \right|$$
$$+ o(\psi_n + n^{-1/2})$$

$$= (2\pi)^{-1} \left| \int_{|t| < T_n} f_n(t)\, e^{-t^2/2} (t - 2\pi\sigma n^{1/2}h^{-1})^{-1}\,dt \right| + o(\psi_n + n^{-1/2}).$$

Hence $\Delta_n \geq cn^{-1/2} + o(\psi_n)$. By (5.39) this implies the lower estimate in Theorem 5.11. □

The function ψ_n is a sum of three summands; it can be represented by (5.38). It is possible to construct the distribution functions $V(x)$ such that any summand in this sum will be dominant.

5.5 Non-uniform estimates

Let us consider two distribution functions $F(x)$ and $G(x)$. We write
$$\Delta(x) = F(x) - G(x).$$
The well-known properties of distribution functions imply the relation $\Delta(x) \to 0$ as $x \to -\infty$ or $x \to +\infty$. If we possess some information about the moments of the distributions, we can obtain estimates of the rates of convergence. For instance, if these distributions have zero means and unit variances, then
$$|F(x) - G(x)| \leq x^{-2}$$
for every $x \neq 0$. This is an easy consequence of Chebyshev's inequality. It follows that $\Delta(x) = O(x^{-2})$ as $|x| \to \infty$.

Let us now consider in more detail the case when $G(x)$ is a normal distribution function. Let $F(x)$ be an arbitrary distribution function satisfying the condition

$$\int_{-\infty}^{\infty} |x|^p \, dF(x) < \infty \tag{5.40}$$

for some $p > 0$. Let $\Phi(x)$ be the standard normal distribution function. We put

$$\Delta = \sup_x |F(x) - \Phi(x)|. \tag{5.41}$$

Lemma 5.4 *Suppose that $0 < \Delta \leq e^{-1/2}$ and (5.40) is fulfilled. Then*

$$|F(x) - \Phi(x)| \leq \frac{c(p)\Delta(\log 1/\Delta)^{p/2} + \lambda_p}{1 + |x|^p} \tag{5.42}$$

for all x, where $c(p)$ is a positive constant depending only on p, and

$$\lambda_p = \left| \int_{-\infty}^{\infty} |x|^p \, dF(x) - \int_{-\infty}^{\infty} |x|^p \, d\Phi(x) \right|. \tag{5.43}$$

Proof Let $a \geq 1$ be such that the points $\pm a$ are points of continuity of $F(x)$. Then

$$\int_{-a}^{a} |x|^p \, dF(x) = \int_{-a}^{a} |x|^p \, d(F(x) - \Phi(x)) + \int_{-a}^{a} |x|^p \, d\Phi(x)$$

$$= a^p(F(a) - \Phi(a)) - a^p(F(-a) - \Phi(-a))$$

$$- p \int_{0}^{a} x^{p-1}(F(x) - \Phi(x)) \, dx + p \int_{-a}^{0} |x|^{p-1}(F(x) - \Phi(x)) \, dx$$

$$+ \int_{-a}^{a} |x|^p \, d\Phi(x).$$

Taking into account (5.41) and (5.43), we obtain

$$\int_{-a}^{a} |x|^p \, dF(x) \geq -4a^p\Delta + \int_{-a}^{a} |x|^p \, d\Phi(x)$$

and

$$\int_{|x| \geq a} |x|^p \, dF(x) \leq \lambda_p + 4a^p\Delta + \int_{|x| \geq a} |x|^p \, d\Phi(x).$$

If $x \geq a$, then

$$\int_{|y| \geq a} |y|^p \, dF(y) \geq x^p(1 - F(x)) \geq x^p(\Phi(x) - F(x)),$$

$$x^p(\Phi(x) - F(x)) \leq \int_{|y| \geq a} |y|^p \, dF(y) \leq \lambda_p + 4a^p\Delta + \int_{|y| \geq a} |y|^p \, d\Phi(y),$$

and
$$x^p(F(x) - \Phi(x)) \le \int_{|y| \ge a} |y|^p \, d\Phi(y).$$

Therefore,
$$|x|^p |F(x) - \Phi(x)| \le \lambda_p + 4a^p \Delta + \int_{|y| \ge a} |y|^p \, d\Phi(y) \qquad (5.44)$$

if $x \ge a$. Similar inequalities hold in the interval $x \le -a$. Accordingly, (5.44) holds for $|x| \ge a$. If $|x| < a$, (5.44) is obviously true. Since $a \ge 1$, we obtain

$$(1 + |x|^p)|F(x) - \Phi(x)| \le \lambda_p + 5a^p \Delta + \int_{|y| \ge a} |y|^p \, d\Phi(y) \qquad (5.45)$$

for all x. The right-hand side of (5.45) is a continuous function of a, therefore, we can remove the requirement that the points $\pm a$ are points of continuity of $F(x)$. Thus the inequality (5.45) is proved for all x and for every $a \ge 1$.

It is easy to find that
$$\int_a^\infty y^p e^{-y^2/2} \, dy \sim a^{p-1} e^{-a^2/2}$$

as $a \to +\infty$. Hence the function
$$I_p(a) = a^{1-p} e^{a^2/2} \int_a^\infty y^p e^{-y^2/2} \, dy$$

is bounded for sufficiently large a. Being continuous in the interval $a \ge 1$, this function is bounded in that interval. We put $K_p = \sup_{a \ge 1} I_p(a)$. Then

$$\int_{|y| \ge a} |y|^p \, d\Phi(y) = 2^{1/2} \pi^{-1/2} \int_a^\infty y^p e^{-y^2/2} \, dy$$
$$\le K_p 2^{1/2} \pi^{-1/2} a^{p-1} e^{-a^2/2} \le K_p 2^{1/2} \pi^{-1/2} a^p e^{-a^2/2}.$$

We may set $a = (2 \log 1/\Delta)^{1/2}$, since the condtion $a \ge 1$ is satisfied in view of the hypothesis $0 < \Delta \le e^{-1/2}$. From (5.45) we obtain

$$(1 + |x|^p)|F(x) - \Phi(x)| \le c(p) \Delta \left(\log \frac{1}{\Delta} \right)^{p/2} + \lambda_p$$

for all x, where $c(p) = 2^{p/2}(5 + K_p 2^{1/2} \pi^{-1/2})$. $\square$

For $p = 2$, Lemma 5.4 implies the following proposition.

Lemma 5.5 *Let the distribution function $F(x)$ satisfy the condition*
$$\int_{-\infty}^\infty x^2 \, dF(x) = 1.$$

If $0 < \Delta \leqslant e^{-1/2} = 0.60653\ldots$, then

$$|F(x) - \Phi(x)| \leqslant \frac{A\Delta \log(1/\Delta)}{1 + x^2}$$

for all x.

In Lemma 5.5 we may put $A = 2(5 + 2\,e^{1/2}\pi^{-1/2}\Gamma(3/2))$, so that $A < 16.5$.

We may choose $F(x)$ to be the distribution function of a normalized sum of independent random variables if these variables have moments of the proper order. The following proposition is an immediate consequence of Lemma 5.5.

Theorem 5.12 *Let $\{X_n\}$ be a sequence of independent random variables with zero means and finite variances. We set*

$$B_n = \sum_{j=1}^{n} \operatorname{Var} X_j, \quad F_n(x) = P\left(B_n^{-1/2} \sum_{j=1}^{n} X_j < x\right),$$

$$\Delta_n = \sup_{x} |F_n(x) - \Phi(x)|.$$

If $0 < \Delta_n \leqslant e^{-1/2}$ for $n \geqslant n_0$ and some n_0, then

$$|F_n(x) - \Phi(x)| \leqslant \frac{A\Delta_n \log(1/\Delta_n)}{1 + x^2}$$

for all x and $n \geqslant n_0$.

Theorem 5.12 provides a non-uniform estimate of the difference $F_n(x) - \Phi(x)$, i.e. the remainder in the central limit theorem. Contrary to the uniform estimates of the previous sections, non-uniform estimates take into account not only the number of summands n but also the value of x. If n is fixed and $|x| \to \infty$, the right-hand sides of non-uniform estimates go to zero.

Non-uniform estimates of the remainder in the central limit theorem are very helpful in proving propositions related to the so-called global form of the central limit theorem. Here is a typical example.

Theorem 5.13 *Let $\{X_n\}$ be a sequence of independent random variables satisfying the conditions of Theorem 5.12 and the condition $\Delta_n \to 0$. Then*

$$\int_{-\infty}^{\infty} |F_n(x) - \Phi(x)|^p \, dx \to 0$$

for every $p > \tfrac{1}{2}$.

This proposition follows immediately from Theorem 5.12. A more general result was proved in Chapter 1 (Theorem 1.12).

In the special case of identical distributions, there are many refined non-uniform estimates of the remainder in the central limit theorem. The proof of the following Osipov's theorem can be found in [342].

Theorem 5.14 *Let $X_1, \ldots, X_n$ be independent identically distributed random variables satisfying the conditions $EX_1 = 0$, $\mathrm{Var}\, X_1 = \sigma^2 > 0$, $E|X_1|^k < \infty$ for some integer $k \geq 3$. Put*

$$V(x) = P(X_1 < x), \quad F_n(x) = P\left(\sigma^{-1}n^{-1/2} \sum_{j=1}^{n} X_j < x\right).$$

Then

$$|F_n(x) - \Phi(x)|$$

$$\leq C(k) \bigg\{ \sigma^{-3} E|X_1|^3 n^{-1/2}(1 + |x|)^{-k-1}$$

$$+ \sigma^{-k} n^{-(k-2)/2} (1 + |x|)^{-k} \int_{|y| \geq \sigma n^{1/2}(1+|x|)} |y|^k \, dV(y)$$

$$+ \sigma^{-k-1} n^{-(k-1)/2} (1 + |x|)^{-k-1} \int_{|y| < \sigma n^{1/2}(1+|x|)} |y|^{k+1} \, dV(y) \bigg\}$$

for all x. Here $C(k)$ is a positive constant depending only on k.

We note that here, as in Theorems 5.4–5.9, n is an arbitrary positive integer.

There are consequences of Theorem 5.14 having much simpler formulation.

Theorem 5.15 *Let $X_1, \ldots, X_n$ be independent identically distributed random variables, $EX_1 = 0$, $\mathrm{Var}\, X_1 = \sigma^2 > 0$, $E|X_1|^r < \infty$ for some $r \geq 3$ (r is not necessarily an integer). Then*

$$|F_n(x) - \Phi(x)| \leq C(r)(1 + |x|)^{-r}(\sigma^{-3} E|X_1|^3 n^{-1/2} + \sigma^{-r} E|X_1|^r n^{-(r-2)/2}).$$

for all x, where $C(r)$ is a positive constant depending only on r.

This theorem follows from Theorem 5.14 with $k = [r]$ and from the inequalities

$$\int_{|y| \geq z} |y|^k \, dV(y) \leq z^{k-r} \int_{|y| \geq z} |y|^r \, dV(y)$$

and

$$\int_{|y| < z} |y|^{k+1} \, dV(y) \leq z^{k+1-r} \int_{|y| < z} |y|^r \, dV(y),$$

where $z > 0$.

In turn, Theorem 5.15 implies the following.

Theorem 5.16 *Let $X_1, \ldots, X_n$ be independent identically distributed random variables, $EX_1 = 0$, $\operatorname{Var} X_1 = \sigma^2 > 0$, $E|X_1|^3 < \infty$. Then*

$$\left| P\left(\frac{1}{\sigma\sqrt{n}} \sum_{j=1}^n X_j < x\right) - \Phi(x) \right| \leq A\sigma^{-3} E|X_1|^3 n^{-1/2}(1 + |x|)^{-3}$$

for all x.

In the case when the summands are not necessarily identically distributed, Theorem 5.16 admits the following generalization.

Theorem 5.17 *Let $X_1, \ldots, X_n$ be independent random variables such that $EX_j = 0$ and $E|X_j|^3 < \infty$ $(j = 1, \ldots, n)$. We put*

$$\sigma_j^2 = \operatorname{Var} X_j, \quad B_n = \sum_{j=1}^n \sigma_j^2, \quad F_n(x) = P\left(B_n^{-1/2} \sum_{j=1}^n X_j < x\right),$$

$$L_n = B_n^{-3/2} \sum_{j=1}^n E|X_j|^3.$$

Then

$$|F_n(x) - \Phi(x)| \leq A L_n (1 + |x|)^{-3}$$

for all x.

Theorems 5.16 and 5.17 are generalizations and strengthenings of Theorems 5.5 and 5.4 respectively. In turn, there are generalizations of Theorems 5.16 and 5.17 under weaker moment conditions.

We shall formulate a consequence of Theorem 5.17 related to the global form of the central limit theorem.

Let $F(x)$ and $G(x)$ be two distribution functions, and let $d_p(F, G)$ be the distance between them in the metric space L_p ($p \geq 1$), that is,

$$d_p(F, G) = \left(\int_{-\infty}^{\infty} |F(x) - G(x)|^p \, dx\right)^{1/p}.$$

Then Theorem 5.17 implies that under the conditions of this theorem we have

$$d_p(F_n, \Phi) \leq A L_n.$$

Therefore, $d_p(F_n, \Phi) \to 0$ for every $p \geq 1$, if $L_n \to 0$.

5.6 Asymptotic expansions in the central limit theorem: formal construction of the expansions

Let $\{X_n; n = 1, 2, \ldots\}$ be a sequence of independent random variables having moments of every order and zero means. We put $B_n = \sum_{j=1}^{n} \operatorname{Var} X_j$. As usual, the trivial case when all X_n have degenerate distributions will be excluded, so that $B_n > 0$ for all sufficiently large n, and we shall consider only such n. Let γ_{vn} denote the cumulant of order v of the random variable X_n, and let $v_j(t)$ and $f_n(t)$ denote the characteristic functions of X_j and $Z_n = B_n^{-1/2} \sum_{j=1}^{n} X_j$ respectively. We note that $\gamma_{1n} = EX_n = 0$, $\gamma_{2n} = \operatorname{Var} X_n$ for all n, and $f_n(t) = \prod_{j=1}^{n} v_j(tB_n^{-1/2})$.

We consider the formal expansion of the function $\log v_j(t)$ in a power series:

$$\log v_j(t) = \sum_{v=2}^{\infty} \frac{\gamma_{vj}}{v!} (it)^v.$$

This implies the following formal equality:

$$\log f_n(t) = \sum_{j=1}^{n} \log v_j(tB_n^{-1/2}) = \sum_{v=2}^{\infty} \frac{\lambda_{vn}}{v!} n^{-(v-2)/2} (it)^v, \qquad (5.46)$$

where

$$\lambda_{vn} = n^{(v-2)/2} B_n^{-v/2} \sum_{j=1}^{n} \gamma_{vj}. \qquad (5.47)$$

Obviously, $\lambda_{2n} = 1$. Hence

$$f_n(t) = e^{-t^2/2} \exp\left\{\sum_{s=1}^{\infty} \frac{\lambda_{s+2,n}}{(s+2)!} n^{-s/2} (it)^{s+2}\right\}.$$

We define $P_{vn}(u)$ as the coefficient of z^v in the formal expansion of the function

$$\exp\left\{\sum_{s=1}^{\infty} \frac{\lambda_{s+2,n}}{(s+2)!} u^{s+2} z^s\right\}$$

in a power series in z. Thus

$$\exp\left\{\sum_{s=1}^{\infty} \frac{\lambda_{s+2,n}}{(s+2)!} u^{s+2} z^s\right\} = 1 + \sum_{v=1}^{\infty} P_{vn}(u) z^v \qquad (5.48)$$

and

$$f_n(t) = e^{-t^2/2} + \sum_{v=1}^{\infty} P_{vn}(it) e^{-t^2/2} n^{-v/2}. \qquad (5.49)$$

To (5.49) there corresponds the expansion of the corresponding distribution function $F_n(x)$ in a power series in $n^{-1/2}$:

$$F_n(x) = \Phi(x) + \sum_{v=1}^{\infty} Q_{vn}(x) n^{-v/2}, \qquad (5.50)$$

so that

$$\int_{-\infty}^{\infty} e^{itx} \, dQ_{vn}(x) = P_{vn}(it) \, e^{-t^2/2}. \tag{5.51}$$

We now find explicit formulae for the functions $P_{vn}(it)$ and $Q_{vn}(x)$, without regard to the convergence of the series in which these functions appear. We need the following elementary proposition.

Lemma 5.6 *Suppose that the functions $y = y(x)$ and $z = z(y)$ have derivatives of order $v \geq 1$. Then*

$$\frac{d^v}{dx^v} z(y(x)) = v! \sum \left. \frac{d^s z(y)}{dy^s} \right|_{y=y(x)} \prod_{m=1}^{v} \frac{1}{k_m!} \left(\frac{1}{m!} \frac{d^m y(x)}{dx^m} \right)^{k_m},$$

where the summation is carried out over all non-negative integer solutions $(k_1, k_2, \ldots, k_v)$ of the equation

$$k_1 + 2k_2 + \cdots + vk_v = v, \tag{5.52}$$

and

$$s = k_1 + k_2 + \cdots + k_v.$$

This proposition (which is contained, for instance, in the book by Goursat [150]) can be proved by induction. We note, incidentally, that from Lemma 5.6 for $y(t) = f(t)$ and $z(y) = \log y$ we can obtain formula (1.13) (Chapter 1), which connects the cumulants and moments of a random variable with the characteristic function $f(t)$.

In Lemma 5.6 we put $z = e^y$. Then we obtain the following proposition: if the function $y = y(x)$ has a derivative of order $v \geq 1$, then

$$\frac{d^v}{dx^v} e^{y(x)} = v! \, e^{y(x)} \sum \prod_{m=1}^{v} \frac{1}{k_m!} \left(\frac{1}{m!} \frac{d^m y(x)}{dx^m} \right)^{k_m}, \tag{5.53}$$

where the summation is carried out over all non-negative integer solutions of eqn (5.52).

If $y(x) = \sum_{s=1}^{\infty} a_s x^s$, we obtain

$$\left. \frac{d^v}{dx^v} \exp\left\{ \sum_{s=1}^{\infty} a_s x^s \right\} \right|_{x=0} = v! \sum \prod_{m=1}^{v} \frac{a_m^{k_m}}{k_m!}.$$

Therefore,

$$P_{vn}(it) = \sum \prod_{m=1}^{v} \frac{1}{k_m!} \left(\frac{\lambda_{m+2,n}(it)^{m+2}}{(m+2)!} \right)^{k_m}, \tag{5.54}$$

where the summation is carried out over all non-negative integer solutions of (5.52).

By (5.54), the function $P_{vn}(it)$ is a polynomial of degree $3v$ in it, with

Rates of convergence in the central limit theorem

coefficients depending on the cumulants of the random variables $X_1, \ldots, X_n$ of order not greater than $v + 2$.

It follows from the equality

$$\int_{-\infty}^{\infty} e^{itx} \, d\Phi(x) = e^{-t^2/2}$$

that

$$\int_{-\infty}^{\infty} e^{itx} \, d\Phi^{(r)}(x) = (-it)^r e^{-t^2/2} \qquad (r = 0, 1, 2, \ldots).$$

Therefore, (5.51) will be satisfied if we define $Q_{vn}(x)$ as $P_{vn}(it)$ after replacing each power $(it)^r$ by $(-1)^r (d^r/dx^r) \Phi(x)$. Then we obtain

$$Q_{vn}(x) = \sum (-1)^{v+2s} \prod_{m=1}^{v} \frac{1}{k_m!} \left(\frac{\lambda_{m+2,n}}{(m+2)!} \right)^{k_m} \frac{d^{v+2s}}{dx^{v+2s}} \Phi(x), \qquad (5.55)$$

where the summation is extended over all non-negative integer solutions of equation (5.52) and $s = k_1 + k_2 + \cdots + k_v$.

Equality (5.55) can be written in a somewhat different way. The Chebyshev–Hermite polynomial of degree m is defined by the equality

$$H_m(x) = (-1)^m e^{x^2/2} \frac{d^m}{dx^m} e^{-x^2/2}.$$

We have

$$H_m(x) = m! \sum_{k=0}^{[m/2]} \frac{(-1)^k x^{m-2k}}{k!(m-2k)! 2^k}$$

for every integer $m \geq 0$. In particular, $H_0(x) = 1$, $H_1(x) = x$, $H_2(x) = x^2 - 1$, $H_3(x) = x^3 - 3x$, $H_4(x) = x^4 - 6x^2 + 3$, $H_5(x) = x^5 - 10x^3 + 15x$.

If m is an arbitrary positive integer, then

$$\frac{d^m}{dx^m} \Phi(x) = (-1)^{m-1} (2\pi)^{-1/2} e^{-x^2/2} H_{m-1}(x).$$

Accordingly,

$$Q_{vn}(x) = -(2\pi)^{-1/2} e^{-x^2/2} \sum H_{v+2s-1}(x) \prod_{m=1}^{v} \frac{1}{k_m!} \left(\frac{\lambda_{m+2,n}}{(m+2)!} \right)^{k_m}, \qquad (5.56)$$

where the summation is extended over all non-negative integer solutions of equation (5.52) and $s = k_1 + k_2 + \cdots + k_v$.

Thus $Q_{vn}(x) = M_{3v-1,n}(x) e^{-x^2/2}$, where $M_{3v-1,n}(x)$ is a polynomial of degree $3v - 1$ in x with coefficients depending only on the cumulants of the random variables $X_1, \ldots, X_n$ up to and including order $v + 2$.

Putting $\alpha_{kj} = EX_j^k$ and using the equalities (5.47) and (5.56), and also (1.13)

(Chapter 1), we find that, in particular,

$$Q_{1n}(x)n^{-1/2} = 6^{-1}(2\pi)^{-1/2}(1-x^2)e^{-x^2/2}B_n^{-3/2}\sum_{j=1}^{n}\alpha_{3j},$$

$$Q_{2n}(x)n^{-1} = -(2\pi)^{-1/2}e^{-x^2/2}\bigg\{(x^5 - 10x^3 + 15x)\bigg(\sum_{j=1}^{n}\alpha_{3j}\bigg)^2\bigg/(72B_n^3)$$

$$+ (x^3 - 3x)\sum_{j=1}^{n}(\alpha_{4j} - 3\alpha_{2j}^2)/(24B_n^2)\bigg\}.$$

We now turn to the special case of independent identically distributed random variables (the i.i.d. case). Suppose, moreover, that the random variables $X_1, X_2, \ldots$ have a positive variance σ^2 and zero mean. In this case the λ_{vn}, defined by (5.47), do not depend on n, and we can denote them simply by λ_v. Thus

$$\lambda_v = \gamma_v/\sigma^v, \tag{5.57}$$

where γ_v is the cumulant of order v of the random variable X_1. In this particular case the formulae (5.54) and (5.56) take the form

$$P_v(it) = \sum \prod_{m=1}^{v}\frac{1}{k_m!}\bigg(\frac{\gamma_{m+2}(it)^{m+2}}{(m+2)!\sigma^{m+2}}\bigg)^{k_m} \tag{5.58}$$

and

$$Q_v(x) = -(2\pi)^{-1/2}e^{-x^2/2}\sum H_{v+2s-1}(x)\prod_{m=1}^{v}\frac{1}{k_m!}\bigg(\frac{\gamma_{m+2}}{(m+2)!\sigma^{m+2}}\bigg)^{k_m}. \tag{5.59}$$

In each of these equalities the summation is extended over all non-negative integer solutions $(k_1, k_2, \ldots, k_v)$ of the equation $k_1 + 2k_2 + \cdots + vk_v = v$, and $s = k_1 + k_2 + \cdots + k_v$.

5.7 Asymptotic expansions in the central limit theorem: the i.i.d. case

We consider a sequence of independent identically distributed random variables $\{X_n; n = 1, 2, \ldots\}$. Suppose that $EX_1 = 0$, $\text{Var } X_1 = \sigma^2 > 0$. Put

$$V(x) = P(X_1 < x), \quad v(t) = Ee^{itX_1}, \quad F_n(x) = P\bigg(\sigma^{-1}n^{-1/2}\sum_{j=1}^{n}X_j < x\bigg).$$

Theorem 5.18 *If* $E|X_1|^k < \infty$ *for some integer* $k \geq 3$, *then the following* inequality holds for all x and n:

$$\left|F_n(x) - \Phi(x) - \sum_{v=1}^{k-2} Q_v(x) n^{-v/2}\right|$$

$$\leqslant c(k)\left\{\sigma^{-k} n^{-(k-2)/2}(1+|x|)^{-k} \int_{|y| \geqslant \sigma n^{1/2}(1+|x|)} |y|^k \, dV(y)\right.$$

$$+ \sigma^{-k-1} n^{-(k-1)/2}(1+|x|)^{-k-1} \int_{|y| < \sigma n^{1/2}(1+|x|)} |y|^{k+1} \, dV(y)$$

$$\left.+ \left(\sup_{|t| \geqslant \delta} |v(t)| + \frac{1}{2n}\right)^n n^{k(k+1)/2}(1+|x|)^{-k-1}\right\}.$$

Here $\delta = \sigma^2/(12 \, E|X_1|^3)$ and $c(k)$ is a positive constant depending only on k. The functions $Q_v(x)$ are defined in Section 5.6.

If the additional condition $\limsup_{|t| \to \infty} |v(t)| < 1$ is satisfied, then $\sup_{|t| \geqslant \delta} |v(t)| < 1$ for every $\delta > 0$, so that the multiplier $(\sup_{|t| \geqslant \delta} |v(t)| + 1/(2n))^n$ is less than n^{-p} for every $p > 0$ and sufficiently large n.

The proof of this theorem of Osipov can be found, for instance, in the book by Petrov [342]. The following propositions are simple consequences of Theorem 5.18.

Theorem 5.19 *Suppose that* $\limsup_{|t| \to \infty} |v(t)| < 1$ *and* $E|X_1|^r < \infty$ *for some* $r \geqslant 3$. *Then there exists a positive function* $\varepsilon(u)$ *such that* $\lim_{u \to +\infty} \varepsilon(u) = 0$ *and*

$$\left|F_n(x) - \Phi(x) - \sum_{v=1}^{[r]-2} Q_v(x) n^{-v/2}\right| \leqslant \frac{\varepsilon(n^{1/2}(1+|x|))}{n^{(r-2)/2}(1+|x|)^r}.$$

Theorem 5.20 *If* $\limsup_{|t| \to \infty} |v(t)| < 1$ *and* $E|X_1|^k < \infty$ *for some integer* $k \geqslant 3$, *then*

$$(1+|x|)^k \left|F_n(x) - \Phi(x) - \sum_{v=1}^{k-2} Q_v(x) n^{-v/2}\right| = o(n^{-(k-2)/2})$$

uniformly in $x \in \mathbb{R}$.

Theorem 5.21 *If the conditions of Theorem 5.20 are satisfied, then*

$$F_n(x) - \Phi(x) - \sum_{v=1}^{k-2} Q_v(x) n^{-v/2} = o(n^{-(k-2)/2})$$

uniformly in $x \in \mathbb{R}$.

The following lemma, which is of interest in itself, plays an important role in the proof of Theorem 5.18.

174 ▌ Limit theorems of probability theory

Lemma 5.7 *Let $F(x)$ be a non-decreasing function, and let $G(x)$ be a differentiable function of bounded variation on the real line. Let $F(-\infty) = G(-\infty)$, $F(+\infty) = G(+\infty)$. We denote the corresponding Fourier–Stieltjes transforms by $f(t)$ and $g(t)$. Suppose that*

$$\int_{-\infty}^{\infty} |x|^s \, |d(F(x) - G(x))| < \infty$$

and $|G'(x)| \leq K(1 + |x|)^{-s}$ for some integer $s \geq 2$ and for all x, where K is a positive constant. Then

$$|F(x) - G(x)| \leq c(s)(1 + |x|)^{-s} \left\{ \int_{-T}^{T} |f(t) - g(t)| \frac{dt}{|t|} + \int_{-T}^{T} |\delta_s(t)| \frac{dt}{|t|} + \frac{K}{T} \right\}$$

for all real x and every $T > 1$. Here

$$\delta_s(t) = \int_{-\infty}^{\infty} e^{itx} \, d\{x^s(F(x) - G(x))\}$$

and $c(s)$ is a positive constant depending only on s.

There is an analogue of Lemma 5.7 in the case when $F(x)$ and $G(x)$ have discontinuities at the points of some countable set. This makes it possible to obtain asymptotic expansions of the distribution function of a sum of independent random variables having a common lattice distribution. For more information, refer to the books of Petrov [342] and Hall [163]. However, we shall prove the following theorem under more general conditions than the conditions of Theorems 5.19–5.21.

Theorem 5.22 *Let $\{X_n\}$ be a sequence of independent random variables with a common non-lattice distribution. Suppose that $EX_1 = 0$, $\mathrm{Var}\, X_1 = \sigma^2 > 0$, $E|X_1|^3 < \infty$. Then*

$$F_n(x) = \Phi(x) + 6^{-1}(2\pi n)^{-1/2}\sigma^{-3}\alpha_3(1 - x^2)\,e^{-x^2/2} + o(n^{-1/2})$$

uniformly in $x \in \mathbb{R}$. Here $\alpha_3 = EX_1^3$, and the function $F_n(x)$ is defined as before.

Proof We shall apply Theorem 5.2, setting $b = 1/\pi$, $F(x) \equiv F_n(x)$, $G(x) \equiv \Phi(x) + 6^{-1}(2\pi n)^{-1/2}\sigma^{-3}\alpha_3(1 - x^2)\,e^{-x^2/2}$, and $T = Kn^{1/2}$, where K is a positive constant. It is easy to show that

$$g(t) = \int_{-\infty}^{\infty} e^{itx} \, dG(x) = e^{-t^2/2}(1 + 6^{-1}n^{-1/2}\sigma^{-3}\alpha_3(it)^3).$$

By Theorem 5.2 we have

$$\sup_x |F_n(x) - G(x)| \leq \pi^{-1} \int_{|t| < Kn^{1/2}} |f_n(t) - g(t)| \frac{dt}{|t|} + ACn^{-1/2}/K,$$

where $f_n(t)$ is the c.f. that corresponds to the d.f. $F_n(x)$, and the constant C is such that $\sup_x |G'(x)| \leqslant C$.

Let ε be an arbitrary positive number. We choose the constant K in such a way that the condition $AC/K < \varepsilon$ is satisfied, and put

$$I = \int_{|t| < Kn^{1/2}} |f_n(t) - g(t)| \frac{dt}{|t|}.$$

The theorem will be proved if we show that $I < \varepsilon n^{-1/2}$ for all sufficiently large n. We have $I = I_1 + I_2$, where

$$I_1 = \int_{|t| < \delta\sigma n^{1/2}} |f_n(t) - g(t)| \frac{dt}{|t|},$$

$$I_2 = \int_{\delta\sigma n^{1/2} \leqslant |t| < Kn^{1/2}} |f_n(t) - g(t)| \frac{dt}{|t|},$$

and δ is a small positive number. We put

$$B = \{t: \delta\sigma n^{1/2} \leqslant |t| < Kn^{1/2}\},$$

$$I_2^{(1)} = \int_B |f_n(t)| \frac{dt}{|t|}, \quad I_2^{(2)} = \int_B |g(t)| \frac{dt}{|t|}.$$

Hence $I_2 \leqslant I_2^{(1)} + I_2^{(2)}$. Since $f_n(t) = v^n(\sigma^{-1} n^{-1/2} t)$, where $v(t)$ is the c.f. of X_1, we have

$$I_2^{(1)} = \int_{\delta \leqslant |u| < K/\sigma} |v(u)| \frac{du}{|u|}.$$

Since X_1 has a non-lattice distribution, there exists a number $b > 0$ such that $|v(t)| \leqslant e^{-b}$ for $\delta \leqslant |t| \leqslant K/\sigma$, by Theorem 1.3 (Chapter 1). Accordingly, $I_2^{(1)} \leqslant C e^{-bn}$, where C is a constant. It follows from the definition of $g(t)$ that $I_2^{(2)} = o(n^{-p})$ for every $p > 0$.

It remains only to show that $I_1 < \varepsilon n^{-1/2}$ for all sufficiently large n. We put $L(t) = \log v(t) + \sigma^2 t^2/2$ for sufficiently small t. It is clear that

$$v(t) = \exp\{L(t) - \sigma^2 t^2/2\}$$

and

$$I_1 = \int_{|t| < \delta\sigma n^{1/2}} e^{-t^2/2} \left| \exp\left\{ nL\left(\frac{t}{\sigma\sqrt{n}}\right)\right\} - 1 - \frac{\alpha_3(it)^3}{6\sigma_3\sqrt{n}} \right| \frac{dt}{|t|}.$$

Making use of the expansion of the function e^u in a power series, we can easily find that the function

$$e^x - 1 - y = (e^{x-y} - 1) e^y + e^y - 1 - y$$

satisfies the inequality

$$|e^x - 1 - y| \leq (|x-y| + |y|^2/2) e^{3z}$$

for all real or complex x and y, where $z = \max(|x|, |y|)$. Hence

$$I_1 \leq \int_{|t| < \delta\sigma n^{1/2}} e^{-t^2/8} \left\{ \left| nL\left(\frac{t}{\sigma\sqrt{n}}\right) - \frac{\alpha_3 (it)^3}{6\sigma^3 \sqrt{n}} \right| + \frac{\alpha_3^2 t^6}{72\sigma^6 n} \right\} \frac{dt}{|t|}.$$

It follows from the definition of the function $L(t)$ that $L(t) = o(t^2)$ and $L(t) = (\alpha_3/6)(it)^3 + o(|t|^3)$ as $t \to 0$. Therefore, $|L(t)| \leq \sigma^2 t^2/8$ and $|L(t) - (\alpha_3/6)(it)^3| \leq \varepsilon_0 |t|^3$ for an arbitrary positive constant ε_0 if $|t| \leq \delta$ and δ is sufficiently small. Thus we have for sufficiently small δ,

$$I_1 \leq \int_{-\infty}^{\infty} e^{-t^2/8} \{\varepsilon_0 |t|^3 \sigma^{-3} n^{-1/2} + \alpha_3^2 t^6 \sigma^{-6} n^{-1}/72\} \frac{dt}{|t|} < \varepsilon n^{-1/2}$$

for all sufficiently large n, if we choose sufficiently small ε_0. □

Theorem 5.22 implies that under the additional condition $EX_1^3 = 0$ the relation $F_n(x) - \Phi(x) = o(n^{-1/2})$ holds uniformly in x. The order of this asymptotic estimate is better than the order of the right-hand side of the Berry–Esseen inequality. Under additional conditions, Theorems 5.19–5.21 can provide even better asymptotic estimates of the remainder in the central limit theorem. A disadvantage of similar asymptotic estimates is that they are useless if we are interested in estimates of the difference $F_n(x) - \Phi(x)$ for a given fixed value n, contrary to the estimates provided by Theorems 5.4–5.9 and 5.14–5.18, which are valid for an arbitrary n.

5.8 Limit theorems for large deviations

Let $\{X_n\}$ be a sequence of independent identically distributed random variables. Suppose that X_1 has zero mean and finite positive variance σ^2. We put

$$S_n = \sum_{j=1}^{n} X_j, \quad Z_n = \sigma^{-1} n^{-1/2} S_n, \quad F_n(x) = P(Z_n < x).$$

By Lévy's theorem, $F_n(x) \to \Phi(x)$ uniformly in x. Therefore, when $x = O(1)$ we have

$$\frac{1 - F_n(x)}{1 - \Phi(x)} \to 1, \quad \frac{F_n(-x)}{\Phi(-x)} \to 1 \tag{5.60}$$

or equivalently,

$$\frac{P(Z_n \geqslant x)}{P(Y \geqslant x)} \to 1, \quad \frac{P(Z_n < -x)}{P(Y < -x)} \to 1, \qquad (5.61)$$

where Y is a random variable having the standard normal distribution. Thus $P(Z_n \geqslant x) \sim P(Y \geqslant x)$ and $P(Z_n < -x) \sim P(Y < -x)$ as $n \to \infty$ if $x = O(1)$. Sometimes we need know what additional conditions guarantee relations (5.60) in the case when x depends on n and tends to $+\infty$ as $n \to \infty$. In particular, we are interested in those conditions for which (5.60) holds in the interval $0 \leqslant x \leqslant \Lambda(n)$, where $\Lambda(n)$ is a non-decreasing function such that $\Lambda(n) \to \infty$. If the relations (5.60) hold in this interval, we call the interval itself a zone of normal convergence.

Thus we shall be concerned with the relative error of the normal approximation to the distribution of a sum of independent random variables. The previous sections of this chapter were devoted to estimates of the absolute error in the central limit theorem. We shall use the corresponding results in what follows. In particular, the following proposition is an easy consequence of the Berry–Esseen inequality.

Lemma 5.8 *If* $\mathrm{E}|X_1|^3 < \infty$, *then the relations* (5.60) *hold in the interval* $0 \leqslant x \leqslant (1 - \varepsilon)(\log n)^{1/2}$, *where* $0 < \varepsilon < 1$.

Proof We need the following elementary facts:

$$1 - \Phi(x) < \frac{1}{x\sqrt{2\pi}} e^{-x^2/2} \quad \text{for every } x > 0 \qquad (5.62)$$

and

$$1 - \Phi(x) = \frac{e^{-x^2/2}}{x\sqrt{2\pi}}\left(1 + O\left(\frac{1}{x^2}\right)\right), \quad 1 - \Phi(x) \sim \frac{1}{x\sqrt{2\pi}} e^{-x^2/2} \quad \text{as } x \to \infty. \qquad (5.63)$$

Both assertions can be easily proved if we integrate by parts in the integral $\int_x^\infty e^{-t^2/2}\,dt = \int_x^\infty (1/t)\,d(-e^{-t^2/2})$.

By $R_n(x)$ we denote the remainder in the central limit theorem, i.e. $R_n(x) = F_n(x) - \Phi(x)$. We have

$$\frac{1 - F_n(x)}{1 - \Phi(x)} = 1 - \frac{R_n(x)}{1 - \Phi(x)} \qquad (5.64)$$

for every real x. Theorem 5.5 implies that $|R_n(x)| \leqslant Cn^{-1/2}$ for all x, where $C = A\sigma^{-3}\mathrm{E}|X_1|^3$ and A is an absolute positive constant. Hence

$$\frac{|R_n(x)|}{1 - \Phi(x)} \leqslant 2(2\pi)^{1/2} Cx\, e^{x^2/2} n^{-1/2}, \qquad (5.65)$$

by (5.63), if $x \geqslant b$ and b is sufficiently large. In the interval $b \leqslant x \leqslant (1-\varepsilon)(\log n)^{1/2}$ the right-hand side of (5.65) tends to zero as $n \to \infty$. It follows from here and (5.64) that $(1 - F_n(x))/(1 - \Phi(x)) \to 1$ in this interval. In the interval $0 \leqslant x \leqslant b$ this relation has already been stated. The first relation in (5.60) is proved. The second relation also follows from our considerations. □

The zone of normal convergence given by Lemma 5.8 is very narrow. It can be shown that imposing the additional conditions of the existence of any power moments of higher order cannot enlarge this zone significantly. This can be done by imposing the following Cramér's condition: there exists a positive constant H such that $\mathrm{E}e^{tX_1} < \infty$ in the interval $|t| < H$.

Equivalent formulations of this condition are given in Lemma 2.2 (Chapter 2).

Theorem 5.23 *If $x \geqslant 0$, $x = o(n^{1/2})$, and if Cramér's condition is satisfied, then*

$$\frac{1-F_n(x)}{1-\Phi(x)} = \exp\left\{\frac{x^3}{\sqrt{n}}\lambda\left(\frac{x}{\sqrt{n}}\right)\right\}\left[1 + O\left(\frac{x+1}{\sqrt{n}}\right)\right], \qquad (5.66)$$

$$\frac{F_n(-x)}{\Phi(-x)} = \exp\left\{-\frac{x^3}{\sqrt{n}}\lambda\left(-\frac{x}{\sqrt{n}}\right)\right\}\left[1 + O\left(\frac{x+1}{\sqrt{n}}\right)\right]. \qquad (5.67)$$

Here $\lambda(t) = \sum_{k=0}^{\infty} c_k t^k$ is a power series with coefficients depending only on the cumulants of the random variable X_1 which converges for sufficiently small values of $|t|$

The series $\lambda(t)$ is called the Cramér series.

Proof Let h be an arbitrary number in the interval $(-H, H)$ so that $\mathrm{E}e^{hX_1} < \infty$. Denoting by $V(x)$ the d.f. of X_1, we introduce a sequence of independent random variables $\{\bar{X}_n\}$ with a common d.f.

$$\bar{V}(x) = \frac{1}{R(h)} \int_{-\infty}^{\infty} e^{hy} \, dV(y), \qquad (5.68)$$

where

$$R(h) = \mathrm{E}e^{hX_1} = \int_{-\infty}^{\infty} e^{hy} \, dV(y).$$

It will be shown later that the distribution $\bar{V}(x)$ has moments of every order. We put

$$v(t) = \mathrm{E}e^{itX_1}, \quad \bar{v}(t) = \mathrm{E}e^{it\bar{X}_1}, \quad \bar{m} = \mathrm{E}\bar{X}_1, \quad \bar{\sigma}^2 = \mathrm{Var}\,\bar{X}_1,$$

$$\bar{S}_n = \sum_{j=1}^{n} \bar{X}_j, \quad W_n(x) = P(S_n < x), \quad \bar{W}_n(x) = P(\bar{S}_n < x),$$

$$w_n(t) = E e^{itS_n}, \quad \bar{w}_n(t) = E e^{it\bar{S}_n}, \quad F_n(x) = P(\bar{S}_n - n\bar{m} < x\bar{\sigma}n^{1/2}).$$

Cramér's condition implies that the function $v(z)$ is analytical in the circle $|z| < H$ and, therefore, in the stripe $|\mathrm{Im}\, z| < H$. We have

$$F_n(x) = W_n(x\sigma\sqrt{n}), \quad \bar{F}_n(x) = \bar{W}_n(n\bar{m} + x\bar{\sigma}n^{1/2}),$$

$$w_n(t) = v^n(t), \quad \bar{w}_n(t) = \bar{v}^n(t), \quad \bar{v}(t) = \int_{-\infty}^{\infty} e^{itx}\, d\bar{V}(x) = \frac{1}{R} v(t - ih),$$

so that

$$v(t) = R\bar{v}(t + ih), \quad w_n(t) = R^n \bar{w}_n(t + ih).$$

The latter equality implies that

$$W_n(x) = R^n \int_{-\infty}^{x} e^{-hy}\, d\bar{W}_n(y)$$

for all x. Consequently,

$$1 - F_n(x) = R^n \int_{x\sigma n^{1/2}}^{\infty} e^{-hy}\, d\bar{W}_n(y)$$

and

$$1 - F_n(x) = R^n e^{-hn\bar{m}} \int_{(x\sigma - \bar{m}n^{1/2})/\bar{\sigma}}^{\infty} e^{-ht\bar{\sigma}\sqrt{n}}\, d\bar{F}_n(t) \qquad (5.69)$$

for all x.

For sufficiently small h we have

$$\log R(h) = \sum_{\nu=1}^{\infty} \frac{\gamma_\nu}{\nu!} h^\nu, \qquad (5.70)$$

where γ_ν is the cumulant of order ν of the random variable X_1 so that $\gamma_1 = EX_1 = 0$ and $\gamma_2 = \mathrm{Var}\, X_1 = \sigma^2$. Since

$$\bar{m} = E\bar{X}_1 = \int_{-\infty}^{\infty} x\, d\bar{V}(x) = \frac{1}{R} \int_{-\infty}^{\infty} x\, e^{hx}\, dV(x)$$

and

$$\bar{\sigma}^2 = \mathrm{Var}\, \bar{X}_1 = E\bar{X}_1^2 - \bar{m}^2 = \frac{1}{R} \int_{-\infty}^{\infty} x^2\, e^{hx}\, dV(x) - \bar{m}^2,$$

we obtain

$$\bar{m} = \frac{d}{dh}\log R(h) = \sum_{\nu=2}^{\infty} \frac{\gamma_\nu}{(\nu-1)!} h^{\nu-1}, \qquad (5.71)$$

$$\bar{\sigma}^2 = \frac{d^2}{dh^2}\log R(h) = \frac{d\bar{m}}{dh} = \sum_{\nu=2}^{\infty} \frac{\gamma_\nu}{(\nu-2)!} h^{\nu-2}. \qquad (5.72)$$

Both series converge in the same circle as the series on the right-hand side of (5.70).

Consider the equation

$$\sigma t = \bar{m}. \qquad (5.73)$$

It follows from (5.71) and (5.72) that $\bar{m}(0) = 0$ and the function $\bar{m}(h)$ has a positive derivative in a neighborhood of the point $h = 0$, since

$$\frac{d\bar{m}}{dh} = \bar{\sigma}^2 = \sigma^2 + \gamma_3 h + \tfrac{1}{2}\gamma_4 h^2 + \cdots > \sigma^2/2 \qquad (5.74)$$

for all sufficiently small h. Therefore, eqn (5.73) has for all sufficiently small t the unique real root h which has the same sign as t and tends to zero when $t \to 0$. Taking account of (5.71), we obtain the following expansion of this root h in a power series in t:

$$h = \frac{t}{\sigma} - \frac{\gamma_3}{2\sigma^4} t^2 + \cdots. \qquad (5.75)$$

This series converges for all sufficiently small t. We choose h to be the unique real root of (5.73). In view of (5.70) and (5.71) we obtain

$$\log R(h) - h\bar{m} = -\frac{t^2}{2} + \frac{\gamma_3}{6\sigma^3} t^3 + \frac{\gamma_4 \sigma^2 - 3\gamma_3^2}{24\sigma^6} t^4 + \cdots$$

or

$$\frac{t^2}{2} + \log R(h) - h\bar{m} = t^3 \lambda(t), \qquad (5.76)$$

where

$$\lambda(t) = \frac{\gamma_3}{6\sigma^3} + \frac{\gamma_4 \sigma^2 - 3\gamma_3^2}{24\sigma^6} t + \cdots \qquad (5.77)$$

The series $\lambda(t)$ was obtained by the substitution of one convergent series in another convergent series and, therefore, is convergent for all sufficiently small values of $|t|$.

Let us turn to equality (5.69). We put $x = \bar{m}\sigma^{-1} n^{1/2}$. Then

$$1 - F_n(\bar{m}\sigma^{-1} n^{1/2}) = \exp\{n \log R - hn\bar{m}\} \int_0^\infty e^{-hy\bar{\sigma} n^{1/2}} \, d\bar{F}_n(y). \qquad (5.78)$$

We put $Q_n(y) = \bar{F}_n(y) - \Phi(y)$. By the Berry–Esseen inequality we have $\sup_y |Q_n(y)| \leq K n^{-1/2}$, where $K = A \, \mathrm{E}|\bar{X}_j - \bar{m}|^3 \bar{\sigma}^{-3}$ and A is an absolute constant. Clearly, $\mathrm{E}|\bar{X}_1 - \bar{m}|^3 \leq 4(\mathrm{E}|\bar{X}_1|^3 + |\bar{m}|^3) \leq 8\,\mathrm{E}|\bar{X}_1|^3$, by Lyapunov's inequality. If $|h| < H$, then

$$\mathrm{E}|\bar{X}_1|^3 = \int_{-\infty}^{\infty} |x|^3 \, d\bar{V}(x) = \frac{1}{R}\left\{\int_{|x|<C_0} + \int_{|x|\geq C_0}\right\} |x|^3 e^{hx} \, dV(x).$$

For a given h in the interval $(-H, H)$ we can choose $\varepsilon > 0$ in such a way that the points $h \pm \varepsilon$ lie in the same interval. Given $\varepsilon > 0$, we choose $C_0 > 0$ such that $|x|^3 e^{hx} \leq e^{(|h|+\varepsilon)|x|}$ for $|x| \geq C_0$. Using Cramér's condition, we obtain $\mathrm{E}|\bar{X}_1|^3 < \infty$ and similarly $\mathrm{E}|\bar{X}_1|^p < \infty$ for every $p > 0$ and all h in the interval $(-H, H)$. Furthermore, $\mathrm{E}|\bar{X}_1|^3 \leq C_1$ for all sufficiently small h, where C_1 does not depend on h. Thus, by (5.74),

$$\sup_y |Q_n(y)| \leq C n^{-1/2} \tag{5.79}$$

for all sufficiently small h, where C does not depend on h and n.

We have

$$\int_0^{\infty} e^{-h\bar{\sigma}n^{1/2}y} \, d\bar{F}_n(y) = (2\pi)^{-1/2} I_1 + I_2,$$

where

$$I_1 = \int_0^{\infty} e^{-h\bar{\sigma}n^{1/2}y - y^2/2} \, dy, \quad I_2 = \int_0^{\infty} e^{-h\bar{\sigma}n^{1/2}y} \, dQ_n(y).$$

Now suppose that $h > 0$, h depends on n in such a way that $h \to 0$ as $n \to \infty$, and

$$hn^{1/2} \geq c \quad \text{for all sufficiently large } n, \tag{5.80}$$

where c is a positive constant. It follows from (5.71) and (5.72) that

$$\sigma^{-1}\bar{m} = \bar{\sigma}h(1 + O(h)). \tag{5.81}$$

In view of the inequality (5.79) we have

$$|I_2| = \left|-Q_n(0) - \int_0^{\infty} Q_n(y) \, d e^{-h\bar{\sigma}n^{1/2}y}\right| \leq C n^{-1/2} \tag{5.82}$$

for all sufficiently large values of n. Furthermore, substituting $u = h\bar{\sigma}n^{1/2}y$, we find that

$$I_1 = (h\bar{\sigma}n^{1/2})^{-1} \int_0^{\infty} \exp\{-u - u^2/(2h^2\bar{\sigma}^2 n)\} \, du$$

and

$$h\bar{\sigma}n^{1/2} I_1 \leq \int_0^{\infty} e^{-u} \, du = 1, \quad h\bar{\sigma}n^{1/2} I_1 > \int_0^{\infty} \exp\{-u - u^2/(c^2\sigma^2)\} \, du$$

for all sufficiently large n, by (5.80). Thus the product $hn^{1/2}I_1$ is bounded from above and below by positive constants for all sufficiently large n, i.e.

$$hn^{1/2}I_1 \asymp 1. \tag{5.83}$$

Making use of (5.82) and (5.83), we conclude that

$$\int_0^\infty e^{-h\bar{\sigma}n^{1/2}y}\,d\bar{F}_n(y) = (2\pi)^{-1/2}I_1 + O(n^{-1/2})$$

$$= (2\pi)^{-1/2}I_1\left(1 + \frac{1}{I_1}O(n^{-1/2})\right) = (2\pi)^{-1/2}I_1(1 + O(h)).$$

$$\tag{5.84}$$

It is easy to express the integrals I_1 and

$$I_3 = \int_0^\infty e^{-\bar{m}\sigma^{-1}n^{1/2}y - y^2/2}\,dy$$

in terms of the so-called Mills ratio

$$\psi(x) = \frac{1 - \Phi(x)}{\Phi'(x)} = e^{x^2/2}\int_x^\infty e^{-y^2/2}\,dy.$$

Clearly,

$$I_1 = \psi(h\bar{\sigma}n^{1/2}), \quad I_3 = \psi(\bar{m}\sigma^{-1}n^{1/2}).$$

We shall prove that

$$I_1 = I_3(1 + O(h)). \tag{5.85}$$

For every $x_1 < x_2$ we have $\psi(x_2) - \psi(x_1) = \psi'(x)(x_2 - x_1)$, where $x_1 < x < x_2$. Furthermore, $|\psi'(x)| = |x\psi(x) - 1| < x^{-2}$ for $x > 0$. The relations (5.81) and (5.83) imply that

$$I_3 = I_1 + O(n^{-1/2}) = I_1\left(1 + \frac{1}{I_1}O(n^{-1/2})\right) = I_1(1 + O(h)).$$

In turn, this implies (5.85).

It follows from (5.78), (5.84), and (5.85) that

$$1 - F_n(\bar{m}\sigma^{-1}n^{1/2}) = (2\pi)^{-1/2}\exp\{n\log R - hn\bar{m}\}I_3(1 + O(h))$$

$$= (2\pi)^{-1/2}\exp\{n\log R - hn\bar{m}\}\psi(\bar{m}\sigma^{-1}n^{1/2})(1 + O(h))$$

$$= \exp\left\{n\left(\frac{\bar{m}^2}{2\sigma^2} + \log R - h\bar{m}\right)\right\}$$

$$\times [1 - \Phi(\bar{m}\sigma^{-1}n^{1/2})](1 + O(h)). \tag{5.86}$$

We note that for $0 \leqslant x \leqslant 1$ the relations (5.66) and (5.67) are consequences of the Berry–Esseen inequality. Let $x > 1$, $x = o(n^{1/2})$ as $n \to \infty$. We put $t = xn^{-1/2}$ and consider the equation (5.73), which takes the form $x = \bar{m}\sigma^{-1}n^{1/2}$. We choose h in (5.78) and (5.86) to be the unique positive root of (5.73) so that $h \to 0$ as $n \to \infty$. By (5.75), $h \sim t/\sigma$ as $t \to 0$, and, therefore, $h > x\sigma^{-1}n^{-1/2}/2$ for all sufficiently large n. Thus the condition (5.80) is satisfied. Taking account of (5.76), we obtain

$$1 - F_n(x) = [1 - \Phi(x)] \exp\left\{ n\left(\frac{x}{\sqrt{n}}\right)^3 \lambda\left(\frac{x}{\sqrt{n}}\right) \right\}\left[1 + O\left(\frac{x}{\sqrt{n}}\right) \right].$$

This implies (5.66).

We can prove (5.67) in a similar way. □

Let us look at some consequences of Theorem 5.23. We shall suppose that $\{X_n\}$ is a sequence of independent identically distributed random variables with zero mean and positive variance and that Cramér's condition is satisfied. If $x \geqslant 0$, $x = O(n^{1/6})$, then we find from (5.66), (5.67), and (5.77) that

$$\frac{1 - F_n(x)}{1 - \Phi(x)} = \exp\left\{ \frac{\gamma_3 x^3}{6\sigma^3 n^{1/2}} \right\}\left[1 + O\left(\frac{x+1}{\sqrt{n}}\right) \right],$$

$$\frac{F_n(-x)}{\Phi(-x)} = \exp\left\{ -\frac{\gamma_3 x^3}{6\sigma^3 n^{1/2}} \right\}\left[1 + O\left(\frac{x+1}{\sqrt{n}}\right) \right].$$

If we replace the condition $x = O(n^{1/6})$ by the stronger condition $x = o(n^{1/6})$, then we obtain the relations (5.60). Taking account of (5.62) and the equality $1 - \Phi(x) = \Phi(-x)$ for all x, we arrive at the following proposition.

Theorem 5.24 *Let the random variable X_1 satisfy Cramér's condition. If $x \geqslant 0$, $x = O(n^{1/6})$, then*

$$1 - F_n(x) = [1 - \Phi(x)] \exp\left\{ \frac{\gamma_3 x^3}{6\sigma^3 n^{1/2}} \right\} + O(n^{-1/2} e^{-x^2/2}),$$

$$F_n(-x) = \Phi(-x) \exp\left\{ -\frac{\gamma_3 x^3}{6\sigma^3 n^{1/2}} \right\} + O(n^{-1/2} e^{-x^2/2}).$$

If $|x| = O(n^{1/6})$ and $EX_1^3 = 0$, then

$$F_n(x) - \Phi(x) = O(n^{-1/2} e^{-x^2/2}).$$

By imposing additional conditions, we can broaden the zone of normal convergence.

Theorem 5.25 *Let the random variable X_1 satisfy Cramér's condition. If $x \geqslant 0$, $x = o(n^{(r+1)/2(r+3)})$ for some positive integer r, and if $\gamma_k = 0$ for $k = 3, \ldots, r+2$, then the relations (5.60) hold.*

We note that the condition $x = o(n^{1/2})$ in Theorem 5.23 cannot be replaced by the weaker conditiion $x = O(n^{1/2})$. In fact, if we consider a sequence of independent random variables having a common symmetric distribution concentrated on an interval $(-C, C)$, then the normed sum $Z_n = S_n \sigma^{-1} n^{-1/2}$ satisfies the inequality $|Z_n| \leq C_1 n^{1/2}$, where $C_1 = C/\sigma$. Therefore, $1 - F_n(x) = 0$ and $F_n(-x) = 0$ for $x > C_1 n^{1/2}$ and all n, so that the relations (5.66) and (5.67) do not hold.

If we consider a sequence of independent not necessarily identically distributed random variables $\{X_n\}$ such that

$$(S_n - b_n)/a_n \xrightarrow{P} 0$$

for some constants $a_n > 0$ and b_n, then probabilities of the form $P(S_n - b_n > a_n)$, $P(S_n - b_n < -a_n)$, and $P(|S_n - b_n| > a_n)$ are called probabilities of large deviations of the sums $S_n = \sum_{j=1}^{n} X_j$. In particular, in the i.i.d. case with zero means, probabilities of the form $P(S_n > x_n \sigma \sqrt{n})$ and $P(S_n < -x_n \sigma \sqrt{n})$, where $x_n \to +\infty$ as $n \to \infty$ and $\sigma^2 = \text{Var } X_1$, are probabilities of large deviations of the sums S_n.

5.9 Bibliographical notes

Estimates of the rate of convergence in the central limit theorem were obtained for the first time by Lyapunov [260].

Theorems 5.1, 5.2 and 5.4 are due to Esseen [119]. Theorem 5.5 was obtained by Berry [32] and Esseen [118]. Theorem 5.3 is a generalization of Esseen's Theorem 5.2 due to Petrov [342]. In the particular case when $F(x)$ and $G(x)$ are distribution functions, a close result was proved by Fainleib [128]. Inequalities of type (5.1) are often called smoothing inequalities. Other smoothing inequalities can be found in Zolotarev [483], Paulauskas [324], Bikelis [36], Sazonov [415], and Bhattacharya and Ranga Rao [34]. Other proofs of the Berry–Esseen inequality can be found in Ho and Chen [193], Sazonov [415], and Barbour and Hall [15].

Theorem 5.6 is a generalization of a result of Katz [212] for the i.i.d. case; this generalization was found by Petrov [330]. In the case when $\delta < 1$, the inequality (5.23) was obtained by Lyapunov [260] without the assertion that the constant A is absolute. Theorem 5.8 was proved by Osipov [317] by another method. Theorem 5.9 was obtained by Osipov and Petrov [322] under the additional assumption about independence. Heyde [186] discovered optimal asymptotic behavior of the estimate given by Theorem 5.9. A strengthening of Theorem 5.9 was obtained by Leslie [250]. Some results close to Theorem 5.9 were later obtained by Feller [131].

The constants in Esseen's inequality and the Berry–Esseen inequality have been computed by Berry [32], Esseen [119], Bergström [22],

Zolotarev [482], and others. Van Beek [460] proved that both inequalities hold with $A = 0.7975$. Using computers, Shiganov [423] obtained the following values of A: 0.7915 for Esseen's inequality and 0.7655 for the Berry–Esseen inequality.

Theorems 5.10 and 5.11 are due to Osipov [319]. Osipov [320] found upper and lower bounds of the remainder in asymptotic expansions connected with the central limit theorem; these bounds have the same order of magnitude. Further results in this direction were obtained by Rozovsky [403, 405], Hall [157, 163], and Heyde and Nakata [190].

Lemma 5.4 is due to Kolodyazhny [236], Theorem 5.12 to Esseen [119], Theorems 5.14 and 5.15 to Osipov [318], Theorem 5.16 to Nagaev [309], and Theorem 5.17 to Bikelis [36]. Information about constants in non-uniform estimates can be found in Michel [296] and Paditz [323]. Generalizations of Theorem 5.17 are proved by Bikelis [36] and Petrov [347].

Asymptotic expansions of the kind considered here were introduced by Chebyshev and later studied by Edgeworth, Cramér and others (see, for instance, Cramér [85], Gnedenko and Kolmogorov [147], and Esseen [119]). The expressions (5.56) and (5.59) for the terms of these expansions were obtained by Petrov [329].

Theorems 5.18–5.20 and Lemma 5.7 are due to Osipov [318, 321], Theorems 5.21 and 5.22 to Esseen [119].

Estimates of the remainder in the central limit theorem and asymptotic expansions take up a significant portion of the books of Gnedenko and Kolmogorov [147], Ibragimov and Linnik [205], Petrov [342, 351], and Hall [163]. These books contain integral and local limit theorems for sums of independent identically distributed random variables [147], [205], and for sums of non-identically distributed summands [163], [342].

The books by Bhattacharya and Ranga Rao [34] and Sazonov [415] are devoted to rates of convergence in the central limit theorems for sums of independent random variables taking on values in k-dimensional Euclidean space and Hilbert space. The latter book uses the convolution method developed by Bergström, Sazonov, and others, instead of the method of characteristic functions.

The central limit theorem for sums of independent random variables taking on values in Banach space is considered in the book by Araujo and Giné [12].

Much attention has been paid to the rate of convergence of distributions of sums of independent random variables to a stable non-normal law. In the i.i.d. case, estimates of the rate of convergence were obtained by Christoph [74, 76], Egorov [111], Paulauskas [326], Hall [158, 160], Mijnheer [301], and others.

Theorem 5.23 is a sharpening of a fundamental result due to Cramér [83], and was obtained by Petrov [327] as a corollary of a more general theorem concerning sequences of independent non-identically distributed random variables. There are a great many investigations on limit theorems for large deviations of sums of independent summands. Surveys of these investigations can be found in Ibragimov and Linnik [205], Sethuraman [420], Petrov [342], Book [49], and Saulis and Statulevicius [411].

5.10 Addenda

5.10.1 Let $F(x)$ be a non-decreasing bounded function, and let $G(x)$ be a function of bounded variation such that $F(-\infty) = G(-\infty)$ and $F(+\infty) = G(+\infty)$. Let $H(x)$ be a distribution function with the density $p(x)$. We put

$$\Delta = \sup_x |F(x) - G(x)|, \quad \Delta_H = \sup_x |(F(x) - G(x)) * H(x)|,$$

$$B(x, \beta) = 2 \int_0^\beta p(x - y) \, dy - 1.$$

If $\beta > 0$, $x \in \mathbb{R}$, and $B(x, \beta) > 0$, then

$$\Delta \leq \frac{1}{B(x, \beta)} \left\{ \Delta_H + \sup_u \int_0^\beta |G(u + y) - G(u)| p(x - y) \, dy \right\}$$

and

$$\Delta \leq \frac{1}{2\pi B(x, \beta)} \int_{-\infty}^\infty \left| (f(t) - g(t)) h\left(\frac{t}{T}\right) \right| \frac{dt}{|t|}$$

$$+ \frac{1}{B(x, \beta)} \sup_u \int_0^\beta \left| G\left(u + \frac{y}{T}\right) - G(u) \right| p(x - y) \, dy$$

for every $T > 0$, where $f(t)$, $g(t)$, and $h(t)$ are the Fourier–Stieltjes transforms of $F(x)$, $G(x)$, and $H(x)$ respectively (Paulauskas [324]).

5.10.2 If $F(x)$ and $G(x)$ are the distribution functions of integer-valued random variables, then

$$\sup_x |F(x) - G(x)| \leq \frac{1}{4} \int_{-\pi}^\pi \left| \frac{f(t) - g(t)}{t} \right| dt,$$

where $f(t)$ and $g(t)$ are the corresponding characteristic functions (Tsaregradskii [455]).

5.10.3 Let $F(x)$ and $G(x)$ be distribution functions, and let $f(t)$ and $g(t)$

be the corresponding characteristic functions. Then

$$\sup_x |F(x) - G(x)| \geq 2^{-1}(2\pi)^{-1/2} \left| \int_{-\infty}^{\infty} (f(t) - g(t)) e^{-t^2/2} \, dt \right|$$

(Matskyavichyus [293]).

In Subsections 5.10.4–5.10.7 we consider independent random variables $X_1, \ldots, X_n$ with zero means and finite variances, an we use the following notation:

$$V_j(x) = P(X_j < x), \quad \sigma_j^2 = \text{Var } X_j, \quad B_n = \sum_{j=1}^{n} \sigma_j^2,$$

$$F_n(x) = P\left(B_n^{-1/2} \sum_{j=1}^{n} X_j < x \right), \quad \Delta_n(x) = |F_n(x) - \Phi(x)|.$$

5.10.4 For every real x,

$$\Delta_n(x) \leq A B_n^{-3/2} (1 + |x|)^{-3} \sum_{j=1}^{n} \int_0^{(1+|x|)B_n^{1/2}} \int_{|u|>v} u^2 \, dV_j(u) \, dv$$

(Bikelis [36]).

5.10.5 If $E|X_j|^{2+\delta} < \infty$ for $j = 1, \ldots, n$ and some positive $\delta \leq 1$, then

$$\Delta_n(x) \leq A B_n^{-1-\delta/2} (1 + |x|)^{-2-\delta} \sum_{j=1}^{n} E|X_j|^{2+\delta}$$

for every real x (Bikelis [36]).

5.10.6 If $EX_j^2 g(X_j) < \infty$ for $j = 1, \ldots, n$ and some function $g(x) \in G$, where G is the same class of functions as in Theorem 5.6, then

$$\Delta_n(x) \leq A B_n^{-1} (1 + |x|)^{-2} (g((1 + |x|)B_n^{1/2}))^{-1} \sum_{j=1}^{n} EX_j^2 g(X_j)$$

for every real x (Petrov [347]).

5.10.7 The following inequality holds:

$$\sup_x \Delta_n(x) \leq A B_n^{-3/2} \left\{ \left| \sum_{j=1}^{n} \int_{|x| \leq B_n^{1/2}} x^3 \, dV_j(x) \right| \right.$$

$$\left. + \sup_{0 < z \leq B_n^{1/2}} z \sum_{j=1}^{n} \int_{|x|>z} x^2 \, dV_j(x) \right\}$$

(Rozovsky [402]; this result is a generalization of a theorem of Esseen [123]).

5.10.8 Let $\{X\}$ be a sequence of independent random variables with a common d.f. $V(x)$. Suppose there exist sequences of constants $\{a_n\}$ and $\{b_n\}$ such that $F_n(x) \to \Phi(x)$, where

$$F_n(x) = P\left(\frac{1}{a_n}\sum_{j=1}^{n} X_j - b_n < x\right).$$

Define

$$r_n = \inf_{a_n, b_n} \sup_{x} |F_n(x) - \Phi(x)|.$$

The relation $r_n = O(n^{-\delta/2})$, where $0 < \delta < 1$, is equivalent to the conditions

$$\int_{-\infty}^{\infty} x^2\, dV(x) < \infty, \quad \int_{|x| \geq z} x^2\, dV(x) = O(z^{-\delta}) \quad \text{as } z \to \infty. \quad (5.87)$$

The relation $r_n = O(n^{-1/2})$ is equivalent to the conditions (5.87) with $\delta = 1$ and

$$\int_{-z}^{z} x^3\, dV(x) = O(1) \quad \text{as } z \to \infty$$

(Ibragimov [202]).

In Subsections 5.10.9–5.10.18 we consider a sequence of independent identically distributed random variables $\{X_n\}$ such that $EX_1 = 0$, $\text{Var } X_1 = 1$, and we put

$$V(x) = P(X_1 < x), \quad F_n(x) = P\left(n^{-1/2}\sum_{j=1}^{n} X_j < x\right),$$

$$\Delta_n(x) = |F_n(x) - \Phi(x)|, \quad \Delta_n = \sup_{x} \Delta_n(x).$$

5.10.9 Define

$$\sigma_n^2 = \int_{|x| < n^{1/2}} x^2\, dV(x) - \left(\int_{|x| < n^{1/2}} x\, dV(x)\right)^2. \quad (5.88)$$

Then the series

$$\sum_{n=1}^{\infty} n^{-1} \sup_{x} \left| P\left(\sigma_n^{-1} n^{-1/2} \sum_{j=1}^{n} X_j < x\right) - \Phi(x) \right|$$

converges (Friedman et al. [139]).

5.10.10 The series $\sum_{n=1}^{\infty} n^{-1} \Delta_n$ converges if and only if $EX_1^2 \log(1 + |X_1|) < \infty$. If $0 < \delta < 1$, the convergence of the series $\sum_{n=1}^{\infty} n^{-1+\delta/2} \Delta_n$ is equivalent to the condition $E|X_1|^{2+\delta} < \infty$ (Heyde [182]).

5.10.11 If $0 < \delta < 1$, the convergence of the series

$$\sum_{n=1}^{\infty} n^{-1+\delta/2} \sup_{x} (1 + x^2) \Delta_n(x)$$

is equivalent to the condition $E|X_1|^{2+\delta} < \infty$ (Maejima [264]).

5.10.12 Let $\{C_n\}$ be a sequence of positive constants,

$$\delta_n(C_n) = \sup_{x} \left| P\left(\sum_{j=1}^{n} X_j < xC_n\right) - \Phi(x) \right|,$$

and $B_n^2 = n\sigma_n^2$, where σ_n^2 is defined by (5.88). Define

$$\nabla_n(C_n) = nP(|X_1| \geq n^{1/2}) + AnB_n^{-3} \int_{|x|<n^{1/2}} |x|^3 \, dV(x)$$

$$+ nB_n^{-1}(2\pi)^{-1/2} \left| \int_{|x|<n^{1/2}} x \, dV(x) \right|$$

$$+ 2^{-1}(2\pi e)^{-1/2} |1 - (B_n^2/C_n^2)| \max(1, C_n^2/B_n^2).$$

(By Theorem 5.9 the inequality $\delta_n(C_n) \leq \nabla_n(C_n)$ holds for every n and some absolute positive constant A). The relation $\delta_n(C_n) \to 0$ is equivalent to the relation $\nabla_n(C_n) \to 0$ (Heyde [186]; in [186] there is a generalization of this result to the case when no assumptions are made about the existence of any moments of the random variable X_1).

5.10.13 Define $G(x) = P(|X_1| < x)$. If $E|X_1|^{2+\delta} < \infty$ for some non-negative $\delta \leq 1$, then

$$\Delta_n(x) \leq Ac_n(\delta)(1 + |x|^{2+\delta})^{-1}$$

for every $x \in \mathbb{R}$, where

$$c_n(\delta) = n^{-1/2} \int_0^{n^{1/2}} x^3 \, dG(x) + n^{-\delta/2} \int_{n^{1/2}}^{\infty} x^{2+\delta} \, dG(x)$$

and A is an absolute positive constant (Heyde [187]).

5.10.14 Let the functions $h_1(x) = x^\alpha(1 - V(x))$ and $h_2(x) = x^\alpha V(-x)$ be slowly varying as $x \to +\infty$ for some $\alpha \geq 2$. We put $K(x) = \int_{|y| \geq x} y^2 \, dV(y)$. If $\alpha = 2$, then

$$8^{-1} \cdot 2^{-1/2} \leq \liminf \Delta_n / K(n^{1/2}) \leq \limsup \Delta_n / K(n^{1/2}) \leq \pi/2.$$

If $2 < \alpha < 3$, then

$$\pi^{1/2}\Gamma((\alpha + 1)/2)/(2^{3/2}\Gamma(\alpha)|\sin(\pi\alpha/2)|)$$
$$\leqslant \liminf\{n[1 - V(n^{1/2}) + V(-n^{1/2})]\}^{-1}\Delta_n$$
$$\leqslant \limsup\{n[1 - V(n^{1/2}) + V(-n^{1/2})]\}^{-1}\Delta_n$$
$$\leqslant \Gamma(\alpha/2)2^{\alpha/2-1}/(\Gamma(\alpha)|\sin \pi\alpha|)$$

(Hall [155]).

5.10.15 Define

$$\Delta_{n,p} = \left(\int_{-\infty}^{\infty} |F_n(x) - \Phi(x)|^p \, dx\right)^{1/p} \quad \text{for } 1 \leqslant p < \infty,$$

$$\Delta_{n,\infty} = \Delta_n = \sup_x |F_n(x) - \Phi(x)|.$$

Suppose that at least one of the following conditions is satisfied:

(a) $x^3 P(|X_1| \geqslant x) \to \infty$ as $x \to \infty$;
(b) $E|X_1|^3 < \infty$ and $EX_1^3 \neq 0$;
(c) X_1 has a lattice distribution and $E|X_1|^3 < \infty$;
(d) $\limsup_{|t| \to \infty} |Ee^{itX_1}| < 1$ and $E|X_1|^r < \infty$, $EX_1^r \neq \int_{-\infty}^{\infty} x^r \, d\Phi(x)$ for some integer $r \geqslant 3$. Then there exist absolute positive constants A_1 and A_2 such that

$$A_1 \leqslant \liminf \Delta_{n,p}/\Delta_{n,p'} \leqslant \limsup \Delta_{n,p}/\Delta_{n,p'} \leqslant A_2$$

for every p and p' satisfying the conditions $p \neq p'$, $1 \leqslant p \leqslant \infty$, $1 \leqslant p' \leqslant \infty$. If condition (a) is satisfied, then $\Delta_{n,p} \asymp \psi_n$ as $n \to \infty$, where

$$\psi_n = EX_1^2 I(|X_1| \geqslant n^{1/2}) + n^{-1/2}|EX_1^3 I(|X_1| < n^{1/2})| + n^{-1} EX_1^4 I(|X_1| < n^{1/2})$$
(5.89)

and $I(B)$ is the indicator of the event B. If (b) or (c) is satisfied, then $\Delta_{n,p} \asymp n^{-1/2}$. If condition (d) is satisfied, then $\Delta_{n,p} \asymp n^{-(r-2)/2}$ (Heyde and Nakata [190].

5.10.16 We set $\kappa = 3\int_{-\infty}^{\infty} x^2 |V(x) - \Phi(x)| \, dx$,

$$\kappa_0 = \int_{-\infty}^{\infty} \max(1, 3x^2)|V(x) - \Phi(x)| \, dx,$$

$$v_0 = \int_{-\infty}^{\infty} \max(1, |x|^3)| \, d(V(x) - \Phi(x))|.$$

For every $n \in \mathbb{N}$ the following inequalities hold:

$$\Delta_n \leqslant A \max(\kappa, \kappa^{n/(3n+1)}) n^{-1/2},$$

$$\Delta_n \leqslant A \max(\kappa_0, \kappa_0^{n/(n+1)}) n^{-1/2}, \quad \Delta_n \leqslant A v_0 n^{-1/2}$$

(Zolotarev [484]).

5.10.17 For every n,

$$\Delta_n \leqslant A n^{-1/2} \int_{-\infty}^{\infty} \max(1, |x|^3) |V(x) - \Phi(x)| \, dx$$

(Sazonov [414]).

5.10.18 Suppose $EX_1^2 g(X_1) < \infty$ for some even function $g(x)$ that satisfies the following conditions: $g(1) = 1$; $g(x)$ is positive and non-decreasing in the interval $x > 0$; $x/g(x)$ is non-decreasing in the interval $x > 0$. Define

$$\lambda = \int_{-\infty}^{\infty} \max(1, |x|g(x)) |V(x) - \Phi(x)| \, dx.$$

Then

$$\Delta_n \leqslant A \max(\lambda^{\alpha_n}, \lambda)/g(n^{1/2}),$$

where $\alpha_0 = 0$, $\alpha_1 = \frac{1}{2}$, and $\alpha_n = (1 + \alpha_{[g^2(n^{1/2})]-1})/2$ for $n \geqslant 2$. (In the case when $g(x) = |x|$, we have $\alpha_n = 1 - 2^{-n}$ for all n.) Moreover,

$$\Delta_n(x) \leqslant A\{g(n^{1/2}) + g(xn^{1/2})x^2\}^{-1} \max(\lambda^{\delta_n}, \lambda)$$

for all $n \in \mathbb{N}$ and $x \in \mathbb{R}$, where $\delta_0 = 0$, $\delta_1 = \frac{1}{2}$, and $\delta_n = \max(1, (n-1)/n + \delta_{[g^2(n^{1/2})]-1})/2$ for $n \geqslant 2$ (Ul'yanov [457]).

Note that in the case when $V(x) \equiv \Phi(x)$ and consequently $\Delta_n = 0$, the right-hand sides of the estimates from Subsections 5.10.16–5.10.18 are equal to zero, in contrast to the Berry–Esseen estimate. Other results of this type can be found in Nagaev and Rotar [313], Sazonov [414, 415], Rotar [400, 401], Paulauskas [325], and Ul'yanov [458].

In Subsections 5.10.19 and 5.10.20 we consider a sequence of independent identically distributed random variables $\{X_n\}$ such that $EX_1 = 0$, $\operatorname{Var} X_1 = \sigma^2 > 0$, and $E|X_1|^3 = \beta_3 < \infty$. We put $\rho = \beta_3/\sigma^3$ and $F_n(x) = P(\sigma^{-1} n^{-1/2} \sum_{j=1}^{n} X_j < x)$.

5.10.19 There exists $\lim_{n \to \infty} n^{1/2} \sup_x |F_n(x) - \Phi(x)|$, not exceeding $[(3 + \sqrt{10})/6\sqrt{2\pi}]\rho$. Equality holds if and only if $X_1 = aX$, where $a \neq 0$ and X takes on the values $-(4 - \sqrt{10})/2$ and $(\sqrt{10} - 2)/2$ with probabilities $(\sqrt{10} - 2)/2$ and $(4 - \sqrt{10})/2$ respectively (Esseen [120]).

5.10.20 We have

$$\limsup_{n\to\infty} \inf_{a,\sigma_1} n^{1/2} \sup_x |F_n(x) - \Phi((x-a)/\sigma_1)| \leq (2\pi)^{-1/2}\rho.$$

Equality holds if X_1 takes on the values -1 and 1 with probability $\frac{1}{2}$ for each (Rogozin [387]).

5.10.21 Let $\{X_n\}$ be a sequence of independent random variables with zero means and finite moments $E|X_n|^3$. We put

$$B_n = \sum_{j=1}^n \operatorname{Var} X_j, \quad L_n = B_n^{-3/2} \sum_{j=1}^n E|X_j|^3,$$

$$F_n(x) = P\left(B_n^{-1/2} \sum_{j=1}^n X_j < x\right), \quad V_j(x) = P(X_j < x),$$

$$Q_n(x) = \sum_{j=1}^n \int_{-\infty}^\infty (\Phi(x) - \Phi(x - B_n^{-1/2}u))\,dV_j(u) - 2^{-1}(2\pi)^{-1/2}xe^{-x^2/2}.$$

The following equality holds for every $x \in \mathbb{R}$:

$$\overline{\lim}_{\substack{l\to 0 \\ L_n = l}} L_n^{-1}|F_n(x) - \Phi(x) + Q_n(x)| = (2\pi)^{-1/2} e^{-x^2/2}$$

(Chistyakov [67]).

5.10.22 Let $\{X_n\}$ be a sequence of independent identically distributed random variables, and let $\{a_n\}$ and $\{b_n\}$ be sequences of numbers, $a_n \uparrow \infty$. We write

$$F_n(x) = P\left(a_n^{-1} \sum_{j=1}^n X_j - b_n < x\right), \quad \Delta_n(x) = |F_n(x) - \Phi(x)|,$$

$$\Delta_n = \sup_x \Delta_n(x), \quad \Delta_{n,p} = \left(\int_{-\infty}^\infty \Delta_n^p(x)\,dx\right)^{1/p}.$$

The following conditions are equivalent:

(A) $EX_1^2 < \infty$;
(B) there exist sequences $\{a_n\}$ and $\{b_n\}$ such that $\Delta_n \to 0$ and $\sum_{n=1}^\infty \Delta_n/n < \infty$;
(C) there exist sequences $\{a_n\}$ and $\{b_n\}$ such that $\Delta_n(x) \to 0$ and $\sum_{n=1}^\infty \Delta_n(x)/n < C/(1+x^2)$ for every $x \in \mathbb{R}$, where C is a positive constant;
(D) there exist sequences $\{a_n\}$ and $\{b_n\}$ such that $\Delta_{n,p} \to 0$ and $\sum_{n=1}^\infty \Delta_{n,p}/n < \infty$ for some p in the interval $1 \leq p \leq 2$ (equivalently, for every p in this interval)

(Egorov [110]; at the same time Heyde [186] proved that (B) ⇒ (A); earlier Friedman *et al.* [139] proved that (A) ⇒ (B); in the case when $p = 1$, Heyde [185] obtained the implication (D) ⇒ (A)).

5.10.23 Let $\{X_n\}$ be a sequence of independent random variables with a common d.f. $V(x)$, and let $\{a_n\}$ and $\{b_n\}$ be sequences of numbers, $a_n > 0$. We write

$$F_n(x) = P\left(a_n^{-1} \sum_{j=1}^n X_j - b_n < x\right), \quad \Delta_n = \inf_{a_n, b_n} \sup_x |F_n(x) - \Phi(x)|,$$

$$R(x) = P(|X_1| \geq x), \quad D(x) = \int_{|y| < x} y^2 \, dV(y).$$

Let $\varepsilon(x)$ be a positive function defined in the interval $x > 0$ and satisfying the following conditions: $\varepsilon(x) \to 0$ as $x \to \infty$; $x^\delta \varepsilon(x)$ does not decrease in the interval $x > x_0$ for some positive $\delta < \tfrac{1}{2}$ and x_0. In order that $\Delta_n = O(\varepsilon(n))$ as $n \to \infty$, it is necessary and sufficient that

$$R(x)x^2/D(x) = O(\varepsilon(x^2/D(x))) \qquad \text{as } x \to \infty.$$

(Rozovsky [406]). In particular, $\Delta_n \to 0$ if and only if $R(x) = o(x^{-2}D(x))$ (see Theorem 3.10, Chapter 3). The article [406] also contains generalizations of some results of Ibragimov and Egorov from Subsections 5.10.8 and 5.10.22.

5.10.24 Let $X_1, \ldots, X_n$ be independent random variables such that $EX_j = 0$, $\text{Var } X_j = \sigma_j^2 < \infty$ ($j = 1, \ldots, n$), $\sum_{j=1}^n \sigma_j^2 = 1$. We put

$$F_n(x) = P\left(\sum_{j=1}^n X_j < x\right), \quad \Delta_{n,p} = \left(\int_{-\infty}^\infty |F_n(x) - \Phi(x)|^p \, dx\right)^{1/p},$$

$$\Delta_{n,\infty} = \sup |F_n(x) - \Phi(x)|,$$

$$\psi_n = \sum_{j=1}^n EX_j^2 I(|X_j| \geq 1) + \sum_{j=1}^n EX_j^4 I(|X_j| < 1) + \left|\sum_{j=1}^n EX_j^3 I(|X_j| < 1)\right|.$$

There exists an absolute constant A such that

$$\psi_n \leq A\left(\Delta_{n,p} + \sum_{j=1}^n \sigma_j^4\right) \tag{5.90}$$

for $1 \leq p \leq \infty$. If $\varepsilon_n > 0$ and $\sum_{j=1}^n EX_j^2 I(|X_j| > \varepsilon_n) \leq 1/8$, then there exists an absolute constant A such that

$$\Delta_{n,p} \leq A\left(\psi_n + \varepsilon_n + \sum_{j=1}^n \sigma_j^4\right)$$

(Heyde and Nakata [190]; the inequality (5.90) for $p = \infty$ was proved by

Hall and Barbour [171]). More complicated results of this type were obtained earlier by Rozovsky [405, 406].

In Subsections 5.10.25–5.10.35 we consider a sequence of independent identically distributed random variables $\{X_n\}$ such that $EX_1 = 0$ and $\operatorname{Var} X_1 = 1$, and we write

$$F_n(x) = P\left(n^{-1/2} \sum_{j=1}^{n} X_j < x\right), \quad \Delta_n = \sup_{x} |F_n(x) - \Phi(x)|.$$

We define ψ_n by the equality (5.89).

5.10.25 The following relation holds:

$$\Delta_n + n^{-1/2} \asymp \psi_n + n^{-1/2}$$

(Osipov [319], Rozovsky [405, 406], and Hall [157]).

5.10.26 The set $\mathscr{S}$ will be said to be convergence-determining to order $n^{-1/2}$ if

$$\sup_{x \in \mathscr{S}} |F_n(x) - \Phi(x)| + n^{-1/2} \asymp \psi_n + n^{-1/2}$$

for every standardized distribution of X_1. In other words, $\mathscr{S}$ is convergence-determining to order $n^{-1/2}$ if the rate of convergence on $\mathscr{S}$ is the same as the rate of convergence on the whole real line, up to terms of order $n^{-1/2}$. If $0 < \alpha < 3^{1/2}$ and if $\alpha \neq 1$, then the pair $(-\alpha, \alpha)$ is convergence-determining to order $n^{-1/2}$. If $\alpha = 0, 1, 3^{1/2}$ or $\alpha \geqslant 2.1242$, then the pair $(-\alpha, \alpha)$ is not convergence-determining to order $n^{-1/2}$ (Hall [166]).

5.10.27 Define

$$G(x, y) = \Phi(x + y) - \Phi(x) - y\varphi(x) - y^2 \varphi'(x)/2,$$

$$H(x_1, x_2, y) = \varphi''(x_1) G(x_2, y) - \varphi''(x_2) G(x_1, y),$$

where $\varphi(x) = (2\pi)^{-1/2} e^{-x^2/2}$. The triplet $\mathscr{T} = \{-\sqrt{3}, 0, \sqrt{3}\}$ is not convergence-determining. If $\{x_1, x_2\}$ is not a subset of $\mathscr{T}$, then $\{x_1, x_2\}$ is convergence-determining to order $n^{-1/2}$ if and only if the only solution in y of the equation $H(x_1, x_2, y) = 0$ is $y = 0$. It follows that no singleton is convergence-determining, since $H(x, x, y) = 0$ for all x and y. If $1 < x_2 < x_1$, or if $x_2 < x_1 < -1$, or if $-1 < x_2 < x_1 < 1$, then the only solution in y of the equation $H(x_1, x_2, y) = 0$ is $y = 0$. Any set of four distinct points is convergence-determining to order $n^{-1/2}$ (Hall and Wightwick [174]).

5.10.28 Suppose $\alpha \geq 0$. We set

$$\Delta_n(\alpha; c, d) = \sup_x (1 + |x|^\alpha) \left| P\left(\sum_{j=1}^n X_j < cx + d \right) - \Phi(x) \right|,$$

$$\Delta_{n1}(\alpha) = \inf_{c > 0, d} \Delta_n(\alpha; c, d),$$

$$\delta_{n1}(\alpha) = n \sup_{x \geq 1} x^\alpha P(|X_1| \geq xn^{1/2}) + n^{-1/2} |EX_1^3 I(|X_1| < n^{1/2})|$$

$$+ n^{-1} EX_1^4 I(|X_1| < n^{1/2}),$$

$$\sigma_n^2 = EX_1^2 I(|X_1| < n^{1/2}), \quad v_n = EX_1 I(|X_1| < n^{1/2}).$$

Let $\beta = \max(0, \alpha/2 - 1)$, $\delta_n(\alpha) = \psi_n$ if $0 \leq \alpha \leq 2$, where ψ_n is defined by (5.89), and $\delta_n(\alpha) = \delta_{n1}(\alpha)$ if $\alpha > 2$. Then

$$\Delta_n(\alpha; \sigma_n n^{1/2}, nv_n) = O(\delta_{n1}(\alpha) + n^{-1/2}(\log n)^\beta)$$

as $n \to \infty$ for every $\alpha \geq 0$, and moreover,

$$\liminf \{\Delta_{n1}(\alpha) + n^{-1/2}(\log n)^\beta\}/\delta_{n1}(\alpha) > 0.$$

Furthermore,

$$\Delta_n(\alpha; n^{1/2}, 0) + n^{-1/2}(\log n)^\beta \asymp \delta_n(\alpha) + n^{-1/2}(\log n)^\beta$$

for every $\alpha \geq 0$. If X_1 has a lattice distribution or satisfies the Cramér condition (C), then it is possible to replace β by zero (Hall [165]).

5.10.29 Let $\mu_1 = 0, \mu_2 = 1, \mu_3, \mu_4, \ldots$ be a given numerical sequence with arbitrary $\mu_3, \mu_4, \ldots$. Let $Q_\nu(x)$ ($\nu = 1, 2, \ldots$) be functions defined by (5.59) with the replacement of the cumulants $\gamma_3, \ldots, \gamma_{\nu+2}$ by $\mu_3, \ldots, \mu_{\nu+2}$. We construct a sequence of numbers $\{\beta_n\}$, where β_n is defined by the numbers $\mu_1, \ldots, \mu_n$ in the same way as the moments are defined by the cumulants, i.e. $\beta_1 = \mu_1, \beta_3 = \mu_2 + \mu_1^2, \ldots$.

In order that

$$F_n(x) = \Phi(x) + \sum_{\nu=1}^{k-2} Q_\nu(x) n^{-\nu/2} + O(n^{-(k-2)/2})$$

hold uniformly in x for some integer $k \geq 3$, it is necessary (and sufficient, for those distributions of X_1 that satisfy the Cramér condition (C)) that the following three conditions be satisfied:

(1) $E|X_1|^{k-1} < \infty$, $EX_1^m = \beta_m$ ($m = 1, \ldots, k-1$);
(2) $\int_{|x| \geq z} |x|^{k-1} dV(x) = o(z^{-1})$ as $z \to \infty$, where $V(x) = P(X_1 < x)$;
(3) $\lim_{z \to +\infty} \int_{-z}^{z} x^k dV(x) = \beta_k$

(Ibragimov [203]).

In order that

$$\sum_{n=1}^{\infty} n^{-1+(k-2+\delta)/2} \sup_x \left| F_n(x) - \Phi(x) - \sum_{v=1}^{k-2} Q_v(x) n^{-v/2} \right| < \infty \quad (5.91)$$

for some integer $k \geq 2$ and some positive $\delta < 1$, it is necessary (and sufficient, for those distributions of X_1 that satisfy the Cramér condition (C)) that

$$E|X_1|^{2+\delta} < \infty, \quad \beta_m = EX_1^m \quad (m = 1, \ldots, k). \quad (5.92)$$

If the conditions (5.92) hold for an even $k \geq 2$ and $\delta = 0$, then in order that (5.91) be satisfied with $\delta = 0$ it is necessary (and sufficient, for distributions satisfying the Cramér condition (C) or in the case when $k = 2$) that $E|X_1|^k \log(1 + |X_1|) < \infty$ (Heyde and Leslie [189]). A generalization of this result and also a generalization of Heyde's theorem from Subsection 5.10.10 can be found in Lifshits [253], where more general weight functions are considered instead of power functions.

5.10.30 Let $k \geq 2$ be an integer, and let $g_v(x)$ ($v = 1, \ldots, k-2$) be functions of bounded variation on $\mathbb{R}$ that satisfy for every odd v at least one of the following conditons: (1) there exists a number $r_v \neq 0$ such that $g_v(r_v) = g_v(-r_v)$; (2) $\liminf_{x \to \infty} x^{v+2} |g_v(x)| = 0$; (3) $\liminf_{x \to 0} |x^{-1}(g_v(x) + g_v(-x))| = 0$. In order that

$$F_n(x) = \Phi(x) + \sum_{v=1}^{k-2} g_v(x) n^{-v/2} + o(n^{-(k-2)/2})$$

uniformly in $x \in \mathbb{R}$, it is necessary (and sufficient, for those distributions that satisfy the Cramér condition (C)) that the following conditions be satisfied:

(i) $E|X_1|^{k-1} < \infty$ and there exists the finite limit $\lim_{z \to +\infty} \int_{-z}^{z} x^k \, dV(x)$, where $V(x) = P(X_1 < x)$;
(ii) $\int_{|x| \geq z} |x|^{k-1} \, dV(x) = o(z^{-1})$ as $z \to +\infty$;
(iii) $g_v(x) \equiv Q_v(x)$ ($v = 1, \ldots, k-2$), where $Q_v(x)$ are defined by (5.59) (Rozovsky [404]). Other results of this type can be found in Rozovsky [403] and Michel [297]; they are generalizations or analogues of Ibragimov's theorem from Subsection 5.10.8.

5.10.31 Suppose X_1 has a non-lattice distribution and $E|X_1|^3 < \infty$. Then

$$(1 + |x|)^3 |F_n(x) - \Phi(x) - 6^{-1}(1 - x^2) e^{-x^2/2} \alpha_3 (2\pi n)^{-1/2}| = o(n^{-1/2})$$

uniformly in $x \in \mathbb{R}$, where $\alpha_3 = EX_1^3$ (Bikelis [36]).

5.10.32 Suppose $E|X_1|^r < \infty$ for some integer $r \geq 3$. Then the following conditions are equivalent:

(1) $EX_1^k = \int_{-\infty}^{\infty} x^k \, d\Phi(x)$ $(1 \leq k \leq r-1)$;
(2) $\int_{-\infty}^{\infty} x^k \, d(F_n(x) - \Phi(x)) = O(n^{-(r-2)/2})$ $(1 \leq k \leq r-1)$;

(3) $\int_{-\infty}^{\infty} e^{itx} d(F_n(x) - \Phi(x)) = O(n^{-(r-2)/2})$ $(t \in \mathbb{R})$;
(4) $\int_{-\infty}^{\infty} f(x+y) d(F_n(x) - \Phi(x)) = O(n^{-(r-2)/2})$ for every function $f \in C_B^r(\mathbb{R})$ uniformly in $y \in \mathbb{R}$;
(5) $\int_{-\infty}^{\infty} f(x) d(F_n(x) - \Phi(x)) = O(n^{-(r-2)/2})$ for every $f \in C_B^r(\mathbb{R})$, where

$$C_B^r(\mathbb{R}) = \{f(x) \in C_B(\mathbb{R}): f^{(k)}(x) \in C_B(\mathbb{R}), 1 \leq k \leq r\}$$

and $C_B(\mathbb{R})$ is the set of all bounded uniformly continuous functions defined on $\mathbb{R}$ (Butzer and Hahn [58]).

Butzer and Hahn [57] and Daugavet [93, 94] obtained bounds for integrals of the type $\int_{-\infty}^{\infty} f(x) d(F_n(x) - F(x))$ for unbounded functions f, where $F_n(x) = P(a_n^{-1} \sum_{j=1}^{n} X_j < x)$, $\{X_n\}$ is a sequence of independent non-identically distributed random variables, and $F(x)$ is a d.f. limiting for $F_n(x)$. In [93] the case when $F(x) \equiv \Phi(x)$ was studied; in [57] and [94] F is a stable d.f.

Asymptotic expansions for integrals $\int_{-\infty}^{\infty} f(x) dF_n(x)$, where $f(x)$ is a smooth function and $F_n(x)$ is the d.f. of a normalized sum of i.i.d. random variables, can be found in von Bahr [463], Bhattacharya and Ranga Rao [34], Hipp [192], Götze and Hipp [149], Hall [168], and Barbour [14].

In Subsections 5.10.33 and 5.10.34 we write

$$Z_n = n^{-1/2} \sum_{j=1}^{n} X_j, \quad \kappa_p = \int_{-\infty}^{\infty} |x|^p d\Phi(x) = \pi^{-1/2} 2^{p/2} \Gamma((p+1)/2) \quad (p > 0).$$

5.10.33 Suppose $E|X_1|^p < \infty$ for some $p > 2$. Then $|E|Z_n|^p - \kappa_p| \leq Cn^{-b}$ for every n, where $b = \frac{1}{2}\min(1, p-2)$ and C is a positive constant (von Bahr [463]).

5.10.34 Suppose $E|X_1|^p < \infty$ for some positive $p < 4$, $p \neq 2$. We write $q = \min(1, (p+1)/2)$,

$$\delta_n(p) = n^{-1} EX_1^4 I(|X_1| < n^{1/2}) + EX_1^2 I(|X_1| \geq n^{1/2}) \quad \text{if } 0 < p < 2,$$

$$\delta_n(p) = n^{-1} EX_1^4 I(|X_1| < n^{1/2}) + n^{1-p/2} E|X_1|^p I(|X_1| \geq n^{1/2}) \quad \text{if } 2 < p < 4.$$

Then

$$\limsup |E|Z_n|^p - \kappa_p|/\{\delta_n(p) + n^{-q}\} < \infty$$

and

$$\liminf \{|E|Z_n|^p - \kappa_p| + n^{-q}\}/\delta_n(p) > 0.$$

A consequence of this result is the equivalence of the conditions $E|X_1|^{2+2t} < \infty$ and $\sum_{n=1}^{\infty} n^{-1+t}|E|Z_n|^p - \kappa_p| < \infty$, where $0 < p < 2$ and $0 < t < q$ (Hall [162]).

5.10.35 Suppose $Ee^{hX_1} < \infty$ for $|h| < h_0$ and some positive h_0. Then there exist positive constants T, γ, and $n_0 \in \mathbb{N}$ such that

$$|F_n^{-1}(y) - \Phi^{-1}(y)| \leq Tn^{-1/2}(1 + (\Phi^{-1}(y))^2)$$

for $n \geq n_0$ and y satisfying the condition

$$(\gamma n^{1/6}\sqrt{2\pi})^{-1} e^{-\gamma^2 n^{1/3}/2} \leq y \leq 1 - (\gamma n^{1/6}\sqrt{2\pi})^{-1} e^{-\gamma^2 n^{1/3}/2}.$$

Here $F^{-1}(y)$ denotes the inverse of $F(y)$ (Khatuntseva [219]). A generalization and strengthening of this result was obtained by Bondarko [45].

5.10.36 Let $\{X_n\}$ be a sequence of independent random variables with a common d.f. that belongs to the domain of attraction of a normal law. Define

$$S_n = \sum_{j=1}^{n} X_j, \quad G(x) = EX_1^2 I(|X_1| < x), \quad \varphi(x) = (2\pi)^{-1/2} e^{-x^2/2},$$

$$a(x) = \sup\{a: a^{-2}G(a)x \geq 1\}, \quad c_n = a(n), \quad \mu_n = EX_1 I(|X_1| < c_n),$$

$$\delta_n = nP(|X_1| \geq c_n) + nc_n^{-4} EX_1^4 I(|X_1| < c_n) + nc_n^{-3}|EX_1^3 I(|X_1| < c_n)|,$$

$$L_n = n\,\mathrm{E}\{\Phi(x - X_1/c_n) - \Phi(x)\} + n\varphi(x)\mu_n/c_n - \varphi'(x)/2,$$

$$D_n(c, d) = \sup_x |P(S_n < cx + d) - \Phi(x)|.$$

Then

$$\liminf \left\{ \inf_{c > 0, d} D_n(c, d) + 1/n \right\} \bigg/ \delta_n > 0$$

and

$$\sup_x |P(S_n < c_n x + n\mu_n) - \Phi(x) - L_n(x)| = O(\delta_n^2 + 1/c_n). \quad (5.93)$$

If the random variable X_1 has a non-lattice distribution, then the right-hand side of (5.93) can be replaced by $O(\delta_n^2) + o(1/c_n)$. Furthermore, $\sup_x |L_n(x)| \leq A\delta_n$ for every n and

$$\liminf \left\{ \sup_x |L_n(x)| \right\} \bigg/ \delta_n > 0$$

(Hall [165]). An approach to rates of convergence connected with the leading term $L_n(x)$ was developed by Hall [163, 169]. This approach was used by Hall and Nakata [173] to derive upper and lower non-uniform bounds for the remainder in Chebyshev–Edgeworth–Cramér asymptotic expansions after the leading term has been accounted for. These upper and lower bounds have the same order of magnitude.

6

Strong limit theorems: the strong law of large numbers

6.1 The Borel–Cantelli lemma

We shall consider a sequence of sets $\{A_n; n = 1, 2, \ldots\}$, each A being a subset of some non-empty set Ω. The upper and lower limits of the sequence $\{A_n\}$ are denoted by lim sup A_n and lim inf A_n respectively, and are defined by the equalities

$$\limsup A_n = \bigcap_{n=1}^{\infty} \bigcup_{k=n}^{\infty} A_k, \quad \liminf A_n = \bigcup_{n=1}^{\infty} \bigcap_{k=n}^{\infty} A_k.$$

Obviously, lim sup A_n is the set of points of Ω that belong to infinitely many A_n, and lim inf A_n is the set of points of Ω that belong to all but a finite number of the A_n. It is also obvious that lim inf $A_n \subset$ lim sup A_n.

If the sets lim inf A_n and lim sup A_n coincide, we write lim A_n = lim inf A_n = lim sup A_n. The set lim A_n is called the limit of the sequence of sets A_n.

Let $(\Omega, \mathscr{A}, P)$ be a probability space, $A_n \in \mathscr{A}$ $(n = 1, 2, \ldots)$. Instead of lim sup A_n we shall sometimes use the abbreviation A_n, i.o.; we shall say that the events A_n occur infinitely often.

Lemma 6.1 (the Borel–Cantelli lemma) *If $\sum_{n=1}^{\infty} P(A_n) < \infty$, then $P(\limsup A_n) = 0$. If $\sum_{n=1}^{\infty} P(A_n) = \infty$ and if $\{A_n\}$ is a sequence of pairwise independent events, then $P(\limsup A_n) = 1$.*

The second part of this lemma was found by Erdös and Rényi. In a traditional formulation of the Borel–Cantelli lemma the condition of pairwise independence is replaced by the stronger condition of the mutual independence of the events $A_1, \ldots, A_n$ for every n.

Proof The first part of the lemma follows from the relations

$$P(\limsup A_n) = P\left(\bigcap_{n=1}^{\infty} \bigcup_{k=n}^{\infty} A_k\right) \leqslant P\left(\bigcup_{k=n}^{\infty} A_k\right) \leqslant \sum_{k=n}^{\infty} P(A_k) \to 0,$$

since the series $\sum_{n=1}^{\infty} P(A_n)$ converges. In order to prove the second part, we need the following proposition.

Lemma 6.2 If $\sum_{n=1}^{\infty} P(A_n) = \infty$ and

$$\liminf \sum_{k=1}^{n} \sum_{l=1}^{n} P(A_k A_l) \Big/ \Big(\sum_{k=1}^{n} P(A_k) \Big)^2 = 1, \qquad (6.1)$$

then $P(\limsup A_n) = 1$.

Proof Let $I_n = I(A_n)$ denote the indicator of the event A_n. We have $EI_n = P(A_n)$ and

$$P\Big(\Big| \sum_{k=1}^{n} I_k - \sum_{k=1}^{n} P(A_k) \Big| \geq \tfrac{1}{2} \sum_{k=1}^{n} P(A_k) \Big) \leq \frac{4 \operatorname{Var}(\sum_{k=1}^{n} I_k)}{(\sum_{k=1}^{n} P(A_k))^2},$$

by Chebyshev's inequality. Since $EI_k I_l = P(A_k A_l)$, we obtain

$$\operatorname{Var}\Big(\sum_{k=1}^{n} I_k \Big) = \sum_{k=1}^{n} \sum_{l=1}^{n} P(A_k A_l) - \Big(\sum_{k=1}^{n} P(A_k) \Big)^2.$$

These relations and (6.1) imply that

$$\liminf P\Big(\Big| \sum_{k=1}^{n} I_k - \sum_{k=1}^{n} P(A_k) \Big| \geq \tfrac{1}{2} \sum_{k=1}^{n} P(A_k) \Big) = 0.$$

Put $B_n = \{ \sum_{k=1}^{n} I_k \leq \tfrac{1}{2} \sum_{k=1}^{n} P(A_k) \}$. Then $\liminf P(B_n) = 0$. Therefore, there exists a strictly increasing sequence of positive integers $\{n_m\}$ such that $\sum_{m=1}^{\infty} P(B_{n_m}) < \infty$. By the first part of the Borel–Cantelli lemma,

$$\sum_{k=1}^{n_m} I_k > \tfrac{1}{2} \sum_{k=1}^{n_m} P(A_k)$$

with probability 1 for all m except for a finite set. The series $\sum_{n=1}^{\infty} P(A_n)$ diverges by the hypothesis, hence $\sum_{k=1}^{\infty} I_k = \infty$ with probability 1. The assertion of the lemma follows. $\square$

The second part of the Borel–Cantelli lemma is a consequence of Lemma 6.2 and the equality

$$\sum_{k=1}^{n} \sum_{l=1}^{n} P(A_k A_l) = \Big(\sum_{k=1}^{n} P(A_k) \Big)^2 + \sum_{k=1}^{n} P(A_k)(1 - P(A_k)), \qquad (6.2)$$

which is valid for pairwise independent events $A_1, \ldots, A_n$.

We can replace the condition of pairwise independence in the second part of the Borel–Cantelli lemma by the weaker condition $P(A_k A_l) \leq P(A_k) P(A_l)$ for $k \neq l$, since these inequalities imply (6.2) with the replacement of the sign '=' by the sign '$\leq$'.

Let $(\Omega, \mathscr{A}, P)$ be a probability space, $A_n \in \mathscr{A}$ ($n = 1, 2, \ldots$).

Lemma 6.3 $P(A_n \text{ i.o.}) = 1$ if and only if $\sum_{n=1}^{\infty} P(A_n B) = \infty$ for every $B \in \mathscr{A}$ such that $P(B) > 0$.

$P(A_n \text{ i.o.}) = 0$ if and only if for every $\varepsilon > 0$ there exists an event $B \in \mathscr{A}$ such that $P(B) > 1 - \varepsilon$ and $\sum_{n=1}^{\infty} P(A_n B) < \infty$.

We shall prove a more general proposition.

Lemma 6.4 Let $0 \leq p \leq 1$. The following conditions are equivalent:
(a) $P(A_n \text{ i.o.}) = p$;
(b) $\sum_{n=1}^{\infty} P(A_n B) = \infty$ for every event $B \in \mathscr{A}$ with $P(B) > 1 - p$; for every $\varepsilon > 0$ there exists an event $B \in \mathscr{A}$ such that $P(B) > 1 - p - \varepsilon$ and $\sum_{n=1}^{\infty} P(A_n B) < \infty$;
(c) for every $B \in \mathscr{A}$ with $P(B) > 1 - p$ the sequence $\{P(A_n B)\}$ contains infinitely many positive numbers; for every $\varepsilon > 0$ there exists an event $B \in \mathscr{A}$ such that $P(B) > 1 - p - \varepsilon$ and $P(A_n B) = 0$ for all sufficiently large n.

Lemma 6.3 is an immediate consequence of Lemma 6.4. In turn, Lemma 6.4 is a consequence of the following proposition.

Lemma 6.5 Let $0 < p \leq 1$. The following conditions are equivalent:
(d) $P(A_n \text{ i.o.}) \geq p$;
(e) $\sum_{n=1}^{\infty} P(A_n B) = \infty$ for every event $B \in \mathscr{A}$ with $P(B) > 1 - p$;
(f) for every $B \in \mathscr{A}$ with $P(B) > 1 - p$ the sequence $\{P(A_n B)\}$ contains infinitely many positive terms.

Proof It is clear that (e) implies (f). We shall show that (d) implies (e). Suppose that $P(A_n \text{ i.o.}) \geq p$ and there exists $B \in \mathscr{A}$ such that $\sum_{n=1}^{\infty} P(A_n B) < \infty$ and $P(B) > 1 - p$. For arbitrary events U and V we have $P(U \cup V) = P(U) + P(V) - P(UV)$, hence $P(U) + P(V) - 1 \leq P(UV)$. Therefore,

$$P(A_n \text{ i.o.}) + P(B) - 1 \leq P(\{A_n \text{ i.o.}\} \cap B) = P(A_n B \text{ i.o.}) = 0,$$

by the Borel–Cantelli lemma. Accordingly, $P(B) \leq 1 - p$. This contradicts our hypothesis. Thus (d) implies (e).

To complete the proof, we need only show that (f) implies (d). Suppose that (f) is satisfied and $P(A_n \text{ i.o.}) < p$, contrary to (d). The inequality $P(A_n \text{ i.o.}) < p$ implies the existence of an integer n_0 such that $P(\bigcup_{k=n_0}^{\infty} A_k) < p$. In fact, if we put $D_n = \bigcup_{k=n}^{\infty} A_k$, then the inequality $P(D_n) \geq p$ for every n together with the relation $D_1 \supset D_2 \supset D_3 \supset \cdots$ implies that

$$\lim P(D_n) = P\left(\bigcap_{n=1}^{\infty} D_n\right) = P\left(\bigcap_{n=1}^{\infty} \bigcup_{k=n}^{\infty} A_k\right) = P(A_n \text{ i.o.}) \geq p,$$

contrary to the condition $P(A_n \text{ i.o.}) < p$. We put $B = \Omega \setminus D_{n_0}$. Then $P(B) > 1 - p$ and $BD_{n_0} = \phi$, hence $P(BA_k) = 0$ for all $k \geq n_0$. This contradicts (f). $\square$

We now introduce a useful characteristic of a random variable. Let X be a random variable such that $P(X = 0) < 1$ and $E|X| < \infty$. We put

$$\Pi(X) = (EX)^2/EX^2. \tag{6.3}$$

Obviously, $\Pi(CX) = \Pi(X)$, if $C \neq 0$ is a constant. Further, $\Pi(X) \leq 1$, by Cauchy's inequality. A strengthening of Lyapunov's inequality given by Theorem 1.2 (Chapter 1) implies that

$$\Pi(X) \leq P(X \neq 0). \tag{6.4}$$

If X and Y are non-negative variables with finite moments of the first order, then

$$\Pi(X + Y) \geq \min\{\Pi(X), \Pi(Y)\}. \tag{6.5}$$

In fact,

$$\Pi(X + Y) = (EX + EY)^2/(EX^2 + EY^2 + 2\,EX\,EY)$$

$$\geq (EX + EY)^2/(EX^2 + EY^2 + 2(EX^2\,EY^2)^{1/2})$$

$$= (EX + EY)^2/((EX^2)^{1/2} + (EY^2)^{1/2})^2)$$

$$\geq (\min\{EX/(EX^2)^{1/2}, EY/(EY^2)^{1/2}\})^2,$$

since $(a + b)/(c + d) \geq \min\{a/c, b/d\}$ for arbitrary positive numbers a, b, c, and d. We note the following consequence of this property: if $X_1, \ldots, X_n$ are non-negative random variables and $\Pi(X_k) \geq p$ $(k = 1, \ldots, n)$, then $\Pi(X_1 + \cdots + X_n) \geq p$.

If $X_1, \ldots, X_n$ are pairwise independent random variables having only two values 0 and 1, and if $EX_1 + \cdots + EX_n \geq M$, then

$$\Pi(X_1 + \cdots + X_n) \geq 1/(1 + 1/M). \tag{6.6}$$

It follows that $\Pi(X_1 + \cdots + X_n) \to 1$ if $\sum_{i=1}^{\infty} EX_i = \infty$.

In order to prove this proposition, we put $E_n = EX_1 + \cdots + EX_n$. We have $\Pi(X_1 + \cdots + X_n) = E_n^2/E(X_1 + \cdots + X_n)^2$. By the hypotheses, $EX_k = EX_k^2 > 0$ for all k and

$$E(X_1 + \cdots + X_n)^2 = \sum_{k=1}^{n} EX_k^2 + \sum_{i \neq j} EX_i\,EX_j$$

$$\leq \sum_{k=1}^{n} EX_k + \sum_{i,j=1}^{n} EX_i\,EX_j = E_n + E_n^2.$$

Therefore, $\Pi(X_1 + \cdots + X_n) \geq E_n^2/(E_n + E_n^2) = 1/(1 + 1/E_n) \geq 1/(1 + 1/M)$.

We shall apply some properties of $\Pi(X)$ to the proof of the following proposition.

Lemma 6.6 *If $\sum_{n=1}^{\infty} P(A_n) = \infty$ and*

$$\limsup \left(\sum_{k=1}^{n} P(A_k)\right)^2 \bigg/ \sum_{k=1}^{n}\sum_{l=1}^{n} P(A_k A_l) \geq p, \tag{6.7}$$

where $0 \leq p \leq 1$, then $P(A_n \text{ i.o.}) \geq p$.

Proof Let X_n be the indicator of the event A_n. We put $S_n = \sum_{k=1}^{n} X_k$. Obviously, $EX_n = P(A_n)$, $ES_n = \sum_{k=1}^{n} P(A_k)$, and $ES_n^2 = \sum_{k=1}^{n}\sum_{l=1}^{n} P(A_k A_l)$. Thus

$$\Pi(S_n) = (ES_n)^2/ES_n^2 = \left(\sum_{k=1}^{n} P(A_k)\right)^2 \bigg/ \sum_{k=1}^{n}\sum_{l=1}^{n} P(A_k A_l). \tag{6.8}$$

Therefore, (6.7) is equivalent to the condition $\limsup \Pi(S_n) \geq p$. It follows that $\limsup \Pi(S_n - S_k) \geq p$ for every positive integer k. If $l > k$, then

$$P\left(\bigcup_{n=k}^{\infty} A_n\right) \geq P\left(\bigcup_{n=k}^{l} A_n\right) \geq P(S_l - S_{k-1} \neq 0) \geq \Pi(S_l - S_{k-1}), \tag{6.9}$$

by the inequality (6.4). Passing to the limit as $l \to \infty$, we obtain

$$P\left(\bigcup_{n=k}^{\infty} A_n\right) \geq \limsup_{l \to \infty} \Pi(S_l - S_{k-1}) \geq p.$$

Hence

$$P(A_n \text{ i.o.}) = \lim_{k \to \infty} P\left(\bigcup_{n=k}^{\infty} A_n\right) \geq p. \qquad \square$$

Taking account of (6.9), we find that

$$P\left(\bigcup_{k=1}^{n} A_k\right) \geq P(S_n \neq 0) \geq \Pi(S_n).$$

Therefore,

$$P\left(\bigcup_{k=1}^{n} A_k\right) \geq \left(\sum_{k=1}^{n} P(A_k)\right)^2 \bigg/ \sum_{k=1}^{n}\sum_{l=1}^{n} P(A_k A_l), \tag{6.10}$$

by (6.8). This inequality is closely related to Bonferroni's inequality:

$$P\left(\bigcup_{k=1}^{n} A_k\right) \geq \sum_{k=1}^{n} P(A_k) - \sum_{k=1}^{n}\sum_{l=1}^{n} P(A_k A_l).$$

Lemma 6.2 is an immediate consequence of Lemma 6.6 for $p = 1$. The following proposition is another corollary of Lemma 6.6.

Lemma 6.7 *Let $\{A_n\}$ be a sequence of events such that $\sum_{n=1}^{\infty} P(A_n) = \infty$ and*

$$\liminf \sum_{k=1}^{n} \sum_{l=1}^{n} P(A_k A_l) \bigg/ \left(\sum_{k=1}^{n} P(A_k) \right)^2 = L.$$

Then

$$P(\limsup A_n) \geq 1/L.$$

6.2 Convergence of series of independent random variables

Let $X, X_1, X_2, \ldots$ be a sequence of random variables defined on a probability space $(\Omega, \mathscr{A}, P)$. The sequence $\{X_n\}$ converges to X almost surely (a.s.) if $X_n(\omega) \to X(\omega)$ everywhere except on a set of points $\omega \in \Omega$ of P-measure zero. Then we write $X_n \to X$ a.s.

Lemma 6.8 *The following conditions are equivalent:*
(A) $X_n \to X$ a.s.;
(B) $P(|X_n - X| \geq \varepsilon \text{ i.o.}) = 0$ for every $\varepsilon > 0$;
(C) $P\left(\bigcup_{m=n}^{\infty} \{|X_m - X| \geq \varepsilon\} \right) \to 0$ for every $\varepsilon > 0$;
(D) $P\left(\sup_{m \geq n} |X_m - X| \geq \varepsilon \right) \to 0$ for every $\varepsilon > 0$.

Proof $X_n \to X$ a.s. if and only if $|X_n(\omega) - X(\omega)| < \varepsilon$ for any $\varepsilon > 0$ at all points ω of Ω except for a set of P-measure zero, provided that $n > n_0(\varepsilon, \omega)$, i.e.

$$P\left(\bigcup_{n=1}^{\infty} \bigcap_{k=0}^{\infty} \{|X_{n+k} - X| < \varepsilon\} \right) = 1$$

for every $\varepsilon > 0$. This is equivalent to the condition that

$$P\left(\bigcap_{n=1}^{\infty} \bigcup_{k=0}^{\infty} \{|X_{n+k} - X| \geq \varepsilon\} \right) = 0,$$

or

$$P(\limsup \{|X_n - X| \geq \varepsilon\}) = 0$$

for every $\varepsilon > 0$. Thus (A) is equivalent to (B).

The sequence of the sets $E_n = \bigcup_{m=n}^{\infty} \{|X_m - X| \geq \varepsilon\}$ is such that $E_1 \supset E_2 \supset \cdots$. Therefore, $\lim P(E_n) = P(\bigcap_{k=1}^{\infty} E_k)$, so that (B) and (C) are equivalent. But (C) and (D) are obviously equivalent, and the lemma is proved. □

Lemma 6.9 *Let $\{X_n\}$ be a sequence of independent random variables. If $\sum_{n=1}^{\infty} \text{Var } X_n < \infty$, then the series $\sum_{n=1}^{\infty} (X_n - EX_n)$ converges a.s.*

Proof We put $S_n = \sum_{k=1}^{n} X_k$. We may assume that $EX_n = 0$ for all n. Let ε be an arbitrary positive number. We have

$$P\left(\sup_{k,n \geq m} |S_k - S_n| > 2\varepsilon\right) \leq P\left(\left\{\sup_{k \geq m} |S_k - S_m| > \varepsilon\right\} \cup \left\{\sup_{n \geq m} |S_n - S_m| > \varepsilon\right\}\right)$$

$$\leq 2P(\sup_{n \geq m} |S_n - S_m| > \varepsilon) \leq 2P\left(\bigcup_{n=m}^{\infty} \{|S_n - S_m| > \varepsilon\}\right)$$

$$= 2 \lim_{M \to \infty} P\left(\bigcup_{n=m}^{M} \{|S_n - S_m| > \varepsilon\}\right)$$

$$= 2 \lim_{M \to \infty} P\left(\max_{m \leq n \leq M} |S_n - S_m| > \varepsilon\right).$$

In order to estimate the last probability we apply Kolmogorov's inequality (2.17) (Chapter 2). Then we obtain

$$P\left(\sup_{k,n \geq m} |S_k - S_n| > 2\varepsilon\right) \leq \frac{2}{\varepsilon^2} \lim_{M \to \infty} \sum_{k=m}^{M} \operatorname{Var} X_k = \frac{2}{\varepsilon^2} \sum_{k=m}^{\infty} \operatorname{Var} X_k \to 0$$

as $m \to \infty$, since the series $\sum \operatorname{Var} X_k$ converges. Furthermore,

$$P(\limsup S_n - \liminf S_n > 2\varepsilon) \leq P\left(\sup_{k,n \geq m} |S_k - S_n| > \varepsilon\right)$$

for every m. Passing to the limit as $m \to \infty$, we conclude that

$$P(\limsup S_n - \liminf S_n > 2\varepsilon) = 0, \quad \text{or} \quad P(\limsup S_n - \liminf S_n \leq 2\varepsilon) = 1$$

for every $\varepsilon > 0$. It follows that $\limsup S_n = \liminf S_n$ with probability 1, i.e. there exists $\lim S_n$ almost surely. This limit is finite, since the previous estimates for $m = 0$ and $X_0 = 0$ imply that

$$P\left(\sup_{n \geq 1} |S_n| > 2\varepsilon\right) \leq \frac{2}{\varepsilon^2} \sum_{k=1}^{\infty} \operatorname{Var} X_k$$

for every $\varepsilon > 0$ and

$$P\left(\sup_{n \geq 1} |S_n| < \infty\right) = \lim_{\varepsilon \to \infty} P\left(\sup_{n \geq 1} |S_n| \leq 2\varepsilon\right) = 1. \quad \square$$

If X is a random variable and c is a positive constant, we write

$$X^c = \begin{cases} X, & \text{if } |X| < c, \\ 0, & \text{if } |X| \geq c. \end{cases}$$

Theorem 6.1 (the three-series theorem) Let $\{X_n; n = 1, 2, \ldots\}$ be a sequence of independent random variables. In order that the series $\sum_{n=1}^{\infty} X_n$ converge almost surely, it is necessary that for every $c > 0$ the three series $\sum_{n=1}^{\infty} \mathrm{E} X_n^c$, $\sum_{n=1}^{\infty} \mathrm{Var}\, X_n^c$, and $\sum_{n=1}^{\infty} P(|X_n| \geq c)$ converge, and it is sufficient that they converge for some $c > 0$.

Proof We first prove the sufficiency. Since the series $\sum P(|X_n| \geq c)$ converges for some $c > 0$, we find from the Borel–Cantelli lemma that $X_n = X_n^c$ for all sufficiently large n with probability 1. Thus we need only prove that the series $\sum X_n^c$ converges a.s. The condition $\sum \mathrm{Var}\, X_n^c < \infty$ and Lemma 6.9 imply that the series $\sum (X_n^c - \mathrm{E} X_n^c)$ converges a.s. Taking into account the condition $\sum \mathrm{E} X_n^c < \infty$, we arrive at the desired conclusion.

We now prove the necessity. If the series $\sum X_n$ converges a.s., then $X_n \to 0$ a.s. and, therefore, $X_n = X_n^c$ for every $c > 0$ and all sufficiently large n. By the second part of the Borel–Cantelli lemma we have $\sum P(|X_n| \geq c) < \infty$. We shall introduce the sequence of the symmetrized random variables $Z_n = X_n^c - Y_n^c$, where X_n^c and Y_n^c are independent random variables with a common distribution. Obviously, $|Z_n| \leq 2c$, $\mathrm{E} Z_n = 0$, and $\mathrm{Var}\, Z_n = 2\,\mathrm{Var}\, X_n^c$. Let $f_n(t)$ be the c.f. of Z_n. The series $\sum X_n^c$ and $\sum Z_n$ converge a.s., therefore, $\prod_{k=1}^{n} f_k(t) \to f(t)$, where $f(t)$ is the c.f. of the symmetric random variable $S = \sum_{n=1}^{\infty} Z_n$. If $0 \leq tc \leq \tfrac{1}{2}$, we obtain

$$f_k(t) = 1 + \sum_{m=2}^{\infty} \frac{(it)^m}{m!} \mathrm{E} Z_k^m \leq 1 - \frac{t^2}{2} \mathrm{E} Z_k^2 \left(1 - \frac{(2tc)^2}{3 \cdot 4} - \cdots \right) \leq 1 - \frac{t^2}{6} \mathrm{E} Z_k^2$$

for every positive integer k. If t is a sufficiently small positive number, then

$$\prod_{k=1}^{n} \left(1 - \frac{t^2}{6} \mathrm{E} Z_k^2 \right) \geq \tfrac{1}{2} f(t) > 0$$

for all sufficiently large n. Accordingly, the series $\sum \mathrm{E} Z_k^2$ converges. It follows that $\sum \mathrm{Var}\, X_n^c < \infty$. By Lemma 6.9, the series $\sum (X_n^c - \mathrm{E} X_n^c)$ converges a.s. Since the series $\sum X_n^c$ also converges a.s., we arrive at the conclusion that the series $\sum \mathrm{E} X_n^c$ converges. $\square$

We shall formulate a simple consequence of Theorem 6.1.

Theorem 6.2 *Let $\{X_n\}$ be a sequence of independent random variables. In order that the series $\sum_{n=1}^{\infty} X_n$ converge absolutely almost surely, it is necessary and sufficient that the series $\sum_{n=1}^{\infty} P(|X_n| \geq c)$ and $\sum_{n=1}^{\infty} \mathrm{E}|X_n^c|$ converge for some $c > 0$ (for every $c > 0$).*

Proof We need only show that if the series $\sum P(|X_n| \geq c)$ and $\sum \mathrm{E}|X_n^c|$ converge, then the series $\sum \mathrm{Var}\, X_n^c$ is also convergent. We have

$$\mathrm{Var}\, X_n^c \leq \int_{|x| \leq c} x^2\, dV_n(x) \leq c \int_{|x| \leq c} |x|\, dV_n(x) = c\, \mathrm{E}|X_n^c|,$$

where $V_n(x)$ is the d.f. of X_n. The desired assertion follows. $\square$

In turn, Theorem 6.2 implies the following.

Theorem 6.3 *If $\{X_n\}$ is a sequence of independent random variables and $\sum_{n=1}^{\infty} E|X_n|^p < \infty$ for some positive $p \leqslant 1$, then the series $\sum_{n=1}^{\infty} X_n$ converges absolutely almost surely.*

The following theorem will be useful in the next sections. Its proof is based on Theorem 6.1.

Theorem 6.4 *Let $\{g_n(x); n = 1, 2, \ldots\}$ be a sequence of even functions, positive and non-decreasing in the interval $x > 0$. Let $\{X_n; n = 1, 2, \ldots\}$ be a sequence of independent random variables. Suppose that for every n at least one of the following conditions is satisfied:*

(a) $x/g_n(x)$ *does not decrease in the interval* $x > 0$;
(b) $x/g_n(x)$ *and* $g_n(x)/x^2$ *do not increase in the interval* $x > 0$, *and also* $EX_n = 0$;
(c) $g_n(x)/x^2$ *does not increase in the interval* $x > 0$, *and moreover*, X_n *has a symmetric distribution.*

Suppose further that $\{a_n\}$ *is a sequence of positive numbers. If*

$$\sum_{n=1}^{\infty} \frac{Eg_n(X_n)}{g_n(a_n)} < \infty, \tag{6.11}$$

then the series $\sum_{n=1}^{\infty} X_n/a_n$ *converges a.s.*

Proof Let $V_n(x)$ be the d.f. of X_n, and let

$$Y_n = \begin{cases} X_n, & \text{if } |X_n| < a_n, \\ 0, & \text{if } |X_n| \geqslant a_n. \end{cases}$$

By Theorem 6.1 it suffices to prove convergence of the corresponding series for the sequence of the random variables $\{X_n/a_n\}$ and $c = 1$, i.e. convergence of the series $\sum P(|X_n| \geqslant a_n)$, $\sum EY_n/a_n$, and $\sum EY_n^2/a_n^2$. We have

$$P(|X_n| \geqslant a_n) \leqslant \int_{|x| \geqslant a_n} \frac{g_n(x)}{g_n(a_n)} \, dV_n(x) \leqslant \frac{Eg_n(X_n)}{g_n(a_n)},$$

since the functions $g_n(x)$ do not decrease. Therefore, (6.11) implies that

$$\sum P(|X_n| \geqslant a_n) < \infty. \tag{6.12}$$

Suppose that the function $g_n(x)$ satisfies condition (a). Then in the interval $|x| < a_n$ we have

$$\frac{x^2}{a_n^2} \leqslant \frac{g_n^2(x)}{g_n^2(a_n)} \leqslant \frac{g_n(x)}{g_n(a_n)}.$$

If condition (b) is satisfied, then in the same interval we obtain $x^2/g_n(x) \leq a_n^2/g_n(a_n)$. This is also true under condition (c). Thus the inequality $x^2/a_n^2 \leq g_n(x)/g_n(a_n)$ holds in the interval $|x| < a_n$ for all n. Hence

$$EY_n^2 = \int_{|x|<a_n} x^2\, dV_n(x) \leq \frac{a_n^2}{g_n(a_n)} \int_{|x|<a_n} g_n(x)\, dV_n(x) \leq \frac{a_n^2}{g_n(a_n)} Eg_n(X_n)$$

and (6.11) implies that

$$\sum EY_n^2/a_n < \infty. \tag{6.13}$$

Furthermore,

$$|EY_n| = \left|\int_{|x|<a_n} x\, dV_n(x)\right| \leq \frac{a_n}{g_n(a_n)} Eg_n(X_n),$$

if condition (a) is satisfied. If (b) is satisfied,

$$|EY_n| = \left|\int_{|x|\geq a_n} x\, dV_n(x)\right| \leq \int_{|x|\geq a_n} |x|\, dV_n(x)$$

$$\leq \frac{a_n}{g_n(a_n)} \int_{|x|\geq a_n} g_n(x)\, dV_n(x) \leq \frac{a_n}{g_n(a_n)} Eg_n(X_n).$$

In the case when (c) holds, we have $EY_n = 0$. Therefore,

$$\sum |EY_n|/a_n < \infty \tag{6.14}$$

Taking account of (6.12)–(6.14), we arrive at the desired assertion. □

6.3 The strong law of large numbers

Let us begin with some simple corollaries of Theorem 6.4. We shall need the following lemmas.

Lemma 6.10 *Let $\{b_n\}$ and $\{x_n\}$ be sequences of real numbers such that $a_n = \sum_{k=1}^n b_k \uparrow \infty$, $x_n \to x$, $|x| < \infty$. Then $1/a_n \sum_{k=1}^n b_k x_k \to x$.*

Proof Let ε be an arbitrary positive number. We have $|x_n - x| < \varepsilon$ for $n \geq n_0$ and some n_0. Therefore,

$$\left|\frac{1}{a_n}\sum_{k=1}^n b_k(x_k - x)\right| \leq \frac{1}{a_n}\sum_{k<n_0} |b_k(x_k - x)| + \varepsilon\frac{a_n + |b_1|}{a_n}$$

for $n \geq n_0$. Passing to the limit as $n \to \infty$ and using the condition $a_n \uparrow \infty$, we obtain $1/a_n \sum_{k=1}^n b_k x_k - x \to 0$. □

Lemma 6.11 (Kronecker's lemma) *If $a_n \uparrow \infty$ and the series $\sum_{n=1}^{\infty} x_n$ converges, then $1/a_n \sum_{k=1}^{n} a_k x_k \to 0$.*

Proof We put

$$s = \sum_{n=1}^{\infty} x_n, \quad s_1 = 0, \quad s_{n+1} = \sum_{k=1}^{n} x_k,$$

$$a_0 = 0, \quad b_k = a_k - a_{k-1}.$$

We have $a_n = \sum_{k=1}^{n} b_k$ and

$$\frac{1}{a_n} \sum_{k=1}^{n} a_k x_k = \frac{1}{a_n} \sum_{k=1}^{n} a_k(s_{k+1} - s_k) = s_{n+1} - \frac{1}{a_n} \sum_{k=1}^{n} b_k s_k \to 0,$$

since $s_{n+1} \to s$ and $(1/a_n) \sum_{k=1}^{n} b_k s_k \to s$ by Lemma 6.10. □

Theorem 6.5 *Let $\{X_n\}$, $\{g_n(x)\}$ and $\{a_n\}$ be sequences satisfying the conditions of Theorem 6.4, and let $a_n \uparrow \infty$. We set $S_n = \sum_{k=1}^{n} X_k$. If (6.11) holds, then $S_n/a_n \to 0$ a.s.*

This theorem is an immediate consequence of Theorem 6.4 and Lemma 6.11.

In Theorem 6.5 we choose $g_n(x) \equiv g(x)$ for all n, and we note certain important consequences arising when we set $g(x) = |x|^p$, $p > 0$.

Theorem 6.6 *Let $\{X_n\}$ be a sequence of independent random variables, and let $a_n \uparrow \infty$. If*

$$\sum_{n=1}^{\infty} \frac{E|X_n|^p}{a_n^p} < \infty \qquad (6.15)$$

for some positive $p \leq 1$, then

$$S_n/a_n \to 0 \text{ a.s.} \qquad (6.16)$$

If the condition (6.15) is satisfied for some p, $1 < p \leq 2$, and if $EX_n = 0$ for all n, then (6.16) holds.

Theorem 6.6 implies a further proposition.

Theorem 6.7 *If $a_n \uparrow \infty$ and $\sum_{n=1}^{\infty} (\text{Var } X_n/a_n^2) < \infty$, then $(S_n - ES_n)/a_n \to 0$ a.s.*

We shall show that when $\sum_{n=1}^{\infty} \text{Var } X_n/a_n^2 = \infty$, the conclusion stated in Theorem 6.7 may not hold. First we formulate an elementary proposition.

Lemma 6.12 Let $\{X_n\}$ be a sequence of pairwise independent random variables, and let $\{a_n\}$ be a sequence of positive numbers. If the relations (6.16) holds and if

$$a_{n-1}/a_n = O(1), \qquad (6.17)$$

then

$$\sum_{n=1}^{\infty} P(|X_n| \geq a_n) < \infty. \qquad (6.18)$$

Proof The relations (6.16) and (6.17) imply that

$$\frac{X_n}{a_n} = \frac{S_n}{a_n} - \frac{a_{n-1}}{a_n} \cdot \frac{S_{n-1}}{a_{n-1}} \to 0 \text{ a.s.} \qquad (6.19)$$

If the series $\sum P(|X_n| \geq a_n)$ diverges, then $P(|X_n| \geq a_n \text{ i.o.}) = 1$, by the Borel–Cantelli lemma, and (6.19) fails. □

Suppose that $a_n \uparrow \infty$ and that $\{\sigma_n^2\}$ is a sequence of non-negative numbers satisfying the condition $\sum \sigma_n^2/a_n^2 = \infty$. We consider the sequence of independent random variables $\{X_n\}$ such that

$$P(X_n = a_n) = P(X_n = -a_n) = \sigma_n^2/(2a_n^2), \quad P(X_n = 0) = 1 - \sigma_n^2/a_n^2$$

for those n for which $\sigma_n^2 \leq a_n^2$, and

$$P(X_n = a_n) = P(X_n = -a_n) = \tfrac{1}{2}$$

for the remaining n. Then $EX_n = 0$ for all n, $\text{Var } X_n = \sigma_n^2$ if $\sigma_n^2 \leq a_n^2$, and $\text{Var } X_n = a_n^2$ otherwise. Therefore, $\sum \text{Var } X_n/a_n^2 = \infty$. Suppose that $S_n/a_n \to 0$ a.s. Then by Lemma 6.12 we obtain (6.18) which contradicts the divergence of the series $\sum P(|X_n| \geq a_n)$. □

Lemma 6.13 Let $\{X_n\}$ be a sequence of independent random variables, and let $\{a_n\}$ be a sequence of positive numbers such that $a_n \uparrow \infty$. We put

$$Z_n = \begin{cases} X_n, & \text{if } |X_n| < a_n, \\ 0, & \text{if } |X_n| \geq a_n. \end{cases}$$

If

$$\sum_{n=1}^{\infty} E \frac{X_n^2}{a_n^2 + X_n^2} < \infty, \qquad (6.20)$$

then

$$\frac{1}{a_n} \sum_{k=1}^{n} (X_k - EZ_k) \to 0 \text{ a.s.}$$

Proof It follows from Lemma 4.6 (Chapter 4) that (6.20) is equivalent to the pair of conditions (6.18) and

$$\sum_{n=1}^{\infty} \frac{1}{a_n^2} \int_{|x|<a_n} x^2 \, dV_n(x) < \infty, \qquad (6.21)$$

where $V_n(x)$ is the d.f. of X_n. Lemma 6.9 and (6.21) imply that the series $\sum (Z_n - EZ_n)/a_n$ converges a.s. and, therefore, $\sum_{k=1}^{n} (Z_k - EZ_k)/a_n \to 0$ a.s. We have $\sum P(X_n \neq Z_n) < \infty$ by (6.18), and accordingly $P(X_n \neq Z_n \text{ i.o.}) = 0$, by the Borel–Cantelli lemma. Hence we conclude that the assertion of the lemma holds. □

Theorem 6.8 *Let $\{X_n\}$ be a sequence of independent of random variables, and let $\{a_n\}$ be a sequence of positive numbers such that $a_n \uparrow \infty$. Let the condition (6.21) be satisfied. Then the relation $1/a_n \sum_{k=1}^{n} X_k \to 0$ a.s. holds if and only if the conditions (6.18) and*

$$\frac{1}{a_n} \sum_{k=1}^{n} \int_{|x|<a_n} x \, dV_k(x) \to 0 \qquad (6.22)$$

are satisfied, where $V_k(x)$ is the d.f. of X_k.

Proof The necessity of the condition (6.18) follows from Lemma 6.12. The necessity of (6.22) follows from Theorem 4.13 (Chapter 4) and the fact that convergence almost surely implies convergence in probability.

By Lemma 6.13, the conditions (6.18), (6.21), and

$$\frac{1}{a_n} \sum_{k=1}^{n} \int_{|x|<a_k} x \, dV_k(x) \to 0 \qquad (6.23)$$

imply the relation $S_n/a_n \to 0$ a.s. Therefore, the sufficiency will be proved if we show that the conditions (6.18) and (6.22) imply (6.23). To prove this, we consider the sum

$$T_{m,n} = \frac{1}{a_n} \sum_{k=m}^{n} \int_{a_k \leq |x| < a_n} x \, dV_k(x).$$

We shall show that $T_{1,n} \to 0$ and, therefore, (6.22) $\Rightarrow$ (6.23). We fix N, and we put $U_n = T_{1,n} - T_{N+1,n}$ for $n > N$. By (6.18) we have for an arbitrary $\varepsilon > 0$

$$|T_{N+1,n}| \leq \sum_{k>N} P(|X_k| \geq a_k) < \varepsilon \qquad \text{if } N \text{ is large enough.}$$

Furthermore,

$$\frac{1}{a_n} \int_{|x|<a_n} |x| \, dV_k(x) = \frac{1}{a_n} \sum_{j=1}^{n} \int_{a_{j-1} \leq |x| < a_j} |x| \, dV_k(x)$$

$$\leq \frac{1}{a_n} \sum_{j=1}^{n} a_j P(a_{j-1} \leq |X_k| < a_j) \to 0$$

for every k, by Kronecker's lemma, since $\sum_{j=1}^{\infty} P(a_{j-1} \leq |X| < a_j) = 1$ for every random variable X. (We put here $a_0 = 0$.) Hence $U_n \to 0$, and we arrive at the desired assertion. □

6.4 The strong law of large numbers: the i.i.d. case

Let us turn to the strong law of large numbers for sequences of independent identically distributed random variables.

Theorem 6.9 *Let $\{X_n\}$ be a sequence of independent random variables with a common distribution function $V(x)$. Let $\{a_n\}$ be a sequence of positive numbers satisfying the conditions $a_n \uparrow \infty$ and*

$$\sum_{k=n}^{\infty} 1/a_k^2 = O(n/a_n^2). \tag{6.24}$$

In order that the relation

$$S_n/a_n \to 0 \text{ a.s.} \tag{6.25}$$

hold, it is necessary and sufficient that

$$\sum_{n=1}^{\infty} P(|X_1| \geq a_n) < \infty \tag{6.26}$$

and

$$\frac{n}{a_n} \int_{|x| < a_n} x \, dV(x) \to 0. \tag{6.27}$$

Proof The necessity of the condition (6.26) follows from Lemma 6.12. The necessity of (6.27) follows from Theorem 4.13 (Chapter 4). It remains to prove the sufficiency. We shall show that the conditions (6.26) and (6.27) imply (6.21). Putting $a_0 = 0$, we have

$$\sum_{n=1}^{\infty} \frac{1}{a_n^2} \int_{|x| < a_n} x^2 \, dV(x) \leq \sum_{n=1}^{\infty} \frac{1}{a_n^2} \sum_{k=1}^{n} a_k^2 P(a_{k-1} \leq |X_1| < a_k)$$

$$= \sum_{k=1}^{\infty} a_k^2 P(a_{k-1} \leq |X_1| < a_k) \sum_{n=k}^{\infty} \frac{1}{a_n^2}$$

$$\leq C \sum_{k=1}^{\infty} k P(a_{k-1} \leq |X_1| < a_k)$$

$$= C \sum_{k=0}^{\infty} P(|X_1| \geq a_k) < \infty,$$

by (6.26). Taking into account the condition (6.27) and making use of Theorem 6.8, we obtain (6.25). □

The condition (6.24) is satisfied in the case when $a_n = n$ and also in the more general case when $\liminf a_{2n}^2/a_n^2 > 2$. We shall consider the condition

$$\sum_{k=n}^{\infty} 1/a_k = O(n/a_n), \qquad (6.28)$$

which is stronger than (6.24).

Theorem 6.10 *Let $\{X_n\}$ be a sequence of identically distributed random variables, and let $\{a_n\}$ be a sequence of positive numbers satisfying the conditions $a_n \uparrow \infty$ and (6.28). Then the condition (6.26) is sufficient for the relation (6.25). If moreover the random variables $X_1, X_2, \ldots$ are pairwise independent, then (6.26) is necessary for (6.25).*

Proof First, we shall prove the sufficiency. If the conditions (6.26) and (6.28) hold, then

$$\sum_{n=1}^{\infty} \frac{1}{a_n} \int_{|x| < a_n} |x| \, dV(x) \leq \sum_{n=1}^{\infty} \frac{1}{a_n} \sum_{k=1}^{n} a_k P(a_{k-1} \leq |X_1| < a_k)$$

$$= \sum_{n=1}^{\infty} a_k P(a_{k-1} \leq |X_1| < a_k) \sum_{n=k}^{\infty} \frac{1}{a_n}$$

$$\leq C \sum_{k=1}^{\infty} k P(a_{k-1} \leq |X_1| < a_k)$$

$$= C \sum_{k=1}^{\infty} P(|X_1| \geq a_{k-1}) < \infty,$$

where $V(x) = P(X_1 < x)$ and $a_0 = 0$. Therefore, the series

$$\sum_{n=1}^{\infty} \frac{1}{a_n} E|X_n| I(|X_n| < a_n)$$

converges, and the series $\sum (1/a_n)|X_n|I(|X_n| < a_n)$ converges a.s. Making use of Kronecker's lemma, we obtain

$$\frac{1}{a_n} \sum_{k=1}^{n} |X_k| I(|X_k| < a_k) \to 0 \text{ a.s.} \qquad (6.29)$$

By the condition (6.26) and the Borel–Cantelli lemma, the series $\sum I(|X_n| \geq a_n)$ converges a.s. and, therefore, $\sum (1/a_n)|X_n|I(|X_n| \geq a_n) < \infty$ a.s. Accordingly, (6.29) implies (6.25). Thus the conditions (6.28) and (6.26) are sufficient for (6.25).

The necessity of the condition (6.26) for (6.25) follows from Lemma 6.12 if the random variables $X_1, X_2, \ldots$ are pairwise independent; we do not refer here to the condition (6.28). □

The condition (6.28) is fulfilled if $a_n = n^p$, where $p > 1$. The sequence $a_n = n$ satisfying the condition (6.24) does not satisfy (6.28). For this sequence the implication (6.26) ⇒ (6.25) may fail without the independence condition. The corresponding example is provided by the sequence $\{X_n\}$ such that $X_n = X_1$ for $n \geq 1$ and $0 < E|X_1| < \infty$.

Lemma 6.14 *Let X be an arbitrary random variable. Then*

$$\sum_{n=1}^{\infty} P(|X| \geq n) \leq E|X| \leq 1 + \sum_{n=1}^{\infty} P(|X| \geq n).$$

Proof We put $V(x) = P(X < x)$. We have

$$\sum_{n=1}^{\infty} P(|X| \geq n) = \sum_{n=1}^{\infty} \sum_{m=n}^{\infty} P(m \leq |X| < m+1)$$

$$= \sum_{m=1}^{\infty} m P(m \leq |X| < m+1)$$

$$\leq \sum_{m=0}^{\infty} \int_{m \leq |x| < m+1} |x| \, dV(x) = E|X|.$$

Furthermore,

$$E|X| \leq \sum_{m=0}^{\infty} (m+1) P(m \leq |X| < m+1)$$

$$= 1 + \sum_{n=1}^{\infty} P(|X| \geq n). \quad \square$$

Theorem 6.11 *Let $\{X_n\}$ be a sequence of pairwise independent identically distributed random variables. There exists a constant b such that*

$$S_n/n \to b \text{ a.s.} \tag{6.30}$$

if and only if

$$E|X_1| < \infty. \tag{6.31}$$

If (6.31) holds, then (6.30) is fulfilled with $b = EX_1$.

Proof If the relation (6.30) holds, we use Lemma 6.12 for $a_n = n$, and we obtain $\sum P(|X_1 - b| \geq n) < \infty$. Hence (6.31) holds, by Lemma 6.14. The necessity is proved.

We shall prove the sufficiency. Suppose that the condition (6.31) is satisfied. For every random variable X we put $X^+ = XI(X \geq 0)$ and $X^- = -XI(X \leq 0)$, so that $X^+ \geq 0$, $X^- \geq 0$ and $X = X^+ - X^-$. The sequences $\{X_n^+\}$ and $\{X_n^-\}$ are sequences of pairwise independent variables having finite absolute moments of the first order, in view of (6.31). If we prove the theorem for these sequences, then we prove it for the sequence $\{X_n\}$. Therefore, we may consider only the case when $X_n \geq 0$ without loss of generality.

We put $Y_n = X_n I(X_n < n)$ for positive integer n and $T_n = \sum_{k=1}^n Y_k$. Let $\beta > 1$ and $k_n = [\beta^n]$. Using the assumption about the pairwise independence and applying Chebyshev's inequality, we obtain

$$\sum_{n=1}^\infty P(|T_{k_n} - ET_{k_n}| \geq \varepsilon k_n) \leq \sum_{n=1}^\infty \operatorname{Var} T_{k_n}/(\varepsilon k_n)^2 = \sum_{n=1}^\infty (\varepsilon k_n)^{-2} \sum_{m=1}^{k_n} \operatorname{Var} Y_m$$

$$\leq C\varepsilon^{-2} \sum_{n=1}^\infty \operatorname{Var} Y_n/n^2$$

for every $\varepsilon > 0$. Since

$$EY_n^2 = \sum_{k=0}^{n-1} \int_k^{k+1} x^2 \, dV(x),$$

where $V(x)$ is the d.f. of X_1, we have

$$\sum_{n=1}^\infty P(|T_{k_n} - ET_{k_n}| \geq \varepsilon k_n) \leq C_1 \varepsilon^{-2} \sum_{k=0}^\infty \frac{1}{k+1} \int_k^{k+1} x^2 \, dV(x)$$

$$\leq C_1 \varepsilon^{-2} \sum_{k=0}^\infty \int_k^{k+1} x \, dV(x) = C_1 \varepsilon^{-2} EX_1 < \infty$$

for every $\varepsilon > 0$. The Borel–Cantelli lemma implies that $(T_{k_n} - ET_{k_n})/k_n \to 0$ a.s. Furthermore,

$$EX_1 = \lim \int_0^n x \, dV(x) = \lim EY_n = \lim ET_{k_n}/k_n.$$

Hence $T_{k_n}/k_n \to EX_1$ a.s. Making use of the definition of Y_n and Lemma 6.14, we obtain

$$\sum_{n=1}^\infty P(X_n \neq Y_n) = \sum_{n=1}^\infty P(X_n \geq n) \leq EX_1 < \infty.$$

Therefore, $P(X_n \neq Y_n \text{ i.o.}) = 0$ and $S_{k_n}/k_n \to EX_1$ a.s. The sum S_n is a non-decreasing function of n. Consequently, the latter relation implies that

$$\frac{1}{\beta} EX_1 \leq \liminf S_n/n \leq \limsup S_n/n \leq \beta EX_1 \text{ a.s.}$$

for every $\beta > 1$. Accordingly, $S_n/n \to EX_1$ a.s. $\square$

6.5 The strong law of large numbers: the necessary and sufficient conditions

We shall return to the sums S_n of independent not necessarily identically distributed random variables and arbitrary norming sequences $\{a_n\}$. So far we have been interested in conditions under which $S_n/a_n \to 0$ a.s. or $(S_n - ES_n)/a_n \to 0$ a.s. (in the case when the means exist). We now consider more general problems.

Let $\{Y_n\}$ be a sequence of random variables. We shall say that the sequence $\{Y_n\}$ is strongly stable if there exists a sequence of constants $\{b_n\}$ such that

$$Y_n - b_n \to 0 \text{ a.s.} \tag{6.32}$$

If (6.32) holds and if $\{b'_n\}$ is another sequence of constants such that $b_n - b'_n \to 0$, then $Y_n - b'_n \to 0$ a.s.

If (6.32) holds, then

$$b_n = mY_n + o(1) \tag{6.33}$$

and

$$Y_n - mY_n \to 0 \text{ a.s.,} \tag{6.34}$$

where mY_n is any median of the random variable Y_n. In fact, (6.32) implies that $P(|Y_n - b_n| < \varepsilon) \to 1$ for every $\varepsilon > 0$ and, therefore, $P(|Y_n - b_n| < \varepsilon) > \frac{1}{2}$ for all sufficiently large n. Thus (6.33) holds. The relations (6.32) and (6.33) imply (6.34).

Let $\{X_n\}$ be a sequence of random variables. We set $S_n = \sum_{j=1}^n X_j$. Let $\{a_n\}$ be a sequence of non-zero real constants. We shall say that the sequence $\{X_n\}$ obeys the strong law of large numbers with the sequence of the norming constants $\{a_n\}$ if the sequence $\{S_n/a_n\}$ is strongly stable.

Taking account of the previous notes, in studying the conditions of applicability of the strong law of large numbers we may limit ourselves to studies of conditions under which $S_n/a_n - m(S_n/a_n) \to 0$ a.s.

We shall need three more lemmas.

Lemma 6.15 *Let $\{A_n\}$ and $\{B_n\}$ ($n = 1, 2, \ldots, N; N \leq \infty$) be two sequences of events. We suppose that for every $n \geq 2$ the events B_n and $A_n \bar{A}_{n-1} \cdots \bar{A}_1$ are independent (here $\bar{A}$ is the complement of A) and that B_1 and A_1 are independent. If $P(B_n) \geq \delta > 0$ for all n, then*

$$P\left(\bigcup_{n=1}^N A_n B_n\right) \geq \delta P\left(\bigcup_{n=1}^N A_n\right).$$

Proof It is clear that

$$\bigcup_{n=1}^{N} A_n B_n = (A_1 B_1) \cup (\overline{A_1 B_1} A_2 B_2) \cup (\overline{A_1 B_1} \, \overline{A_2 B_2} A_3 B_3) \cup \cdots$$

$$\supset (A_1 B_1) \cup (\overline{A}_1 A_2 B_2) \cup (\overline{A}_1 \overline{A}_2 A_3 B_3) \cup \cdots.$$

Therefore,

$$P\left(\bigcup_{n=1}^{N} A_n B_n\right) \geq P(A_1)P(B_1) + P(\overline{A}_1 A_2)P(B_2) + P(\overline{A}_1 \overline{A}_2 A_3)P(B_3)$$

$$+ \cdots \geq \delta P\left(\bigcup_{n=1}^{N} A_n\right). \qquad \square$$

If Y is a random variable, we denote by Y^s the symmetrized random variable as usual.

Lemma 6.16 *Let $\{Y_n\}$ be a sequence of random variables, and let $\{b_n\}$ be a sequence of real numbers. For every $\varepsilon > 0$ and every $k \geq 1$ the following inequalities (the symmetrization inequalities) hold:*

$$P\left(\sup_{n \geq k}(Y_n - mY_n) \geq \varepsilon\right) \leq 2P\left(\sup_{n \geq k} Y_n^s \geq \varepsilon\right), \qquad (6.35)$$

$$P\left(\sup_{n \geq k}|Y_n - mY_n| \geq \varepsilon\right) \leq 2P\left(\sup_{n \geq k}|Y_n^s| \geq \varepsilon\right)$$

$$\leq 4P\left(\sup_{n \geq k}|Y_n - b_n| \geq \varepsilon/2\right). \qquad (6.36)$$

Proof We consider a sequence of random variables $\{Z_n\}$ such that for every n the random variables Z_n and Y_n have the same distribution, and Z_n and $(Y_1, \ldots, Y_n)$ are independent. We put

$$A_n = \{Y_n - mY_n \geq \delta\}, \quad B_n = \{Z_n - mZ_n \leq 0\}, \quad C_n = \{Y_n^s \geq \delta\}$$

for every $\delta > 0$. We also put $mY_n = mZ_n$ so that $A_n B_n \subset C_n$. Lemma 6.15 and the inequality $P(B_n) \geq \frac{1}{2}$ imply that

$$P\left(\bigcup_{n=k}^{\infty} A_n\right) \leq 2P\left(\bigcup_{n=k}^{\infty} A_n B_n\right) \leq 2P\left(\bigcup_{n=k}^{\infty} C_n\right).$$

From this we obtain (6.35). If we replace Y_n by $-Y_n$ in (6.35), we obtain the first of the inequalities (6.36). It remains to note that

$$P\left(\sup_n |Y_n^s| \geq \varepsilon\right) = P\left(\sup_n |Y_n - b_n - (Z_n - b_n)| \geq \varepsilon\right)$$

$$\leq P\left(\sup_n |Y_n - b_n| \geq \varepsilon/2\right) + P\left(\sup_n |Z_n - b_n| \geq \varepsilon/2\right)$$

$$= 2P\left(\sup_n |Y_n - b_n| \geq \varepsilon/2\right). \qquad \square$$

Lemma 6.17 *Let $\{Y_n\}$ be a sequence of random variables, and let $\{b_n\}$ be a sequence of real numbers. If $Y_n - b_n \to 0$ a.s., then $Y_n^s \to 0$ a.s. and $b_n - mY_n \to 0$. If $Y_n^s \to 0$ a.s., then $Y_n - b_n \to 0$ a.s. for every sequence of numbers $\{b_n\}$ satisfying the condition $b_n - mY_n \to 0$.*

This lemma is a corollary of Lemmas 6.8 and 6.16.

Let $\{a_n\}$ be a sequence of non-zero real numbers. We fix an arbitrary number $c > 1$. Let i_n be the largest integer for which $|a_{i_n}| \leq c^n$, if there are finitely many such integers; otherwise we put $i_n = 0$. Moreover, we put $a_0 = 0$ and $S_0 = 0$.

Theorem 6.12 *In order that the sequence of independent random variables $\{X_n\}$ obey the strong law of large numbers with the sequence of the norming constants $\{a_n\}$, it is necessary and sufficient that the following conditions be satisfied:*

(A) $|a_n| \to \infty$ or all X_n are degenerate,

(B) $\sum_{n=1}^{\infty} P(|S_{i_n} - S_{i_{n-1}} - m(S_{i_n} - S_{i_{n-1}})| \geq \varepsilon c^n) < \infty$ for every $\varepsilon > 0$.

Proof We begin with the proof of the necessity. If the sequence $\{X_n\}$ obeys the strong law of large numbers with the norming sequence $\{a_n\}$, then $S_n^s/a_n \to 0$ a.s. by Lemma 6.17. Suppose that the relation $|a_n| \to \infty$ does not hold. Then there exists an infinite sequence of integers $\{k_n\}$ such that $\sup |a_{k_n}| < \infty$. We have $S_{k_n}^s \to 0$ a.s. Therefore, the series $\sum U_n$, where $U_n = X_{k_{n-1}}^s + \cdots + X_{k_n}^s$ converges a.s. (We put $k_0 = 0$.) Moreover, the sum of this series equals zero. Consequently, for every $n \geq 1$ the series $\sum_{k \neq n} U_k$ converges a.s. and its sum equals $-U_n$ a.s. The latter series and $-U_n$ are independent random variables, and this equality implies the degeneracy of the distributions of the random variables. Thus all the random variables U_n are degenerate. The condition (A) follows. Taking account of the inequality $|a_{i_n}| \leq c^n$, we obtain $S_{i_n}^s/c^n \to 0$ a.s. and $(S_{i_n}^s - S_{i_{n-1}}^s)/c^n \to 0$ a.s. Applying Lemma 6.17 and the Borel–Cantelli lemma, we obtain (B).

Let us prove the sufficiency. The case when all X_n have degenerate distributions is trivial. Let the conditions (B) and $|a_n| \to \infty$ be satisfied. Making use of Lemma 6.17 and Lévy's inequality (inequality (2.7), Chapter 2), we find that

$$\sum_{n=1}^{\infty} P\left(\max_{i_n < j \le i_{n+1}} |S_j^s - S_{i_n}^s| \ge \varepsilon c^n \right) < \infty \qquad (6.37)$$

for every $\varepsilon > 0$, by (B). We put

$$T_k = c^{-k} \max_{i_k < j \le i_{k+1}} |S_j^s - S_{i_k}^s|.$$

It follows from (6.37), the Borel–Cantelli lemma, and Lemma 6.8 that $T_n \to 0$ a.s. In turn, this implies by Lemma 6.10 that $c^{-n} \sum_{k=0}^{n-1} c^k T_k \to 0$ a.s. Therefore, $c^{-n} S_{i_n}^s \to 0$ a.s. If $i_k < n \le i_{k+1}$, then $|a_n| > c^k$ and

$$|S_n^s/a_n| \le (|S_n^s - S_{i_k}^s| + |S_{i_k}^s|)/|a_n| \le T_k + c^{-k}|S_{i_k}^s| \to 0 \text{ a.s.}$$

as $k \to \infty$. Thus $S_n^s/a_n \to 0$ a.s. Applying Lemma 6.17 once more, we conclude that the sequence $\{X_n\}$ obeys the strong law of large numbers with the norming sequence $\{a_n\}$. □

6.6 Estimates of the growth of sums of independent random variables in terms of sums of their moments

We shall consider a sequence of independent random variables $\{X_n; n = 1, 2, \ldots\}$. We put $S_n = \sum_{k=1}^{n} X_k$. Up to now we have sought those conditions under which $S_n = o(a_n)$ a.s. for a given sequence of numbers $\{a_n\}$ such that $a_n \to \infty$. It is interesting to find the 'true' growth rate of the sums S_n. In the next chapter we shall impose some rather strong conditions on the distributions of our random variables and we shall investigate relations of the type $\limsup |S_n|/c_n = 1$ a.s., where $c_n \to \infty$. Nevertheless, even in this chapter, and with more elementary arguments, and with less severe restrictions, we shall obtain estimates of the growth order of the sums S_n that are optimal in some sense. These estimates will be expressed in terms of the sums of moments.

Let us introduce some additional notation. We shall denote by Ψ_c (or, respectively, Ψ_d) the set of functions $\psi(x)$ such that each $\psi(x)$ is positive and non-decreasing in the interval $x > x_0$ for some x_0 and the series $\sum \dfrac{1}{n\psi(n)}$ converges (respectively, diverges).

Here, and in what follows, $\sum f(n)$ denotes the summation over all positive integers n for which $f(n)$ is defined and non-negative. In the definitions of the sets Ψ_c and Ψ_d the value x_0 need not be the same for different functions ψ.

For example, $\psi(x) = x^\alpha \in \Psi_c$ for every $\alpha > 0$; if $\psi(x) = (\log x)^{1+\delta}$, then $\psi(x) \in \Psi_c$ for $\delta > 0$ and $\psi(x) \in \Psi_d$ for $\delta = 0$.

Theorem 6.13 *Let $g(x)$ be an even continuous function, positive and strictly increasing in the interval $x > 0$, and such that $g(x) \to \infty$ as $x \to \infty$. Suppose that at least one of the two following conditions is satisfied:*

(A) $x/g(x)$ *is non-decreasing in the interval $x > 0$;*
(B) $x/g(x)$ *and $g(x)/x^2$ do not increase in the interval $x > 0$.*

Suppose further that
$$Eg(X_n) < \infty \quad (n = 1, 2, \ldots) \tag{6.38}$$
and
$$M_n \to \infty \tag{6.39}$$
where
$$M_n = \sum_{k=1}^{n} Eg(X_k). \tag{6.40}$$

Finally, if condition (B) holds, suppose that
$$EX_n = 0 \quad (n = 1, 2, \ldots). \tag{6.41}$$
Then
$$S_n = o(g^{-1}(M_n \psi(M_n))) \text{ a.s.} \tag{6.42}$$
for every function $\psi(x) \in \Psi_c$. Here g^{-1} is the inverse of g.

The proof is based on Theorem 6.4 and the following elementary proposition.

Lemma 6.18 *Let $\{a_n\}$ be a sequence of non-negative numbers, $A_n = \sum_{k=1}^{n} a_k$, $A_n \to \infty$. Then the series $\sum a_n/(A_n \psi(A_n))$ converges for every $\psi \in \Psi_c$.*

Proof Let n_0 be such that $A_n > 0$ and $\psi(A_{n_0}) > 0$. The series $\sum 1/(n\psi(n))$ converges, and, therefore, so does the integral
$$I = \int_{A_{n_0}}^{\infty} \frac{dx}{x\psi(x)}.$$

By the mean-value theorem we have
$$\int_{A_{n-1}}^{A_n} \frac{dx}{x\psi(x)} = (A_n - A_{n-1})c_n$$
for $n > n_0$, where
$$\frac{1}{A_n \psi(A_n)} \le c_n \le \frac{1}{A_{n-1} \psi(A_{n-1})}.$$

Since $A_n - A_{n-1} = a_n$ and

$$I = \sum_{n=n_0+1}^{\infty} \int_{A_{n-1}}^{A_n} \frac{dx}{x\psi(x)},$$

we obtain the assertion of the lemma. $\square$

We shall now complete the proof of Theorem 6.13. Suppose $\psi \in \Psi_c$. By the hypotheses of the theorem, there exists a positive inverse $g^{-1}(x)$ in the interval $x > 0$. We put $a_n = g^{-1}(M_n\psi(M_n))$. Then $a_n \uparrow \infty$.

Making use of (6.39) and Lemma 6.18, we obtain

$$\sum \frac{Eg(X_n)}{M_n\psi(M_n)} < \infty.$$

We note that $g(a_n) = M_n\psi(M_n)$. By Theorem 6.4 the series $\sum X_n/a_n$ converges a.s. Hence

$$\frac{S_n}{g^{-1}(M_n\psi(M_n))} \to 0 \text{ a.s.,}$$

by Lemma 6.11. $\square$

If instead of $\psi \in \Psi_c$ we take a more slowly increasing function $\psi \in \Psi_d$, (6.42) may fail, as we see from the following theorem.

Theorem 6.14 *Let $g(x)$ be an even continuous function, positive and strictly increasing in the interval $x > 0$, with $g(0) = 0$, $g(x) \to \infty$ as $x \to \infty$. Corresponding to an arbitrary function $\psi(x) \in \Psi_d$ there exists a sequence of independent symmetric bounded random variables $\{X_n\}$, satisfying the conditions (6.38) and (6.39) but not satisfying (6.42).*

Proof We consider the sequence of independent random variables $\{X_n\}$ such that X_n for $n > n_1$ takes on the values $\pm g^{-1}(n\psi(n))$, each with probability $(2n\psi(n))^{-1}$ and the value zero with probability $1 - (n\psi(n))^{-1}$, while for $n \leq n_1$ it takes on the values $\pm g^{-1}(1)$ with probability $\frac{1}{2}$ each. Here $\psi \in \Psi_d$ and n_1 is such that $n_1\psi(n_1) > 1$.

Clearly, $EX_n = 0$ and $Eg(X_n) = 1$ for all n. By (6.40) we have $M_n = n$. Thus the conditions (6.38), (6.39), and (6.41) are satisfied. Furthermore,

$$\sum_{n>n_1} P(|X_n| = g^{-1}(n\psi(n))) = \sum_{n>n_1} \frac{1}{n\psi(n)}.$$

The latter series diverges, since $\psi \in \Psi_d$. In view of Lemma 6.12 and the equality $M_n = n$ we conclude that the relation (6.42) does not hold. $\square$

Theorems 6.13 and 6.14 have much simpler consequences in the case when $g(x) = |x|^p$, where $0 < p \leq 2$.

Theorem 6.15 *Suppose that*
$$E|X_n|^p < \infty \qquad (6.43)$$
for all n and some positive $p \leq 2$. Suppose also that
$$M_n = \sum_{k=1}^{n} E|X_k|^p \to \infty. \qquad (6.44)$$
Then if $0 < p \leq 1$ we have
$$S_n = o(M_n \psi(M_n))^{1/p}) \text{ a.s.} \qquad (6.45)$$
for every $\psi(x) \in \Psi_c$, and if $1 < p \leq 2$ the estimate (6.45) is valid under the additional condition (6.41).

Theorem 6.16 *For every function $\psi \in \Psi_d$ and every positive number p there exists a sequence of independent symmetric bounded random variables $\{X_n\}$ satisfying the conditions (6.43) and (6.44), but not (6.45).*

We now consider a sequence of independent random variables $\{X_n\}$ with finite variances. We set $B_n = \sum_{k=1}^{n} \text{Var } X_k$.

Theorem 6.17 *Suppose that $B_n \to \infty$. Then*
$$S_n - ES_n = o((B_n \psi(B_n))^{1/2}) \text{ a.s.} \qquad (6.46)$$
for every function $\psi \in \Psi_c$.

Theorem 6.18 *For every $\psi \in \Psi_d$ there exists a sequence of independent random variables $\{X_n\}$ with zero means and finite variances such that $B_n = \sum_{k=1}^{n} \text{Var } X_k \to \infty$ and (6.46) does not hold.*

Theorem 6.17 implies that the sums S_n of independent random variables with finite variances and unboundedly increasing variance of the sum, $B_n = \text{Var } S_n$, admit the following estimates of their growth rate (in order of increasing severity): for every $\varepsilon > 0$,

$$S_n - ES_n = o(B_n^{1/2 + \varepsilon}) \text{ a.s.,}$$
$$S_n - ES_n = o(B_n^{1/2} (\log B_n)^{1/2 + \varepsilon}) \text{ a.s.,}$$
$$S_n - ES_n = o(B_n^{1/2} (\log B_n)^{1/2} (\log \log B_n)^{1/2 + \varepsilon}) \text{ a.s.,}$$

and so on. In view of Theorem 6.18 we cannot replace ε by zero without introducing additional assumptions.

We shall apply these results to the strong law of large numbers with the classical norming constants, namely of the form

$$(S_n - ES_n)/n \to 0 \text{ a.s.} \qquad (6.47)$$

According to Theorem 4.16 (Chapter 4), Markov's condition

$$B_n = o(n^2), \qquad (6.48)$$

where $B_n = \text{Var } S_n$, is sufficient for the applicability of the weak law of large numbers, i.e. for the convergence of $(S_n - ES_n)/n$ to zero in probability. With the help of Theorem 6.17 we may suggest a sharpening of the condition (6.48) that is sufficient for the applicability of the strong law of large numbers.

Theorem 6.19 *If*

$$B_n = O(n^2/\psi(n)) \qquad (6.49)$$

for some function $\psi \in \Psi_c$, then (6.47) holds. On the other hand, for every $\psi \in \Psi_d$ such that $n/\psi(n)$ is non-decreasing in the domain $n > n_0$ for some n_0 there exists a sequence of independent symmetric random variables $\{X_n\}$ with finite variances and satisfying (6.49) but not (6.47)).

Proof We shall prove the first assertion. Let (6.49) hold for some $\psi \in \Psi_c$. We note that every function $\psi \in \Psi_c$ satisfies the condition $\psi(n) \to \infty$.

We may assume that $B_n \to \infty$, since otherwise the series $\sum \text{Var } X_n$ converges and, consequently, $\sum \text{Var } X_n/n^2 < \infty$; then Theorem 6.7 immediately yields the desired assertion.

For every $f \in \Psi_c$ we have $S_n - ES_n = o((B_n f(B_n))^{1/2})$ a.s., by Theorem 6.17. Therefore, (6.49) implies that

$$S_n - ES_n = o(n(f(B_n)/\psi(n))^{1/2}) \text{ a.s.} \qquad (6.50)$$

Using (6.49) again, and using the fact that functions of the set Ψ_c are non-decreasing, we conclude that

$$S_n - ES_n = o(n(f(n^2)/\psi(n))^{1/2}) \text{ a.s.} \qquad (6.51)$$

Now we choose f so that $f(n^2)/\psi(n)$ is bounded for all sufficiently large n. To this end, we put $f(n^2) = \psi(n)$, and for values of n that are not squares of integers we choose $f(x)$ in such a way as to make it non-decreasing. We shall show that $\sum 1/(nf(n))$ converges, and this will prove that $f \in \Psi_c$.

The series $\sum 1/(nf(n))$ can be written in the form

$$\sum_k \sum_{k^2 \leq n < (k+1)^2} \frac{1}{nf(n)}.$$

Obviously,

$$\sum_{k^2 \leq n < (k+1)^2} \frac{1}{nf(n)} \leq \frac{2k+1}{k^2 f(k^2)}.$$

Since $f(n^2) = \psi(n)$ and $\sum 1/(n\psi(n)) < \infty$, we conclude that the series $\sum 1/(nf(n))$ converges.

With this choice of the function f, (6.51) implies (6.47).

We now turn to the proof of the second assertion of the theorem. Suppose that $\psi(x) \in \Psi_d$ and $n/\psi(n)$ does not decrease. We consider a sequence of independent random variables $\{X_n\}$ such that for $n > n_1$ the variable X_n takes on the values 0, n, and $-n$ with probabilities $1 - (1/n\psi(n))$, $1/2n\psi(n)$, and $1/2n\psi(n)$ respectively. Here n_1 satisfies the condition $n_1\psi(n_1) > 1$. For $n \leqslant n_1$, let X_n take on the values 1 and -1, each with probability $\frac{1}{2}$. Then $EX_n = 0$ for all n and

$$\text{Var } X_n = \begin{cases} 1 & (n \leqslant n_1) \\ n/\psi(n) & (n > n_1). \end{cases}$$

Therefore, $B_n = O(n^2/\psi(n))$.

Since $\sum 1/n\psi(n) = \infty$, we have

$$P(|X_n| = n \text{ i.o.}) = 1, \tag{6.52}$$

by the Borel–Cantelli lemma. If our sequence were to satisfy (6.47), we would have $S_n/n \to 0$ a.s. and, therefore, $X_n/n \to 0$ a.s., which contradicts (6.52). □

It follows from Theorem 6.19 that the condition $B_n = O(n^2/(\log n)^{1+\delta})$ for some $\delta > 0$ is sufficient for (6.47). If $B_n = O(n^2/\log n)$, or even $B_n = O(n^2/(\log n)(\log \log n))$, then the relation (6.47) may fail.

In terms of the classes Ψ_c and Ψ_d it is convenient to describe the behaviour of $\liminf b(n)|S_n|$ where $b(n) \uparrow \infty$ is a given function and S_n is the sum of n independent identically distributed random variables. We shall present two theorems of this kind.

Theorem 6.20 *Let $\{X_n\}$ be a sequence of independent random variables with a common distribution function. Suppose that the characteristic function $f(t) = Ee^{itX_1}$ satisfies the Cramér condition (C),*

$$\limsup_{|t| \to \infty} |f(t)| < 1.$$

Then for every function $\psi \in \Psi_c$,

$$\lim \sqrt{n\psi(n)}|S_n| = \infty \text{ a.s.} \tag{6.53}$$

Theorem 6.21 *Let $\{X_n\}$ be a sequence of independent identically distributed random variables with the characteristic function $f(t)$ satisfying the Cramér condition (C). Suppose that $EX_1 = 0$ and $\text{Var } X_1 < \infty$. Then*

$$\liminf \sqrt{n\psi(n)}|S_n| = 0 \text{ a.s.} \tag{6.54}$$

for every function $\psi \in \Psi_d$.

We shall prove only Theorem 6.20, referring to Petrov [346] for the proof of Theorem 6.21.

There are no moment assumptions in Theorem 6.20. The proof of this theorem is based on some inequalities for the concentration functions.

We define the concentration function $Q(X; \lambda)$ of a random variable X by the equality

$$Q(X; \lambda) = \sup_x P(x \leq X \leq x + \lambda).$$

Lemma 1.16 (Chapter 1) implies that

$$Q(X_1; \lambda) \leq (\tfrac{96}{95})^2 \lambda \int_{|t| \leq 1/\lambda} |f(t)|\, dt$$

for every $\lambda > 0$. Therefore,

$$Q(S_n; \lambda) \leq A\lambda \int_{|t| \leq 1/\lambda} |f(t)|^n\, dt \tag{6.55}$$

for every $\lambda > 0$, where $A = (96/95)^2$.

Suppose that $\psi \in \Psi_c$. We put $\lambda = \lambda_n = L/\sqrt{n\psi(n)}$, where L is a positive constant. By Lemma 1.4 (Chapter 1) and the accompanying note, Cramér's condition (C) implies that for every $\delta > 0$ there exists a positive number c such that $|f(t)| \leq e^{-c}$ for $|t| \geq \delta$. Let

$$I_1 = \int_{|t| \leq \delta} |f(t)|^n\, dt, \quad I_2 = \int_{\delta \leq |t| \leq 1/\lambda} |f(t)|^n\, dt.$$

Taking (6.55) into account, we get

$$Q(S_n; \lambda) \leq A\lambda(I_1 + I_2). \tag{6.56}$$

In view of Cramér's condition (C) the random variable X_1 has a non-degenerate distribution. Therefore, there exist positive constants δ and ε such that $|f(t)| \leq e^{-\varepsilon t^2}$ for $|t| \leq \delta$, according to Lemma 1.5 (Chapter 1). Taking the δ in the definitions of the integrals I_1 and I_2 to satisfy the last condition, we obtain

$$I_1 \leq \int_{|t| \leq \delta} e^{-n\varepsilon t^2}\, dt \leq \frac{1}{\sqrt{n\varepsilon}} \int_{-\infty}^{\infty} e^{-u^2}\, du = \sqrt{\frac{\pi}{n\varepsilon}}$$

for all n. Furthermore, $I_2 \leq 2 e^{-cn}/\lambda$ for all n. The inequality (6.56) implies that

$$Q(S_n; \lambda) \leq A\lambda \left(\sqrt{\frac{\pi}{n\varepsilon}} + \frac{2}{\lambda} e^{-cn} \right) \leq C_1 \left(\frac{\lambda}{\sqrt{n}} + e^{-cn} \right)$$

for all n, where C_1 is a positive constant. Hence

$$P\left(|S_n| \leq \frac{L}{2\sqrt{n\psi(n)}}\right) \leq Q\left(S_n; \frac{L}{\sqrt{n\psi(n)}}\right) \leq C_1 \left(\frac{L}{n\psi(n)} + e^{-cn} \right).$$

It is clear that
$$\sum \left(\frac{L}{n\psi(n)} + e^{-cn} \right) < \infty,$$
since $\psi \in \Psi_c$. By the Borel–Cantelli lemma,
$$P\left(|S_n| \leq \frac{L}{2\sqrt{n\psi(n)}} \text{ i.o.} \right) = 0.$$
Since the number L is arbitrary, the relation (6.53) follows from this. □

6.7 Bibliographical notes

Borel [50] proved that if $\{A_n\}$ is a sequence of independent events, then $P(A_n \text{ i.o.})$ is 0 or 1 according as the series $\sum P(A_n)$ converges or diverges. Without the independence condition, the implication $\sum P(A_n) < \infty \Rightarrow P(A_n \text{ i.o.}) = 0$ was proved by Hausdorff [178] and later by Cantelli [60].

The second part of Lemma 6.1 and Lemma 6.2 are due to Erdös and Rényi [114]. Lemma 6.3 was obtained by Shuster [425]. Lemmas 6.4 and 6.5 can be found in Martikainen and Petrov [291]. The equivalence of (a) and (b) in Lemma 6.4 as well as the equivalence of (d) and (e) was also proved by O'Brien and Tomkins [personal communication]. The inequality (6.10) is due to Chung and Erdös [81]. Lemmas 6.6 and 6.7 are due to Kochen and Stone [228] and Spitzer [433]. Better lower estimates for $P(A_n \text{ i.o.})$ were found by Móri and Székely [308] and Martikainen and Petrov [291].

Theorem 6.1 is due to Kolmogorov [231] and Theorem 6.4 for $g_n(x) = g(x)$ to Chung [77]. Theorems 6.2 and 6.3 can be found in van Kampen [461].

Theorem 6.7 for $a_n = n$ and Theorem 6.11 for a sequence of independent random variables were obtained by Kolmogorov [230, 231], Lemma 6.13 by Heyde [183], and Theorems 6.9 and 6.10 by Martikainen and Petrov [290]. Theorem 6.11 is due to Etemadi [125], Theorem 6.12 to Martikainen [282]. The latter theorem is a generalization of some results of Martikainen and Petrov [289] and Volodin and Nagaev [462] for a non-decreasing sequence of the norming constants. The necessary and sufficient conditions for the strong law of large numbers with an arbitrary sequence of the norming constants were also obtained by Buldygin [55].

Theorems 6.13–6.19 were proved by Petrov [334]. Their generalizations can be found in Egorov [109]. Theorems 6.20 and 6.21 are due to Petrov [345, 346]; they are generalizations and strengthening of some results of Chung and Erdös [79].

Surveys on the strong law of large numbers for sequences of independent random variables can be found in the books of Csörgő and Révész [87], Révész [382, 383], and Stout [440].

6.8 Addenda

When we consider a sequence of random variables $\{X_n\}$, we put $S_n = \sum_{k=1}^{n} X_k$.

6.8.1 Let $\{X_n\}$ be a sequence of independent random variables such that $EX_n = 0$ for all n and ess sup $X_n = O(n/\log \log n)$. The relation $S_n/n \to 0$ a.s. holds if and only if $\sum_{k=1}^{\infty} \exp\{-\varepsilon/H_k\} < \infty$ for every $\varepsilon > 0$, where $H_k = 2^{-2k} \sum_{j=2^k+1}^{2^{k+1}} \text{Var } X_j$ (Prohorov [369]).

6.8.2 For every positive integer r there exist sequences of independent random variables $\{X_n\}$ and $\{Y_n\}$ such that $EX_n^p = EY_n^p < \infty$ ($1 \leq p \leq r$), $\{X_n\}$ obeys, and $\{Y_n\}$ does not obey the strong law of large numbers with the sequence of norming constants $a_n = n$ (Nagaev [310]).

6.8.3 Let $\{X_n\}$ be a sequence of random variables such that $\liminf E|X_n| < \infty$. Then there exist a sub-sequence $\{X_{n_k}\}$ and a random variable X satisfying the condition $E|X| < \infty$ and

$$N^{-1} \sum_{k=1}^{N} X_{n_k} \to X \text{ a.s. as } N \to \infty \qquad (6.57)$$

(Komlós [237]; this proposition is a strengthening of an earlier result of Révész [381]). Other proofs were given by Hall and Heyde [172] and Trautner [454].

In the assertion of the previous proposition it is possible to replace (6.57) by the relation

$$N^{-1} \sum_{k=1}^{N} Y_{\sigma(k)} \to X \text{ a.s. as } N \to \infty$$

for every permutation $\sigma(1), \sigma(2), \ldots$ of the sequence $1, 2, \ldots$, where $Y_m = X_{n_m}$ for all m (Berkes [25]).

6.8.4 Let $\{X_n\}$ be a sequence of random variables such that $\liminf E|X_n|^p < \infty$ for some positive $p < 2$.

(A) If $0 < p < 1$, then there exists a sub-sequence $\{X_{n_k}\}$ such that $N^{-1/p} \sum_{k=1}^{N} X_{n_k} \to 0$ a.s. as $N \to \infty$.

(B) If $1 \leq p < 2$, then there exists a sub-sequence $\{X_{n_k}\}$ and a random variable X such that $E|X|^p < \infty$ and $N^{-1/p} \sum_{k=1}^{N} X_{n_k} \to X$ a.s. (Chatterjee [61] and Gaposhkin [143]).

A general sub-sequence principle for weak and strong limit theorems is given by Aldous [3]. A simplified approach to the sub-sequence theorems of Aldous and various extensions were provided by Berkes and Péter [28].

6.8.5 $\{X_n\}$ be a sequence of independent random variables, and let $a_n \uparrow \infty$. Suppose that there exist a sub-sequence $\{a_{n_k}\}$ and constants $c_1 > 1$ and c_2 such that $c_1 \leqslant a_{n_{k+1}}/a_{n_k} \leqslant c_2$ for all sufficiently large k. We write $S_{n_0} = 0$, $T_k = (S_{n_k} - S_{n_{k-1}})/a_{n_k}$. The following conditions are equivalent:

(I) $(S_n - mS_n)/a_n \to 0$ a.s.;
(II) $T_k - mT_k \to 0$ a.s. as $k \to \infty$;
(III) $\sum_{k=1}^{\infty} P(|T_k - mT_k| \geqslant \varepsilon) < \infty$ for every $\varepsilon > 0$ (Prohorov [364] for $a_n = n$ and Loève [256]).

6.8.6 Let $\{a_n\}$ be a sequence of non-zero real numbers, and let $\{X_n\}$ be a sequence of independent symmetric random variables. Let $c > 1$ be a fixed number, and let i_n be the greatest integer satisfying the condition $|a_{i_n}| \leqslant c^n$ if there are finitely many integers with this property; otherwise we put $i_n = 0$. In order that the sequence $\{X_n\}$ satisfy the strong law of large numbers with the norming sequence $\{a_n\}$, it is necessary that for every $r \geqslant 2$ (and sufficient that for some $r \geqslant 2$) the following conditions be satisfied:

(A) $|a_n| \to \infty$ or X_n are degenerate for all n;

(B) $\sum_{n=1}^{\infty} \sum_{k=i_n+1}^{i_{n+1}} P(|X_k| \geqslant \varepsilon c^n) < \infty$ for every $\varepsilon > 0$;
(C) for every $\varepsilon > 0$ there exists a sequence of positive constants $\{h_n\}$ (possibly depending on ε) such that

$$\sum_{n=1}^{\infty} \exp\{-r\varepsilon h_n c^n\} \prod_{k=i_n+1}^{i_{n+1}} \int_{|x| < \varepsilon c^n} e^{h_n x} \, dV_k(x) < \infty,$$

where $V_k(x) = P(X_k < x)$.

If $a_n \uparrow \infty$, it is possible to replace (B) by the condition $\sum_{n=1}^{\infty} P(|X_n| \geqslant \varepsilon a_n) < \infty$ for every $\varepsilon > 0$ (Martikainen [282]).

More complicated necessary and sufficient conditions for the relation $S_n/n \to 0$ a.s. were obtained earlier by Nagaev [310].

In Subsections 6.8.7–6.8.10 we consider a sequence of independent random variables $\{X_n\}$.

6.8.7 A sequence $\{X_n\}$ obeys the strong law of large numbers with an arbitrary norming sequence of non-zero constants $\{a_n\}$ if and only if it obeys the strong law of large numbers with the norming sequence $c_n = \inf_{k \geqslant n} |a_k|$ (Martikainen [282]). Therefore, we do not diminish the generality of considerations connected with the conditions of the applicability of the strong law of large numbers if we restrict ourselves by non-decreasing sequences of norming constants.

6.8.8 Suppose that $a_n \uparrow \infty$. We put $V_n(x) = P(X_n < x)$,

$$\sigma_n^2(x) = \int_{|y|<x} y^2 \, dV_n(y) - \left(\int_{|y|<x} y \, dV_n(y)\right)^2, \quad B_T(x) = \sum_{k \in T} \sigma_k^2(x).$$

Let $c > 1$, and let i_n be the greatest integer satisfying the condition $a_{i_n} \leq c^n$. We write $I_n = \{k \in \mathbb{N} : i_{n-1} < k \leq i_n\}$. Suppose that $mX_n = 0$ for all n and

$$\sum_{n=1}^{\infty} a_n^{-3} \int_{|x|<\delta a_n} |x|^3 \, dV_n(x) < \infty$$

for some $\delta > 0$. In order that the sequence $\{X_n\}$ satisfy the strong law of large numbers with the norming sequence $\{a_n\}$, it is necessary and sufficient that the following conditions be satisfied:

(I) $\sum_{n=1}^{\infty} P(|X_n| \geq \delta a_n) < \infty$;

(II) $\sum_{n=1}^{\infty} \Phi(-\varepsilon c^n / B_{I_n}(\delta c^n)) < \infty$ for every $\varepsilon > 0$, where $\Phi(x)$ is the standard normal d.f. (Martikainen [282]).

6.8.9 Let $a_n \to \infty$. In order that

$$-\infty < \liminf (S_n - mS_n)/a_n \leq \limsup (S_n - mS_n)/a_n < \infty, \quad (6.58)$$

it is necessary and sufficient that condition (B) of Theorem 6.12 be satisfied (Martikainen [286]). Other criteria for (6.58) can be found in [286] and in Kruglov [242].

6.8.10 If $\sum_{n=1}^{\infty} n^{-1} P(|S_n/n - m(S_n/n)| \geq \varepsilon) < \infty$ for every $\varepsilon > 0$, then $S_n/n - m(S_n/n) \to 0$ a.s. (The converse is not true.) If $|X_n| < n$ for all n and $\sum_{n=1}^{\infty} n^{-1} P(|S_n - ES_n| \geq n\varepsilon) < \infty$ for every $\varepsilon > 0$, then $(S_n - ES_n)/n \to 0$ a.s. (Baum and Katz [16]).

6.8.11 Let $\{X_n\}$ be an arbitrary sequence of random variables such that $E|X_n|^p < \infty$ for some positive $p \leq 1$ and for all n. We write $M_n = \sum_{k=1}^{n} E|X_k|^p$. If $M_n \to \infty$, then $S_n = o((M_n \psi(M_n))^{1/p})$ a.s. for every function $\psi \in \Psi_c$, where the class Ψ_c is defined in Section 6.6 (Petrov [336]).

6.8.12 Let $\{X_n\}$ be a sequence of non-negative random variables such that $\sup_{n \geq 1} EX_n < \infty$, $EX_n X_m \leq EX_n EX_m$ for $n \neq m$, and

$$\sum_{n=1}^{\infty} \text{Var } X_n/n^2 < \infty. \quad (6.59)$$

Then $(S_n - ES_n)/n \to 0$ a.s. (Etemadi [126]).

6.8.13 Let $\{X_n\}$ be a sequence of pairwise independent random variables satisfying the conditions (6.59) and

$$n^{-1} \sum_{k=1}^{n} E|X_k - EX_k| = O(1). \tag{6.60}$$

Then $(S_n - ES_n)/n \to 0$ a.s. One cannot omit the condition (6.60) in this proposition (Csörgő et al. [90]).

6.8.14 Let $\{X_n\}$ be a sequence of orthogonal random variables (this means that $EX_n^2 < \infty$ for all n and $EX_n X_m = 0$ for $n \neq m$). We put $B_n = \sum_{k=1}^{n} EX_k^2$. If $B_n \to \infty$, then $S_n = o((B_n \psi(B_n))^{1/2} \log n)$ a.s. for every function $\psi \in \Psi_c$. On the other hand, for every function $\psi \in \Psi_d$ there exists a sequence of orthogonal random variables $\{X_n\}$ such that $B_n \to \infty$ and

$$\limsup |S_n|/((B_n \psi(B_n))^{1/2} \log n) = \infty \text{ a.s.}$$

Petrov [343]).

6.8.15 Let $\{X_n\}$ be a sequence of independent random variables, and let $\{a_n\}$ be a sequence of positive constants satisfying the conditions $a_n \uparrow \infty$ and $\sum_{k=n}^{\infty} a_k^{-2} = O(na_n^{-2})$ as $n \to \infty$. Suppose there exist a random variable X and a positive constant C such that $\sum_{n=1}^{\infty} P(|X| \geq a_n) < \infty$ and

$$P(|X_n| \geq x) \leq CP(|X| \geq x)$$

for all $x > 0$ and all n. If at least one of the following conditions is satisfied: (A) X_n is symmetric for every n, (B) $a_k/a_n \leq Ck/n$ for $k \geq n$ and $EX_n = 0$ for every n, then $S_n/a_n \to 0$ a.s. (Petrov [356]). This is a generalization of Feller's theorem [130] (see also Petrov [342], Chapter 9, Theorem 17).

6.8.16 Let $\{X_n\}$ be a sequence of independent variables, and let $\{a_n\}$ be a sequence of positive constants such that $\liminf a_{n+1}/a_n > 1$. Then $S_n/a_n \to 0$ a.s. if and only if $\sum P(|X_n| \geq \varepsilon a_n) < \infty$ for every $\varepsilon > 0$ (Rozovsky [408]).

From now on we consider a sequence of independent identically distributed random variables $\{X_n\}$. We put $V(x) = P(X_1 < x)$.

6.8.17 If $0 < p < 1$, then the relation $S_n/n^{1/p} \to 0$ a.s. is equivalent to the condition $E|X_1|^p < \infty$. If $1 < p < 2$, then the latter condition is equivalent to the relation $(S_n - ES_n)/n^{1/p} \to 0$ a.s. (Marcinkiewicz and Zygmund [275]).

6.8.18 If $EX_1^+ = \infty$ and $a \in \mathbb{R}$, then the following conditions are equivalent:

(1) $\limsup S_n > -\infty$ a.s.;
(2) $\limsup S_n/n > -\infty$ a.s.;
(3) $\limsup S_n/n = +\infty$ a.s.;

(4) $\limsup S_n = +\infty$ a.s.;
(5) $\sum_{n=1}^{\infty} n^{-1} P(S_n > na) = \infty$

(Kesten [217]).

6.8.19 If $EX_1^+ = EX_1^- = +\infty$, then one of the following assertions holds:

(1) $\lim S_n/n = +\infty$ a.s.;
(2) $\lim S_n/n = -\infty$ a.s.;
(3) $\liminf S_n/n = -\infty$ and $\limsup S_n/n = +\infty$ a.s.

(Kesten [217]). Another proof can be found in Tanny [446]. Erickson [116] found the necessary and sufficient conditions for each assertion. Chow and Zhang [73] obtained extensions of these results to other normalizing sequences of constants.

Earlier, Katz [213] proved that if $S_n/n \to 0$ in probability but not a.s., then assertion (3) holds.

6.8.20 Let $E|X_1| = \infty$, and let $\{a_n\}$ be an arbitrary sequence of constants. Then either $\limsup |S_n/a_n| = \infty$ a.s. or $\liminf |S_n/a_n| = 0$ a.s. (Chow and Robbins [72]).

6.8.21 If X_1 has a non-degenerate distribution, then either

$$\limsup S_n/n^{1/2} = \infty \quad \text{or} \quad \lim S_n/n^{1/2} = -\infty \text{ a.s.}$$

An immediate consequence of this result: if $P(S_n \geq 0 \text{ i.o.}) > 0$, then

$$\limsup S_n/n^{1/2} = \infty \text{ a.s.}$$

(Stone [439]).

6.8.22 Suppose that $E|X_1| = \infty$ and $\{S_n/n\}$ has a finite strong limit point b (i.e. there exists a non-random sequence $\{n_k\}$ such that $S_{n_k}/n_k \to b$ a.s. as $k \to \infty$). Then the sequence $\{S_n/n\}$ is dense in $\mathbb{R}$ (Erickson and Kesten [117]). The possible structures for the set of strong limit points of S_n/n^α are determined in Kesten [217] and Erickson and Kesten [117].

6.8.23 Let $\{a_n\}$ be a sequence of positive numbers satisfying the conditions $a_n \uparrow \infty$ and $\sum_{k=n}^{\infty} a_k^{-3} = O(na_n^{-3})$. We put

$$\sigma^2(x) = \int_{|y|<x} y^2 \, dV(y) - \left(\int_{|y|<x} y \, dV(y) \right)^2.$$

Let $\delta > 0$ and $0 < x < 1 < y$. Let $j_n(x, y)$ be the number of the integers k satisfying the inequalities $xa_n \leq a_k \leq ya_n$. The sequence $\{X_n\}$ obeys the strong law of large numbers with the norming sequence of constants $\{a_n\}$ if and only if $\sum_{n=1}^{\infty} P(|X_1| \geq a_n) < \infty$ and

$$\sum_{n=1}^{\infty} \Phi(-\varepsilon a_n/(j_n(x,y)\sigma^2(\delta a_n))^{1/2})/j_n(x,y) < \infty \qquad (6.61)$$

for every $\varepsilon > 0$. Under the additional condition $\sup_{n \geq 1} a_{2n}/a_n < \infty$ it is possible to replace (6.61) by the condition $\sum_{n=1}^{\infty} n^{-1}\Phi(-\varepsilon a_n/(n\sigma^2(\delta a_n))^{1/2}) < \infty$ for every $\varepsilon > 0$ (Martikainen [286]).

6.8.24 Let $\delta > 0$ and $c > 1$. The relation $\lim \sup(S_n - mS_n)/n = 0$ a.s. is equivalent to the fulfilment of the following conditions:

(A) $\int_0^{\infty} x \, dV(x) < \infty$;

(B) $nP(X_1 \leq -n) \to 0$;

(C) $\sum_{n=1}^{\infty} \exp\{-\varepsilon c^n/v_{[c^n]}(\varepsilon/2, \delta)\} < \infty$ for every $\varepsilon > 0$, where $v_n(a, \delta)$ is the least solution of the inequalities $\delta \leq v_n \leq \delta n$ and $\int_{-\delta n}^{-v} |x| \, dV(x) \leq a$ (Martikainen [285]). The same paper contains the necessary and sufficient conditions for the relation $\lim \sup(S_n - mS_n)/a_n = \alpha$ a.s., where α is a constant and $a_n \uparrow \infty$.

6.8.25 Let $EX_1 = 0$. The condition $E|X_1|^t < \infty$, where $1 \leq t \leq 2$, is equivalent to the fulfilment of the relation $n^{-1/t} \sum_{k=1}^{n} a_{nk} X_k \to 0$ a.s. for every sequence of real numbers $\{a_{nk}; k = 1, \ldots, n; n = 1, 2, \ldots\}$ such that

$$\lim \sup \sum_{k=1}^{n} a_{nk}^2 < \infty. \qquad (6.62)$$

The condition $Ee^{hX_1} < \infty$ for every real h is equivalent to the relation $(\log n)^{-1} \sum_{k=1}^{n} a_{nk} X_k \to 0$ a.s. for every sequence of real numbers $\{a_{nk}\}$ satisfying the condition (6.62) (Chow and Lai [69]). For general double arrays $\{a_{nk}; k = 1, \ldots, k_n; n = 1, 2, \ldots\}$ where $k_n \to \infty$ or $k_n = \infty$ and i.i.d. random variables $\{X_n\}$, Teicher [448] obtained conditions under which $\sum_{k=1}^{k_n} a_{nk} X_k \to 0$ a.s.

6.8.26 Let X_1 have a non-degenerate distribution. If f is a bounded measurable function such that $f(x) \to \alpha$ as $|x| \to \infty$ and $|\alpha| < \infty$, then $n^{-1} \sum_{k=1}^{n} f(S_k) \to \alpha$ a.s.

Let X_1 have a non-lattice distribution. If f is a periodic function with period p and Riemann-integrable in the interval $(0, p)$, then

$$n^{-1} \sum_{k=1}^{n} f(S_k) \to p^{-1} \int_0^p f(t) \, dt \text{ a.s.}$$

(Robbins [386]).

6.8.27 Let the d.f. of X_1 have an absolutely continuous component, and let $EX_1 = a > 0$. Suppose that f is a bounded measurable function. Then

$$n^{-1}\left\{\sum_{k=1}^{n} f(S_k) - \int_0^n f(ax)\,dx\right\} \to 0 \text{ a.s.}$$

(Mejlijson [294]). If additionally $E|X_1|^{2+\delta} < \infty$ for some positive $\delta \leq 1$, then

$$n^{-1}\left\{\sum_{k=1}^{n} Ef(S_k) - \int_0^n f(ax)\,dx\right\} = o(n^{-(\delta/2)+\varepsilon})$$

for every $\varepsilon > 0$ (Bingham and Goldie [40]). In [40] one can find estimates of rates of convergence in the relation

$$n^{-\alpha}\sum_{k=1}^{n}\{f(S_k) - Ef(S_k)\} \to 0 \text{ a.s.},$$

where $\alpha > \tfrac{1}{2}$.

6.8.28 Let $|B|$ denote the number of elements in the finite set B. A sequence $\{(n)\}$ of finite sets of natural numbers satisfying the condition $|(n)| = n$ for $n \geq 1$ is called a rule and denoted by $(\ )$. A rule satisfying the condition

$$m, n \in \mathbb{N}, \quad m \neq n \Rightarrow (m) \cap (n) = \varnothing$$

is denoted by $\langle\ \rangle$; a rule satisfying the condition $(n) = \{1, 2, \ldots, n\}$ for every $n \in \mathbb{N}$ is denoted by $(\ )_0$. The set of all rules is denoted by T. A sequence $\{S_{(n)}\}$ (where $S_B = \sum_{k \in B} X_k$ for any set B of natural numbers and $\{X_k\}$ is a sequence of independent identically distributed random variables) is called a sequence of ruled sums corresponding to the rule $(\ )$. (This is a generalization of the cumulative sums $S_n = \sum_{k=1}^{n} X_k$; we have $S_n = S_{(n)_0}$.)
If $(\ ) \in T$ and $a_n \uparrow \infty$, then

$$S_{\langle n \rangle}/a_n \to 0 \text{ a.s.} \Rightarrow S_{(n)}/a_n \to 0 \text{ a.s.} \Rightarrow S_n/a_n \to 0 \text{ a.s.}$$

Furthermore, the following conditions are equivalent:

(I) $S_{(n)}/n \to 0$ a.s. for all $(\ ) \in T$;
(II) $EX_1^2 < \infty$ and $EX_1 = 0$ (Baum et al. [17]).

6.8.29 The following conditions are equivalent:

(a) $\limsup S_{(n)}/n \leq 0$ a.s. for all $(\ ) \in T$;
(b) $\sum_{n=1}^{\infty} P(S_n \geq n\varepsilon) < \infty$ for every $\varepsilon > 0$.

If $EX_1 = 0$, then (a) and (b) are equivalent to the condition $\int_0^\infty x^2\,dV(x) < \infty$, where $V(x)$ is the d.f. of X_1 (Petrov [339]).

6.8.30 Let $(\ ) \in T$ and $a_n = |\bigcup_{k=1}^{n}(k)|$. The sequence $\{S_{(n)}/a_n\}$ converges a.s. to a finite limit if and only if $E|X_1| < \infty$ and there exists the limit $\lim n\, EX_1/a_n$. If these hypotheses are satisfied, then $\lim S_{(n)}/a_n = \lim n\, EX_1/a_n$ a.s. (Martikainen [278]).

6.8.31 A sequence $\{B_n\}$ of finite sets of natural numbers will be called a rule if $|\bigcup_{k=1}^{n} B_k| \to \infty$. We put $b_n = |\bigcup_{k=1}^{n} B_k|$. Suppose $b_1 > 0$ and $\limsup b_{n+1}/b_n < \infty$. Then the following conditions are equivalent:

(1) $E|X_1| < \infty$;
(2) $\limsup S_{B_n}/b_n \leq EX_1^+$ a.s., $\liminf S_{B_n}/b_n \geq -EX_1^-$ a.s.;
(3) $\limsup |S_{B_n}|/b_n < \infty$ a.s.;
(4) $\sum_{n=1}^{\infty} P(|S_{a_n}| \geq cb_n) < \infty$ for some (for every) $c > 0$, where $a_n = |B_n \setminus \bigcup_{k=1}^{n-1} B_k|$ (Martikainen [278]). In [279] there are estimates of the growth of ruled sums of dependent random variables satisfying the condition $E|X_n|^p < \infty$ for some positive $p \leq 1$; these results generalize Petrov's theorem from Subsection 6.8.11.

Newer results on the strong law of large numbers for ruled sums can be found in Skovoroda and Mikosch [429].

6.8.32 If X_1 is a positive random variable, then the conditions $EX_1 = \infty$ and $\limsup X_n/S_{n-1} = \infty$ a.s. are equivalent (Kesten [217]). If X_1 is symmetric, then the conditions $EX_1^2 = \infty$ and

$$\limsup |X_n|/\max_{1 \leq j \leq n-1} |S_j| = \infty \text{ a.s.}$$

are equivalent (Wittmann [468]).

6.8.33 Let $u_k = P(2^k < |X_1| \leq 2^{k+1} | |X_1| > 2^k)$. Let $X_n^{(1)}$ be the term of maximum modulus among $X_1, \ldots, X_n$. Then $X_n^{(1)}/S_n \to 1$ a.s. if and only if $\sum u_k^2 < \infty$. If $r > 1$ is an integer, then $\liminf X_n^{(1)}/S_n = r^{-1}$ a.s. if and only if $\sum_k u_k^r = \infty$. If $\sum_k u_k^r = \infty$ for all r, then $\liminf |X_n^{(1)}/S_n| = 0$ a.s. (Pruitt [373]. Earlier, Darling [92] showed that with non-negative summands, $E|X_n^{(1)}/S_n - 1| \to 0$ if and only if $P(X_1 > x)$ is a slowly varying function.

6.8.34 Let $R(t) = Ee^{tX_1} < \infty$ for some $t > 0$, and let X_1 have a non-degenerate distribution. Define

$$\rho(x) = \inf_t e^{-tx} R(t) \tag{6.63}$$

(the Chernoff function) and

$$T_n = (S_{n+f(n)} - S_n)/f(n), \tag{6.64}$$

where $f(n)$ is a non-decreasing function having only positive integer values. Let $a = \text{ess sup } X_1 \leq \infty$, and let r denote the radius of convergence of the series $\sum_n x^{f(n)}$. Then $\limsup T_n = \alpha_r < \infty$ a.s., where $\alpha_r = a$ if $r < \rho(a)$ and α_r is the unique solution of the equation $\rho(a) = r$ if $r \geqslant \rho(a)$ (Shepp [422]).

6.8.35 Let X_1 be a zero mean random variable with a non-degenerate distribution. Suppose that

$$Ee^{tX_1} < \infty \quad \text{for } |t| < b \text{ and some } b > 0. \tag{6.65}$$

We put $K = [c \log n]$ and

$$U_n = \max_{0 \leqslant i \leqslant n - K} (S_{i+K} - S_i). \tag{6.66}$$

Then for every $c > 0$,

$$U_n/K \to \alpha(c) \text{ a.s.},$$

where $\alpha(c) = \sup\{x: \rho(x) \geqslant e^{-1/c}\}$ and $\rho(x)$ is defined by (6.63). The function $\alpha(c)$ uniquely determines the d.f. of X_1 (Erdös and Rényi [115]).

Generalizations and extensions of the Erdös–Rényi theorem were obtained by Book [47, 48], Deheuvels and Steinebach [100], Deheuvels and Devroye [98], Komlós and Tusnády [239], and many others. Surveys and extensions can be found in Csörgö [88], Csörgö and Révész [87] and Steinebach [436]. Most papers are devoted to the i.i.d. case. Book [48], Frolov [141], and Lin [254] considered non-identically distributed random variables. Frolov [140, 141] and Lin [254] obtained extensions of the Erdös–Rényi theorem in the case when the Cramér condition (6.65) or its one-sided analogue is violated.

6.8.36 Suppose that $Ee^{tX_1} = \infty$ for every $t > 0$. Then $\limsup U_n/K = \infty$ a.s. for every $c > 0$, where U_n and K are the same as in Subsection 6.8.35 (Steinebach [435]).

6.8.37 Suppose that $Ee^{tX_1} = \infty$ for every $t > 0$. Define T_n by the equality (6.64). Then $\limsup T_n = \infty$ a.s. for all sequences $\{f(n)\}$ for which $r < 1$. Here r is the radius of convergence of the series $\sum_n x^{f(n)}$ (Lynch [262]).

6.8.38 Let X_1 have a non-degenerate distribution with mean zero. Let $R(t) = Ee^{tX_1}$, $t_0 = \sup\{t: R(t) < \infty\} > 0$, $m(t) = R'(t)/R(t)$, $A = \lim_{t \uparrow t_0} m(t)$, $K = [c \log n]$. Define U_n by (6.66). Then for every $\alpha \in (0, A)$ and c such that $e^{-1/c} = \inf_t e^{-t\alpha} R(t)$ we have

$$\limsup (U_n - \alpha K)/\log K = 1/(2t^*) \text{ a.s.}$$

and

$$\liminf (U_n - \alpha K)/\log K = -1/(2t^*) \text{ a.s.},$$

where t^* is defined by $m(t^*) = \alpha$ (Deheuvels et al. [99]).

6.8.39 Let $1 < r < 2$, $EX_1 = 0$, $EX_1^2 = 1$, and let the condition (6.65) be satisfied. We put

$$M_r(n, L) = \max_{0 \leq j \leq n-L} L^{-1/r}(S_{j+L} - S_j).$$

Then

$$M_r(n, [(2\lambda^{-2} \log n)^{r/(2-r)}]) \to \lambda \text{ a.s.}$$

for every $\lambda > 0$ (Book [46]).

6.8.40 Let $V(x)$ be a d.f. satisfying the conditions

$$\int_{-\infty}^{\infty} x \, dV(x) = 0, \quad \int_{-\infty}^{\infty} x^2 \, dV(x) = 1. \tag{6.67}$$

Then there exist a probability space $(\Omega, \mathscr{A}, P)$, a sequence of i.i.d random variables $\{X_n\}$ with the d.f. $V(x)$, and a sequence of i.i.d. random variables $\{Y_n\}$ with the standard normal distribution (both sequences are defined on Ω), such that the sums

$$S_n = \sum_{k=1}^{n} X_k, \quad T_n = \sum_{k=1}^{n} Y_k \tag{6.68}$$

satisfy the relation $S_n - T_n = o((n \log \log n)^{1/2})$ a.s. (Strassen [442]). Another proof is given by Major [267]).

6.8.41 Let $f(n)$ be an arbitrary positive function such that $f(n) \to \infty$. Then there exists a d.f. $V(x)$ satisfying the conditions (6.67) and the following condition: for every pair of sequences of i.i.d. random variables $\{X_n\}$ and $\{Y_n\}$ with the distribution functions $V(x)$ and $\Phi(x)$ respectively we have

$$\limsup |S_n - T_n|(n \log \log n)^{-1/2} f(n) = \infty \text{ a.s.}$$

(Major [267]). Here S_n and T_n are defined by (6.68).

6.8.42 Let $V(x)$ be a d.f. satisfying the conditions (6.67) and

$$\int_{-\infty}^{\infty} e^{tx} \, dV(x) < \infty \quad \text{for } |t| < b \text{ and some } b > 0.$$

Then one can construct sequences of i.i.d. random variables $\{X_n\}$ and $\{Y_n\}$ on a common probability space having the distribution functions $V(x)$ and $\Phi(x)$ respectively and satisfying the condition

$$S_n - T_n = O(\log n) \text{ a.s.,} \tag{6.69}$$

where S_n and T_n are defined by (6.68) (Komlós et al. [238]).

6.8.43 Let $V(x)$ be a d.f. satisfying the conditions (6.67) and

$$\int_{-\infty}^{\infty} |x|^r \, dV(x) < \infty$$

for some $r > 3$. Then the assertion of Subsection 6.8.42 holds with the replacement of the relation (6.69) by $S_n - T_n = o(n^{1/r})$ a.s. (Komlós et al. [238]). This result is a strengthening of a result of Strassen [444]. Various generalizations and strengthenings were obtained by Sahanenko [409, 410] for sums of independent non-identically distributed random variables.

6.8.44 Let $V(x)$ be a d.f. satisfying the conditions (6.67) and

$$\int_{-\infty}^{\infty} x^2 g(|x|) \, dV(x) < \infty,$$

where $g(x)$ is such that $x/g(x)$ and $g(x)/x^\varepsilon$ do not decrease in the interval $x > 0$ for some $\varepsilon > 0$. Then the assertion of Subsection 6.8.42 holds with the replacement of the relation (6.69) by the inequality

$$\limsup \frac{|S_n - T_n|}{h(n)} \leq C \text{ a.s.,}$$

where C is a constant and $h(x)$ is the inverse of $x^2 g(x)$. In particular, if $2 < r \leq 3$, then the conditions (6.67) and $\int_{-\infty}^{\infty} |x|^r \, dV(x) < \infty$ imply the relation $S_n - T_n = o(n^{1/r})$ a.s. (Major [266]).

6.8.45 Let $V(x)$ be a d.f. satisfying the conditions (6.67) and

$$\int_{-\infty}^{\infty} e^{|x|^r} \, dV(x) < \infty, \quad \int_{-\infty}^{\infty} e^{|x|^{2r}} \, dV(x) = \infty$$

for some positive $r < 1$. Then there exist sequences of i.i.d. random variables $\{X_n\}$ and $\{Y_n\}$ on a common probability space such that X_1 has the d.f. $V(x)$, Y_1 has the d.f. $\Phi(x)$, and $S_n - T_n = O((\log n)^{1/r})$ a.s. Here S_n and T_n are defined by (6.68). One cannot replace O by o in the latter relation (Shao [421]).

6.8.46 Let $V(x)$ be a d.f. satisfying the conditions (6.67). We put

$$\sigma_k^2 = \int_{|x| < 2^{n/2}} x^2 \, dV(x) - \left(\int_{|x| < 2^{n/2}} x \, dV(x) \right)^2$$

if $2^n \leq k < 2^{n+1}$. Then one can construct a sequence of independent random variables $\{X_n\}$ with the d.f. $V(x)$ and a sequence of independent normally distributed random variables $\{Y_n\}$ such that $EY_k = 0$, $\text{Var } Y_k = \sigma_k^2$ for all k, and $S_n - T_n = o(n^{1/2})$ a.s. (Major [269]).

Almost-sure approximations of sums of i.i.d. random variables in the domain of attraction of a stable non-normal law by sums of i.i.d. stable random variables were obtained by Stout [44], Mijnheer [300], Christoph [75], and Berkes and Dehling [26]. Weak approximations of this type are due to Simons and Stout [426] and Berkes and Dehling [26].

Surveys on strong approximations of sums of independent random variables can be found in Major [268], Csörgő and Révész [87], Csörgő and Hall [89], and Steinebach [436].

6.8.47 Let $EX_1 = 0$ and $Var\, X_1 = 1$. Then

$$\frac{1}{\log n} \sum_{k=1}^{n} k^{-1} I(k^{-1/2} S_k < x) \to \Phi(x) \text{ a.s.} \qquad (6.70)$$

for every real x, where I is the indicator function (Lacey and Philipp [246] and Fisher [134]). Under some stronger moment conditions this result was proved by Brosamler [52] and Schatte [417]. Berkes and Dehling [27] obtained extensions to the case when the distributions of X_1 belongs to the domain of attraction of an arbitrary stable law.

6.8.48 Let $\{X_n\}$ be a sequence of random variables. Suppose that there exists a sequence of i.i.d. random variables $\{Y_n\}$ having the standard normal distribution such that

$$\sum_{k=1}^{n} X_k - \sum_{k=1}^{n} Y_k = o(n^{1/2}) \text{ a.s.}$$

Then the relation (6.70) holds (Lacey and Philipp [246]).

6.8.49 Let $\{A_n\}$ be a sequence of events on a probability space $(\Omega, \mathscr{A}, P)$. We say that $\{A_n\}$ is a Borel–Cantelli sequence of events if either $\sum P(A_n) < \infty$ or $\sum P(A_n) = \infty$ and $P(A_n \text{ i.o.}) = 1$. The sequence of events $\{A_n\}$ is a Borel–Cantelli sequence if it satisfies one of the following conditions:

(a) A_n are pairwise independent;
(b) the relation (6.1) holds;
(c) $\lim(P(A_n B) - P(A_n)P(B)) = 0$ for every $B \in \mathscr{A}$ and $P(A_n) \not\to 0$ (Fishler [135]);
(d) the series $\sum (P(A_n B) - P(A_n) P(B))$ converges for every $B \in \mathscr{A}$;
(e) there exists a sequence of independent events $\{B_n\}$, $B_n \in \mathscr{A}$, such that $A_n \subset B_n$ for $n \geq 1$, $P(A_n)/P(B_n) \to 1$, and $\sum P(B_n) = \infty$ (Anděl and Dupač [8]). It is possible to replace the independence condition by the weaker condition of pairwise independence of the events B_n (Martikainen and Petrov [291]).

7

Strong limit theorems: the law of the iterated logarithm

Let $\{X_n; n = 1, 2, \ldots\}$ be a sequence of independent random variables, and let $\{b_n; n = 1, 2, \ldots\}$ be a sequence of real numbers. We put $S_n = \sum_{k=1}^{n} X_k$. We shall say that the sequence $\{b_n\}$ belongs to the lower class if $P(S_n > b_n \text{ i.o.}) = 1$, and to the upper class if $P(S_n > b_n \text{ i.o.}) = 0$. We shall be investigating the conditions under which relations of the type

$$\limsup \frac{S_n}{a_n} = 1 \text{ a.s.}$$

or

$$\limsup \frac{S_n}{a_n} \leq 1 \text{ a.s.}$$

hold, where $a_n \to \infty$. The first equality is equivalent to the statement that for every $\delta > 0$ the sequence $\{(1 - \delta)a_n\}$ belongs to the lower class and the sequence $\{(1 + \delta)a_n\}$ to the upper class.

7.1 Kolmogorov's theorem

In this section we shall consider a sequence of independent random variables $\{X_n\}$ with zero means and finite variances. We put

$$\sigma_n^2 = \text{Var } X_n, \quad B_n = \sum_{k=1}^{n} \sigma_k^2.$$

The following fundamental result is due to Kolmogorov.

Theorem 7.1 *Suppose $B_n \to \infty$. Suppose also that there exists a sequence of positive constants $\{M_n\}$ such that*

$$M_n = o\left(\left(\frac{B_n}{\log \log B_n}\right)^{1/2}\right) \tag{7.1}$$

and

$$|X_n| \leq M_n \text{ a.s.} \tag{7.2}$$

Then

$$\limsup \frac{S_n}{(2B_n \log \log B_n)^{1/2}} = 1 \text{ a.s.} \quad (7.3)$$

Obviously, if the sequence $\{X_n\}$ satisfies the conditions of this theorem, so does the sequence $\{-X_n\}$. Accordingly, the conditions of Theorem 7.1 imply that

$$\liminf \frac{S_n}{(2B_n \log \log B_n)^{1/2}} = -1 \text{ a.s.} \quad (7.4)$$

and, therefore,

$$\limsup \frac{|S_n|}{(2B_n \log \log B_n)^{1/2}} = 1 \text{ a.s.} \quad (7.5)$$

We shall need a few lemmas.

Lemma 7.1 *We put $q_n(x) = P(S_n \geq x)$. If $0 \leq xM_n \leq B_n$, then*

$$q_n(x) \leq \exp\left\{-\frac{x^2}{2B_n}\left(1 - \frac{xM_n}{2B_n}\right)\right\}. \quad (7.6)$$

If $xM_n \geq B_n$, we have

$$q_n(x) \leq \exp\left\{-\frac{x}{4M_n}\right\}. \quad (7.7)$$

Proof Suppose $t > 0$ and $tM_n \leq 1$. The condition (7.2) implies that $|EX_n^k| \leq M_n^{k-2}\sigma_n^2$ for every $k \geq 2$. Therefore,

$$Ee^{tX_n} = 1 + \sum_{k=2}^{\infty} \frac{t^k}{k!} EX_n^k$$

$$\leq 1 + \frac{t^2}{2}\sigma_n^2\left(1 + \frac{t}{3}M_n + \frac{t^2}{12}M_n^2 + \cdots\right)$$

$$\leq 1 + \frac{t^2}{2}\sigma_n^2\left(1 + \frac{t}{2}M_n\right) \leq \exp\left\{\frac{t^2}{2}\sigma_n^2\left(1 + \frac{t}{2}M_n\right)\right\}.$$

Without loss of generality we may suppose that the sequence $\{M_n\}$ is non-decreasing. We have

$$Ee^{tS_n} = \prod_{k=1}^{n} Ee^{tX_k} \leq \exp\left\{\frac{t^2}{2}B_n\left(1 + \frac{t}{2}M_n\right)\right\}.$$

Consequently,

$$q_n(x) \leq e^{-tx} Ee^{tS_n} \leq \exp\left\{-tx + \frac{t^2}{2}B_n\left(1 + \frac{t}{2}M_n\right)\right\}.$$

Putting here $t = x/B_n$ when $xM_n \leq B_n$ and $t = 1/M_n$ when $xM_n \geq B_n$, we obtain the inequalities (7.6) and (7.7) respectively. □

Lemma 7.2 *If $x > 0$, $xM_n/B_n \to 0$, and $x^2/B_n \to \infty$, then for every fixed $\mu > 0$ and all sufficiently large n we have*

$$q_n(x) \geq \exp\left\{-\frac{x^2}{2B_n}(1 + \mu)\right\}. \tag{7.8}$$

Proof The inequality $1 + x \geq \exp\{x(1-x)\}$ holds for every $x \geq 0$. If $0 \leq tM_n \leq 1$, we have

$$Ee^{tX_n} \geq 1 + \frac{t^2}{2}\sigma_n^2\left(1 - \frac{t}{3}M_n - \frac{t^2}{12}M_n^2 - \cdots\right)$$

$$\geq 1 + \frac{t^2}{2}\sigma_n^2\left(1 - \frac{t}{2}M_n\right) \geq \exp\left\{\frac{t^2}{2}\sigma_n^2\left(1 - \frac{t}{2}M_n - \frac{t^2}{2}\sigma_n^2\right)\right\}$$

$$\geq \exp\left\{\frac{t^2}{2}\sigma_n^2(1 - tM_n)\right\},$$

$$Ee^{tS_n} \geq \exp\left\{\frac{t^2}{2}B_n(1 - tM_n)\right\}.$$

We put $t = x/(1-\delta)B_n$, where δ is a small positive number to be chosen later. Then $tM_n \to 0$, and for every fixed $\alpha > 0$ we have

$$Ee^{tS_n} \geq \exp\left\{\frac{B_n}{2}t^2(1 - \alpha)\right\} \tag{7.9}$$

for all sufficiently large n. Furthermore,

$$Ee^{tS_n} = -\int_{-\infty}^{\infty} e^{ty}\,dq_n(y) = t\int_{-\infty}^{\infty} e^{ty}q_n(y)\,dy = t\sum_{k=1}^{5} I_k, \tag{7.10}$$

where $I_1, \ldots, I_5$ are the integrals of $e^{ty}q_n(y)$ over the intervals $(-\infty, 0)$, $(0, t(1-\delta)B_n)$, $(t(1-\delta)B_n, t(1+\delta)B_n)$, $(t(1+\delta)B_n, 8tB_n)$, and $(8tB_n, \infty)$ respectively. Clearly,

$$tI_1 \leq t\int_{-\infty}^{0} e^{ty}\,dy = 1.$$

If $yM_n \geq B_n$, we have $q_n(y) \leq \exp\{-y/4M_n\} \leq \exp\{-2ty\}$ for all sufficiently large n, by Lemma 7.1. In the interval $8tB_n \leq y \leq B_n/M_n$ the same lemma implies that $q_n(y) \leq \exp\{-y^2/4B_n\} \leq \exp\{-2ty\}$. Hence

$$tI_5 \leq t\int_{8tB_n}^{\infty} e^{-ty}\,dy < 1.$$

Taking account of (7.9), we obtain

$$tI_1 + tI_5 < 2 < \tfrac{1}{4} E e^{tS_n} \qquad (7.11)$$

for sufficiently large n.

To estimate the integrals I_2 and I_4 we use (7.6) and the condition $xM_n/B_n \to 0$. Then we get

$$q_n(y) \leq \exp\left\{-\frac{y^2}{2B_n}(1-\beta)\right\}$$

for every fixed $\beta > 0$ and sufficiently large n. Thus we arrive at the inequality

$$tI_2 + tI_4 \leq t\int_D e^{\psi(y)}\,dy$$

for large n, where

$$\psi(y) = ty - \frac{y^2}{2B_n}(1-\beta)$$

and

$$D = (0, t(1-\delta)B_n) \cup (t(1+\delta)B_n, 8tB_n).$$

The function $\psi(y)$ has a maximum at the point $y = tB_n/(1-\beta)$, which lies in the interval $(t(1-\delta)B_n, t(1+\delta)B_n)$, if β is chosen to be small enough. Therefore,

$$\sup_{y \in D} \psi(y) = \max\{\psi(t(1-\delta)B_n), \psi(t(1+\delta)B_n)\}.$$

Furthermore,

$$\psi(t(1 \pm \delta)B_n) = \frac{t^2}{2}(1-\delta^2+\beta(1\pm\delta)^2) \leq \frac{t^2 B_n}{2}\left(1-\frac{\delta^2}{2}\right),$$

if $\beta < \delta^2/2(1+\delta)^2$. Therefore,

$$tI_2 + tI_4 \leq 8t^2 B_n \exp\left\{\frac{t^2 B_n}{2}\left(1-\frac{\delta^2}{2}\right)\right\}.$$

Making use of the relations $t^2 B_n = x^2/(1-\delta)^2 B_n \to \infty$ and (7.9), we obtain

$$32t^2 B_n \leq \exp\left\{\frac{t^2 B_n}{8}\delta^2\right\}$$

and

$$tI_2 + tI_4 \leq \tfrac{1}{4}\exp\left\{\frac{t^2 B_n}{2}\left(1-\frac{\delta}{4}\right)^2\right\} \leq \tfrac{1}{4} E e^{tS_n} \qquad (7.12)$$

for sufficiently large n.

The function $q_n(y)$ does not increase. Taking account of the equality $x = (1-\delta)tB_n$, we get

$$tI_3 \leq 2\delta t^2 B_n \exp\{t^2 B_n(1+\delta)\} q_n(x).$$

In view of (7.10)–(7.12) we have $tI_3 > \frac{1}{2} E e^{tS_n}$. Using (7.9), we find that

$$q_n(x) \geq \frac{1}{2t^2 B_n} \exp\left\{-\frac{t^2 B_n}{2}(1+\alpha+2\delta)\right\}$$

$$\geq \exp\left\{-\frac{x^2}{2B_n(1-\delta)^2}\left(1+\alpha+2\delta+\frac{\delta^2}{4}\right)\right\}$$

for sufficiently large n, if $\delta < 1/2$.

Let μ be an arbitrary positive number. We choose the positive numbers δ and α so that

$$\left(1+\alpha+2\delta+\frac{\delta^2}{4}\right)(1-\delta)^{-2} < 1+\mu.$$

Then (7.8) holds for sufficiently large n. □

In what follows we put $\chi(n) = (2B_n \log \log B_n)^{1/2}$ for sufficiently large n.

Lemma 7.3 *If the conditions of Theorem 7.1 are satisfied, then for all positive constants b and μ and for all sufficiently large n the following inequalities hold:*

$$(\log B_n)^{-(1+\mu)b^2} \leq P(S_n \geq b\chi(n)) \leq (\log B_n)^{-(1-\mu)b^2} \qquad (7.13)$$

Proof The right-hand inequality follows from Lemma 7.1 for $x = b\chi(n)$. We have $xM_n/B_n \to 0$ and $x^2/B_n \to \infty$, so that we can apply Lemma 7.2 and the inequality (7.6). The left-hand inequality in (7.13) follows from Lemma 7.2. □

We shall prove that for every $\varepsilon > 0$ the sequence $(1+\varepsilon)\chi(n)$ belongs to the upper class, i.e. that

$$P(S_n > (1+\varepsilon)\chi(n) \text{ i.o.}) = 0. \qquad (7.14)$$

Taking account of the relations $B_n \to \infty$ and (7.1), we obtain

$$\frac{B_n}{B_{n+1}} = 1 - \frac{\sigma_{n+1}^2}{B_{n+1}} = 1 + o\left(\frac{1}{\log \log B_{n+1}}\right) \to 1.$$

For every $\tau > 0$ there exists a non-decreasing sequence of integers $\{n_k\}$ such that $n_k \to \infty$ as $k \to \infty$ and

$$B_{n_k-1} \leq (1+\tau)^k < B_{n_k} \qquad (k=1,2,\ldots). \qquad (7.15)$$

(We set $B_0 = 0$.) Hence
$$\frac{B_{n_k} - \sigma^2_{n_k}}{B_{n_k}} \leq \frac{(1+\tau)^k}{B_{n_k}} < 1$$
for all k. Therefore,
$$B_{n_k} \sim (1+\tau)^k \qquad (7.16)$$
and
$$B_{n_k} - B_{n_{k-1}} = B_{n_k}\left(1 - \frac{B_{n_{k-1}}}{B_{n_k}}\right) \sim B_{n_k} \frac{\tau}{1+\tau} \qquad (7.17)$$
as $k \to \infty$.

We write $\bar{S}_{n_k} = \max_{n \leq n_k} S_n$. We shall show that
$$\sum_k P(\bar{S}_{n_k} > (1+\gamma)\chi(n_k)) < \infty \qquad (7.18)$$
for every $\gamma > 0$. Using inequality (2.13) (Chapter 2), we obtain
$$P(\bar{S}_{n_k} > (1+\gamma)\chi(n_k)) \leq 2P(S_{n_k} > (1+\gamma)\chi(n_k) - \sqrt{2B_{n_k}})$$
$$\leq 2P(S_{n_k} > (1+\gamma_1)\chi(n_k))$$
for every positive $\gamma_1 < \gamma$ and sufficiently large k. Lemma 7.3 and (7.15) imply that
$$P(S_{n_k} > (1+\gamma_1)\chi(n_k)) \leq (\log B_{n_k})^{-(1-\mu)(1+\gamma_1)^2} \leq (k \log(1+\tau))^{-(1-\mu)(1+\gamma_1)^2}$$
for every positive μ and γ_1 and sufficiently large k. If we choose μ to be so small that $(1-\mu)(1+\gamma_1)^2 > 1$, we obtain (7.18) from our estimates.

Further, for every $\varepsilon > 0$ we have
$$P(S_n > (1+\varepsilon)\chi(n) \text{ i.o.}) \leq P\left(\max_{n_{k-1} \leq n \leq n_k} S_n > (1+\varepsilon)\chi(n_{k-1}) \text{ i.o.}\right)$$
$$\leq P(\bar{S}_{n_k} > (1+\varepsilon)\chi(n_{k-1}) \text{ i.o.}).$$
It follows from (7.16) that $\chi(n_k)/\chi(n_{k-1}) < (1+2\tau)^{1/2}$ for sufficiently large k. Hence
$$P(S_n > (1+\varepsilon)\chi(n) \text{ i.o.}) \leq P(\bar{S}_{n_k} > (1+\varepsilon)(1+2\tau)^{-1/2}\chi(n_k) \text{ i.o.})$$
$$\leq P(\bar{S}_{n_k} > (1+\varepsilon/2)\chi(n_k) \text{ i.o.})$$
if τ is sufficiently small. Using (7.18) and the Borel–Cantelli lemma, we obtain (7.14).

Replacing S_n by $-S_n$, we get
$$P(-S_n > (1+\varepsilon)\chi(n) \text{ i.o.}) = 0$$

and, therefore,
$$P(|S_n| > (1 + \varepsilon)\chi(n) \text{ i.o.}) = 0 \tag{7.19}$$
for every $\varepsilon > 0$.

To complete the proof of Theorem 7.1 we have to show that
$$P(S_n > (1 - \varepsilon)\chi(n) \text{ i.o.}) = 1 \tag{7.20}$$
for every $\varepsilon > 0$. We write
$$\psi(n_k) = (2(B_{n_k} - B_{n_{k-1}}) \log \log(B_{n_k} - B_{n_{k-1}}))^{1/2}.$$
Making use of (7.16), we find that
$$\log(B_{n_k} - B_{n_{k-1}}) < \log B_{n_k} < 2k \log(1 + \tau)$$
for sufficiently large k. Furthermore, $\psi(n_k)/\chi(n_{k-1}) \sim \tau^{1/2}$ by (7.17). If A and B are any events, then $P(A \cap B) = P(A) - P(A \cap \bar{B}) \geq P(A) - P(\bar{B})$. Therefore, for every positive $\gamma < 1$ we have

$P(S_{n_k} - S_{n_{k-1}} > (1 - \gamma)\psi(n_k))$
$$\geq P([S_{n_k} > (1 - \gamma/2)\psi(n_k)] \cap [S_{n_{k-1}} < \gamma\psi(n_k)/2])$$
$$\geq P(S_{n_k} > (1 - \gamma/2)\psi(n_k)) - P(S_{n_{k-1}} \geq \gamma\psi(n_k)/2)$$
$$\geq P(S_{n_k} > (1 - \gamma/2)\chi(n_k)) - P(S_{n_k} \geq \gamma\tau^{1/2}\chi(n_{k-1})/3)$$
$$\geq (\log B_{n_k})^{-(1+\mu)(1-\gamma/2)^2} - (\log B_{n_{k-1}})^{-\gamma^2\tau/10},$$

by Lemma 7.3. For sufficiently large k and τ the last difference is greater than
$$C[k^{-(1+\mu)(1-\gamma/2)^2} - k^{-\gamma^2\tau/10}] > \frac{C}{2} k^{-(1+\mu)(1-\gamma/2)^2}$$

Here C is a positive constant not dependent on k. If we choose μ to be so small that $(1 + \mu)(1 - \gamma/2)^2 < 1$, then
$$\sum_k P(S_{n_k} - S_{n_{k-1}} > (1 - \gamma)\psi(n_k)) = \infty.$$

Applying the Borel–Cantelli lemma once again, we obtain
$$P(S_{n_k} - S_{n_{k-1}} > (1 - \gamma)\psi(n_k) \text{ i.o.}) = 1 \tag{7.21}$$
for every positive $\gamma < 1$.

Furthermore,
$$(1 - \gamma)\psi(n_k) - 2\chi(n_{k-1}) \sim [(1 - \gamma)\tau^{1/2}(1 + \tau)^{-1/2} - 2(1 + \tau)^{-1/2}]\chi(n_k)$$

as $k \to \infty$. Let Ω be the space of elementary events on which the random variables of the sequence $\{X_n\}$ are defined. It follows from (7.19) that $|S_n(\omega)| \leq 2\chi(n)$ for $n > n_0(\omega)$ and for all $\omega \in \Omega$ except for a set of points having probability measure zero. If ε is an arbitrary positive number, we choose the positive numbers γ and τ in such a way that

$$(1-\gamma)\tau^{1/2}(1+\tau)^{-1/2} - 2(1+\tau)^{-1/2} > 1 - \varepsilon.$$

Taking (7.21) into account, we then obtain

$$P(S_{n_k} > (1-\varepsilon)\chi(n_k) \text{ i.o.}) \geqslant P(S_{n_k} > (1-\gamma)\psi(n_k) - 2\chi(n_{k-1}) \text{ i.o.})$$
$$\geqslant P(S_{n_k} - S_{n_{k-1}} > (1-\gamma)\psi(n_k) \text{ i.o.}) = 1.$$

This implies (7.20). □

Theorem 7.1 is in some sense as sharp as possible. Namely, if we replace o by O in the condition (7.1), the assertion of the theorem may fail.

Using the arguments that we employed in the proof of Theorem 7.1, we can prove the following proposition on the connection between an estimate of the remainder term in the central limit theorem and the law of the iterated logarithm, i.e. the relation (7.3).

Let $\{X_n\}$ be a sequence of independent random variables with zero means and finite variances. We put

$$S_n = \sum_{k=1}^n X_k, \quad B_n = \sum_{k=1}^n \text{Var } X_k,$$

$$\Delta_n = \sup_x |P(S_n < xB_n^{1/2}) - \Phi(x)|,$$

where $\Phi(x)$ stands for the standard normal distribution function.

Theorem 7.2 *If*

$$B_n \to \infty, \tag{7.22}$$

$$B_{n+1}/B_n \to 1, \tag{7.23}$$

and

$$\Delta_n = ((\log B_n)^{-1-\delta}) \quad \text{for some } \delta > 0, \tag{7.24}$$

then the relation (7.3) holds.

Proof We put $\chi(n) = (2B_n \log \log B_n)^{1/2}$ for sufficiently large n. The theorem will be proved if we prove (7.14) and (7.20) for every $\varepsilon > 0$.

Since $1 - \Phi(x) \sim (1/x\sqrt{2\pi}) e^{-x^2/2}$ as $x \to \infty$ by the relation (5.63) (Chapter 5), we have

$$1 - \Phi(b(2 \log \log B_n)^{1/2}) = \frac{1 + o(1)}{2b(\pi \log \log B_n)^{1/2}(\log B_n)^{b^2}}$$

for every positive constant b. It follows from (7.24) that

$$P(S_n \geq b\chi(n)) = \frac{1}{\sqrt{2\pi}} \int_{b(2\log\log B_n)^{1/2}}^{\infty} e^{-t^2/2} \, dt + O((\log B_n)^{-1-\delta}).$$

In turn, this implies that

$$(\log B_n)^{-(1+c)b^2} \leq P(S_n \geq b\chi(n)) \leq (\log B_n)^{-b^2}$$

for all positive constants c and $b < (1 + \delta)^{1/2}$ and for all sufficiently large n.

We now use the arguments that were employed in the proof of (7.14) and (7.20). In the proof of (7.18) we set $\mu = 0$ and choose the positive number $\gamma_1 < \gamma$ in such a way that $\gamma_1(\gamma_1 + 2) < \delta$, where δ is the constant in (7.24). The rest of the argument is unchanged. $\square$

It it not hard to prove that Theorem 7.2 is still true if we replace (7.24) by the weaker condition

$$\Delta_n = O((\log B_n)^{-1}(\log\log B_n)^{-1-\delta})$$

for some $\delta > 0$. However, Theorem 7.2 fails if $\delta = 0$ in (7.24).

We shall formulate a consequence of Theorem 7.2.

Theorem 7.3 *If the sequence $\{X_n\}$ satisfies the conditions*

$$\liminf B_n/n > 0 \tag{7.25}$$

and

$$\limsup \frac{1}{n} \sum_{k=1}^{n} EX_k^2 |\log |X_k||^{1+\delta} < \infty \tag{7.26}$$

for some $\delta > 0$, then (7.3) holds.

Proof For the Lindeberg ratio

$$\Lambda_n(\varepsilon) = B_n^{-1} \sum_{k=1}^{n} \int_{|x| \geq \varepsilon B_n^{1/2}} x^2 \, dV_k(x),$$

where $V_k(x)$ is the d.f. of X_k, we have the estimates

$$\Lambda_n(\varepsilon) \leq B_n^{-1} \sum_{k=1}^{n} \int_{|x| \geq \varepsilon B_n^{1/2}} \frac{x^2 (\log |x|)^{1+\delta}}{(\log(\varepsilon B_n^{1/2}))^{1+\delta}} \, dV_k(x)$$

$$\leq B_n^{-1} (\log(\varepsilon B_n^{1/2}))^{-1-\delta} \sum_{k=1}^{n} EX_k^2 |\log |X_k||^{1+\delta}$$

$$\leq C(\log(\varepsilon B_n^{1/2}))^{-1-\delta}$$

248 | Limit theorems of probability theory

for sufficiently large n by (7.25) and (7.26). Using (7.25) again, we obtain $\Lambda_n(\varepsilon) \to 0$ for every fixed $\varepsilon > 0$. By Theorem 4.7 (Chapter 4) the relation (7.23) holds. If (7.26) is satisfied, it is also satisfied when $|\log |X_k||^{1+\delta}$ is replaced by $(\log(3 + |X_k|))^{1+\delta}$. The function $g(x) = (\log(3 + |x|))^{1+\delta}$ satisfies all the conditions of Theorem 5.6 (Chapter 5). Therefore, the estimate (7.24) holds for positive δ satisfying (7.26). It remains to refer to Theorem 7.2. □

One can formulate conditions that are sufficient for the hypotheses of Theorem 7.2 to be satisfied but do not require the existence of moments of order higher than two, contrary to Theorem 7.3. The inequality (5.26) (Chapter 5) implies that for (7.24) it is sufficient that

$$\Lambda_n(\gamma(\log B_n)^{-1-\delta}) = O((\log B_n)^{-1-\delta}) \tag{7.27}$$

for some positive constants γ and δ. Here $\Lambda_n(\varepsilon)$ is the Lindeberg ratio. Thus if $B_n \to \infty$ and (7.27) holds, then the sequence $\{X_n\}$ obeys the law of the iterated logarithm, i.e. the relation (7.3) holds.

7.2 The Hartman–Wintner theorem

Another basic result is the Hartman–Wintner theorem. We shall formulate it in Strassen's form.

Theorem 7.4 *Let $\{X_n\}$ be a sequence of independent identically distributed random variables with zero mean and finite variance σ^2. We put $S_n = \sum_{k=1}^{n} X_k$, $a_n = (2n \log \log n)^{1/2}$. Then*

$$\limsup S_n/a_n = \sigma \text{ a.s.}, \tag{7.28}$$

$$\liminf S_n/a_n = -\sigma \text{ a.s.} \tag{7.29}$$

and every point of the closed interval $[-\sigma, \sigma]$ is the limit point (in the sense of almost sure convergence) for the sequence $\{S_n/a_n\}$.

Proof We set $LLn = \log \log n$ for $n \geq 3$, $LLn = 1$ for $n = 1$ and $n = 2$. In order that the sequence $\{a_n\}$ be defined for all n and be positive, we set $a_n = (2n\,LLn)^{1/2}$ for all positive integers n.

Lemma 7.4 *Let $\{X_n\}$ be a sequence of independent identically distributed random variables with finite variance. Then there exists a sequence of positive numbers $\{\tau_n\}$ such that $\tau_n \downarrow 0$, $\tau_n(n/LLn)^{1/2} \uparrow \infty$, and*

$$U_n/a_n \to 0 \text{ a.s.}, \tag{7.30}$$

where $U_n = \sum_{k=1}^{n} Z_k$, $Z_k = X_k I(|X_k| \geq \tau_k (k/LLk)^{1/2})$.

Proof It suffices to show that the series $\sum E|Z_n|/a_n$ converges, since this implies the convergence a.s. of the series $\sum Z_n/a_n$ and consequently the relation (7.30) by Kronecker's lemma.

The condition $EX_1^2 < \infty$ implies the existence of an even non-negative function $f(x)$ such that $f(x)$ does not decrease in the interval $x > 0$, $f(x) \to \infty$ as $x \to \infty$. and $EX_1^2 f(X_1) < \infty$. We may also assume that

$$f(n^{1/3})\left(\frac{LLn}{n}\right)^{1/2} \downarrow 0, \quad f(n) \leqslant n^{1/3}.$$

We put $\tau_n = 1/f(n^{1/3})$, $b_n = \tau_n(n/LLn)^{1/2}$. We have

$$\sum_{n=1}^{\infty} E|Z_n|/a_n = \sum_{n=1}^{\infty} a_n^{-1} \sum_{m=n}^{\infty} E|X_n|I(b_m \leqslant |X_n| < b_{m+1})$$

$$\leqslant \sum_{n=1}^{\infty} a_n^{-1} \sum_{m=n}^{\infty} b_{m+1} P(b_m \leqslant |X_1| < b_{m+1})$$

$$= \sum_{m=1}^{\infty} b_{m+1} P(b_m \leqslant |X_1| < b_{m+1}) \sum_{n=1}^{m} a_n^{-1}.$$

Furthermore, $\sum_{n=1}^{m} a_n^{-1} \leqslant cb_m/\tau_m$ and $b_{m+1} \leqslant cb_m$ for every m. Therefore,

$$\sum_{n=1}^{\infty} \frac{E|Z_n|}{a_n} \leqslant c_1 \sum_{m=1}^{\infty} \frac{b_m^2}{\tau_m} P(b_m \leqslant |X_1| < b_{m+1}) < \infty,$$

since $b_m^2/\tau_m = b_m^2 f(m^{1/3})$ and $EX_1^2 f(X_1) < \infty$. $\square$

If $\sigma = 0$, the assertions of the theorem are obvious. Suppose that $\sigma > 0$. Let $\{\tau_n\}$ be a sequence of numbers from Lemma 7.4. We put

$$X_n' = X_n I(|X_n| < \tau_n(n/LLn)^{1/2}), \quad Y_n = X_n' - EX_n',$$

$$Z_n = X_n - X_n', \quad S_n' = \sum_{k=1}^{n} X_k', \quad T_n = \sum_{k=1}^{n} Y_k, \quad U_n = \sum_{k=1}^{n} Z_k.$$

Using the equality $EX_n' = -EZ_n$ and the results stated in the proof of Lemma 7.4, we obtain

$$\sum_{n=1}^{\infty} \frac{|EX_n'|}{a_n} \leqslant \sum_{n=1}^{\infty} \frac{|EZ_n|}{a_n} < \infty$$

and, therefore,

$$\frac{1}{a_n}|ES_n'| \leqslant \frac{1}{a_n} \sum_{k=1}^{n} |EX_k'| \to 0, \qquad (7.31)$$

by Kronecker's lemma.

Further, $EY_n^2 \to \sigma^2$, and $1/n \sum_{k=1}^n EY_k^2 \to \sigma^2$. Applying Theorem 7.1 to the sequence $\{Y_n\}$, we get

$$\limsup T_n/a_n = \sigma \text{ a.s.}, \quad \liminf T_n/a_n = -\sigma \text{ a.s.} \qquad (7.32)$$

From (7.31) and Lemma 7.4 we conclude that

$$\limsup |S_n - T_n|/a_n \leq \limsup |ES_n'|/a_n + \limsup |U_n|/a_n = 0 \text{ a.s.}$$

Taking (7.32) into account, we arrive at the equalities (7.28) and (7.29). We shall show that

$$\frac{S_{n+1}}{a_{n+1}} - \frac{S_n}{a_n} \to 0 \text{ a.s.} \qquad (7.33)$$

Obviously,

$$\frac{S_{n+1}}{a_{n+1}} - \frac{S_n}{a_n} = \frac{X_{n+1}}{a_{n+1}} + S_n\left(\frac{1}{a_{n+1}} - \frac{1}{a_n}\right). \qquad (7.34)$$

In view of the condition $EX_1^2 < \infty$ and Lemma 6.14 (Chapter 6), we have $\sum P(X_1^2 \geq n) < \infty$ and, therefore, $\sum P(|X_n| \geq \varepsilon a_n) < \infty$ for every $\varepsilon > 0$. Accordingly, $X_n/a_n \to 0$ a.s. by the Borel–Cantelli lemma.

We introduce the function $a(x) = (2x \log \log x)^{1/2}$ for $x \geq 3$. Using the relations $a'(x) \sim ((\log \log x)/(2x))^{1/2}$ as $x \to \infty$ and

$$\frac{1}{a_n} - \frac{1}{a_{n+1}} = \frac{1}{a_n a_{n+1}} \int_n^{n+1} a'(x) \, dx \sim \frac{1}{a_n^2}\left(\frac{\log \log n}{2n}\right)^{1/2}$$

$$= (2n)^{-3/2}(\log \log n)^{-1/2},$$

we conclude that $S_n(1/a_n - 1/a_{n+1}) \to 0$ a.s. by the strong law of large numbers. Therefore, (7.34) implies (7.33).

To complete the proof of Theorem 7.4, it suffices to show that every point of the set $[-\sigma, \sigma]$ is the limit point for the sequence $\{S_n/a_n\}$. This assertion follows from (7.28), (7.29), (7.33), and the following elementary proposition, which is easy to prove: if $\{x_n\}$ is a sequence of numbers such that $x_n - x_{n-1} \to 0$, then the set of the limit points of the sequence $\{x_n\}$ coincides with the set $[\liminf x_n, \limsup x_n]$. □

Theorem 7.4 implies that the set of the limit points for the sequence $S_n/(2n \log \log n)^{1/2}$ coincides with the set $[-\sigma, \sigma]$.

7.3 The generalized law of the iterated logarithm

In this section we do not suppose the existence of any moments of the random variables under consideration.

Let $\{X_n; n = 1, 2, \ldots\}$ be a sequence of independent random variables, and let $\{a_n; n = 1, 2, \ldots\}$ be a non-decreasing sequence of positive numbers such that $a_n \to \infty$.

We set $S_n = \sum_{k=1}^{n} X_k$ for $n \geq 1$, $S_0 = 0$, and $a_0 = 0$.

We introduce the following condition (the condition (B)): for every $\delta > 0$ there exists $\lambda > 0$ such that

$$P(S_n - S_j \geq -\delta a_n) \geq \lambda$$

for all non-negative $j \leq n$ and all sufficiently large n. This condition (B) is satisfied if the random variables $X_1, X_2, \ldots$ are symmetric and also in the case when $S_n/a_n \xrightarrow{P} 0$.

For every fixed number $c > 1$ and for every integer $n \geq 0$ we define $i_n = i_n(c)$ as the greatest integer satisfying the condition $a_{i_n} \leq c^n$.

Theorem 7.5 *Suppose that the condition (B) is satisfied. Then the relation*

$$\limsup S_n/a_n \leq 1 \text{ a.s.} \tag{7.35}$$

is equivalent to any one of the following conditions:
(D_1) *for every $\varepsilon > 0$ and every non-decreasing sequence of positive integers $k_n \to \infty$,*

$$\sum_{n=2}^{\infty} P(S_{k_n} - S_{k_{n-1}} > (1 + \varepsilon) a_{k_n}) < \infty; \tag{7.36}$$

(E_1) *for every $\varepsilon > 0$, every integer $r \geq 1$ and every $c > 1$,*

$$\sum_{n=r}^{\infty} P(S_{i_n} - S_{i_{n-r}} > (1 + \varepsilon) c^n) < \infty. \tag{7.37}$$

The proof of Theorem 7.5 makes use of the following lemma in which $\{A_n; n = 1, 2, \ldots\}$ and $\{B_n; n = 1, 2, \ldots\}$ are sequences of events defined on a common probability space.

Lemma 7.5 *Suppose that $P(B_n) \geq \alpha$ for all n, where α is a positive constant. If the following pairs of events are independent for every n: A_n and B_n, A_n and $B_n A_{n-1} B_{n-1}$, A_n and $B_n \overline{A_{n-1} B_{n-1}} \overline{A_{n-2} B_{n-2}}, \ldots$, and if $\sum_{n=1}^{\infty} P(A_n) = \infty$, then $P(A_n B_n \text{ i.o.}) \geq \alpha$.*

Proof Suppose that the assertion of the lemma does not hold, i.e.

$$P(A_n B_n \text{ i.o.}) < \alpha.$$

Then there exist $\delta > 0$ and $m \in \mathbb{N}$ such that the condition $k > m$ implies the inequality

$$P\left(\bigcup_{n=m}^{k-1} A_n B_n\right) < \alpha - \delta. \tag{7.38}$$

252 | Limit theorems of probability theory

We have
$$\bigcup_{n=m}^{\infty} D_n = D_m \cup \bar{D}_m D_{m+1} \cup \bar{D}_m \bar{D}_{m+1} D_{m+2} \cup \cdots$$

for every sequence of events $\{D_n\}$ and every m. Putting $D_n = A_n B_n$, we obtain

$$P\left(\bigcup_{n=m}^{\infty} A_n B_n\right) = P(A_m)P(B_m) + P(A_{m+1})P(B_{m+1}\overline{A_m B_m})$$
$$+ P(A_{m+2})P(B_{m+2}\overline{A_{m+1}B_{m+1}}\,\overline{A_m B_m}) + \cdots$$

For all events A and B we have $P(AB) = P(A) - P(A\bar{B})$. Therefore,

$$P\left(\bigcup_{n=m}^{\infty} A_n B_n\right) = P(A_m)P(B_m) + P(A_{m+1})\{P(B_{m+1}) - P(B_{m+1}A_m B_m)\}$$
$$+ P(A_{m+2})\{P(B_{m+2}) - P(B_{m+2} \cap (A_m B_m \cup A_{m+1}B_{m+1}))\} + \cdots.$$

It follows from the condition $P(B_n) \geq \alpha$ and (7.38) that

$$P\left(\bigcup_{n=m}^{\infty} A_n B_n\right) \geq \sum_{n=m}^{\infty} P(A_n)\left(\alpha - P\left(B_n \cap \bigcup_{k=m}^{n-1} A_k B_k\right)\right)$$
$$\geq \sum_{n=m}^{\infty} P(A_n)(\alpha - (\alpha - \delta)).$$

This contradicts the hypothesis that the series $\sum P(A_n)$ diverges. □

We continue the proof of Theorem 7.5. Suppose that (7.35) holds. Then
$$P(S_n > (1 + \varepsilon)a_n \text{ i.o.}) = 0 \qquad (7.39)$$

for every $\varepsilon > 0$. For every non-decreasing sequence of positive integers $k_n \to \infty$ and every $\varepsilon > 0$ we set

$$A_n = \{S_{k_n} - S_{k_{n-1}} > (1 + 2\varepsilon)a_{k_n}\}, \quad B_n = \{S_{k_{n-1}} > -\varepsilon a_{k_n}\}$$

Obviously, $P(A_n B_n \text{ i.o.}) \leq P(S_n > (1 + \varepsilon)a_n \text{ i.o.}) = 0$ by (7.39). Therefore, the series $\sum P(A_n)$ converges by Lemma 7.5 and the condition (D_1) is satisfied. Thus $(7.35) \Rightarrow (D_1)$.

If (D_1) is satisfied, then we conclude, putting $k_n = i_{rn+j}$ for $j = 0, 1, \ldots, r-1$, that (E_1) holds. To complete the proof of Theorem 7.5, it remains to show that $(E_1) \Rightarrow (7.35)$. We put

$$T_k = \max_{i_{mk+j} < r \leq i_{mk+j+m}} (S_r - S_{i_{mk+j}}). \qquad (7.40)$$

For every $n \in \mathbb{N}$ we have

$$\max_{i_{mn+j} < i \leq i_{mn+j+m}} S_i \leq \sum_{k=1}^{n} T_k + \max_{r \leq i_{m+j}} S_r. \qquad (7.41)$$

We shall show that

$$\limsup \frac{T_n}{c^{mn+j+m}} \leq 1 \text{ a.s.} \tag{7.42}$$

By the condition (B), for every $\delta > 0$ there exists a positive constant λ such that

$$P(S_{i_{mk+j+m}} - S_r \geq -\delta c^{mk+j+m}) \geq \lambda$$

for $i_{mk+j} < r \leq i_{mk+j+m}$ and for all sufficiently large k. Therefore, (7.40) and Theorem 2.3 (Chapter 2) imply that

$$P(T_k > (1+\varepsilon)c^{mk+j+m}) \leq \frac{1}{\lambda} P\left(S_{i_{mk+j+m}} - S_{i_{mk+j}} > (1+\varepsilon)c^{mk+j+m} - \frac{\varepsilon}{2} c^{mk+j+m}\right)$$

for every $\varepsilon > 0$ and all sufficiently large k. Taking (E_1) into account, we obtain

$$\sum_{n=1}^{\infty} P(T_n > (1+\varepsilon)c^{mn+j+m}) < \infty.$$

Accordingly, $P(T_n > (1+\varepsilon)c^{mn+j+m} \text{ i.o.}) = 0$ for every $\varepsilon > 0$, and (7.42) follows.

Lemma 7.6 *Let $\{b_n\}$ and $\{x_n\}$ be sequences of numbers, and let $\limsup x_n \leq x$, $\sum_{k=1}^{n} b_k \uparrow \infty$. Then*

$$\limsup \frac{b_1 x_1 + \cdots + b_n x_n}{b_1 + \cdots + b_n} \leq x.$$

This lemma is the one-sided analogue of Lemma 6.10 (Chapter 6). It can be proved by the same method.

It is clear that

$$\sum_{k=1}^{n} c^{mk+j} \leq c^{mn+j} \frac{c^m}{c^m - 1}.$$

In view of (7.41), (7.42), and Lemma 7.6, we have

$$\limsup c^{-(mn+j+m-1)} \max_{i_{mn+j+m-1} < i \leq i_{mn+j+m}} S_i \leq \frac{c^m}{c^m - 1} \text{ a.s.}$$

Therefore,

$$\limsup \frac{S_n}{a_n} \leq \max_{0 \leq j \leq m-1} \limsup \max_{i_{mn+j+m-1} < i \leq i_{mn+j+m}} \frac{S_i}{a_i} \leq \frac{c^{m+1}}{c^m - 1} \text{ a.s.}$$

Let ε be an arbitrary positive number. We choose $c > 1$ such that $c < \sqrt{1+\varepsilon}$. Moreover, we choose $m \in \mathbb{N}$ sufficiently large in order to satisfy the inequality $c^m/(c^m - 1) < \sqrt{1+\varepsilon}$. Then $\limsup S_n/a_n < 1 + \varepsilon$ a.s. This implies (7.35). □

Theorem 7.6 *Suppose that the condition (B) is satisfied. Then the relation*

$$\limsup S_n/a_n = 1 \text{ a.s.} \qquad (7.43)$$

is equivalent to any one of the following conditions:

(D_2) *for every $\varepsilon > 0$ and every non-decreasing sequence of positive integers $k_n \to \infty$ the condition (7.36) holds; for every $\varepsilon > 0$ there exists a non-decreasing sequence of positive numbers $k_n \to \infty$ such that*

$$\sum_{n=2}^{\infty} P(S_{k_n} - S_{k_{n-1}} > (1-\varepsilon)a_{k_n}) = \infty; \qquad (7.44)$$

(E_2) *for every $\varepsilon > 0$, every integer $r \geq 1$, and every $c > 1$, (7.37) holds; for every $\varepsilon > 0$ there exists $c > 1$ and an integer $r \geq 1$ such that*

$$\sum_{n=r}^{\infty} P(S_{i_n} - S_{i_{n-r}} > (1-\varepsilon)c^n) = \infty. \qquad (7.45)$$

Proof In view of Theorem 7.5. the implication (7.43) $\Rightarrow$ (E_2) will be proved if we prove the following proposition: if (7.43) holds, then for every positive $\varepsilon < 1$ there exist $c > 1$ and an integer $r \geq 1$ such that (7.45) holds. Suppose that (7.43) is satisfied and there exists a number $\varepsilon_0 \in (0,1)$ satisfying the condition $Q(c, \varepsilon_0) < \infty$ for all $c > 1$ and $r \geq 1$, where

$$Q(c, \varepsilon) = \sum_{n=r}^{\infty} P(S_{i_n} - S_{i_{n-r}} > (1-\varepsilon)c^n).$$

We shall apply Theorem 7.5 to the sequence of independent random variables $\{X_n\}$ and the sequence of numbers $\{(1-\varepsilon_0)a_n\}$. Since $(1+\varepsilon)(1-\varepsilon_0) > 1 - \varepsilon_0$ for every $\varepsilon > 0$, we obtain

$$\limsup S_n/((1-\varepsilon_0)a_n) \leq 1 \text{ a.s.}$$

Hence $\limsup S_n/a_n < 1$ a.s. This inequality contradicts the condition (7.43). Thus (7.43) $\Rightarrow$ (E_2).

It is easy to prove that (E_2) $\Rightarrow$ (D_2). If we suppose that (E_2) holds but (D_2) does not hold, we shall get a contradiction. To complete the proof of Theorem 7.6, it suffices to show that (D_2) $\Rightarrow$ (7.43). By Theorem 7.5 we have (D_2) $\Rightarrow$ (E_1) $\Rightarrow$ (7.35). Let τ be an arbitrary number such that $\tau > 1$. The condition (D_2) implies the existence of a number $\varepsilon \in (0,1)$ and a sequence of numbers $k_n \uparrow \infty$ satisfying the conditions $(1-\varepsilon)\tau > 1$ and (7.44). Applying Theorem 7.5 to the sequence of random variables $\{X_n\}$ and the numerical sequence $\{a_n/\tau\}$, we conclude that $\limsup \tau S_n/a_n \geq 1$ a.s. Since $\tau > 1$ is an arbitrary number, we obtain $\limsup S_n/a_n \geq 1$ a.s. The relation (7.43) follows from the latter inequality and (7.35). □

7.4 Bibliographical notes

The first result on the law of the iterated logarithm was obtained by Khintchine [220] for a sequence of two-valued i.i.d. random variables. The much more general Theorem 7.1 is due to Kolmogorov [229]. Marcinkiewicz and Zygmund [276] have constructed an example of a sequence of independent random variables having non-identical symmetric distributions with two values, for which $B_n \to \infty$, $|X_n| \leq M_n = O((B_n/\log \log B_n)^{1/2})$, but $\limsup S_n/(2B_n \log \log B_n)^{1/2} < 1$ a.s.

Theorems 7.2 and 7.3 were obtained by Petrov [331].

Theorem 7.4 is the Hartman–Wintner theorem in Strassen's form [442]. Hartman and Wintner [176] obtained sufficient conditions for the applicability of the law of the iterated logarithm to sequences of independent non-identically distributed random variables. In the i.i.d. case these conditions reduce to finiteness of variance. Other proofs of the Hartman–Wintner theorem were found by Strassen [442], Heyde [185], Csörgő and Révész [87], and de Acosta [97]. The proof of Theorem 7.4 in Section 7.2 was suggested by Martikainen.

Theorems 7.5 and 7.6 were obtained by Martikainen and Petrov [289] with a more complicated proof. Tomkins [451] proved analogous results in different terms. Martikainen [280] found generalizations of Theorems 7.5 and 7.6 for not necessarily increasing sequences of norming constants. Lemma 7.5 is due to Baum et al. [17].

Necessary and sufficient conditions for the law of the iterated logarithm without assumptions about the existence of moments were obtained by Martikainen [281], and sufficient conditions by Klass and Tomkins [227]. These conditions are expressed in terms of the distributions of the summands.

Surveys of the results of the iterated logarithm type can be found in Bingham [39] and also in the books by Csörgő and Révész [87], Révész [383], and Stout [440].

7.5 Addenda

In Subsections 7.5.1–7.5.25 we consider a sequence of independent identically distributed random variables $\{X_n\}$, and we write $S_n = \sum_{k=1}^{n} X_k$.

7.5.1 If $EX_1^2 = \infty$, then $\limsup |S_n|/(n \log \log n)^{1/2} = \infty$ a.s. (Strassen [443]).

7.5.2 Let $f(n)$ be an arbitrary function such that $\lim f(n) = \infty$. Then there exists a sequence of independent identically distributed random variables $\{X_n\}$ satisfying the conditions $EX_1 = 0$, $\text{Var } X_1 = \infty$, and $\lim S_n/(f(n)(n \log \log n)^{1/2}) = 0$ a.s. (Berkes [23]).

7.5.3 If $\limsup S_n/(2n \log \log n)^{1/2} = 1$ a.s., then $EX_1 = 0$ and $\text{Var } X_1 = 1$ (Martikainen [283], Rosalsky [391], and Pruitt [371]).

7.5.4 Suppose that X_1 has a non-degenerate distribution with zero mean. Define $K(y)$ as the unique positive solution of the equation

$$y E\{(X_1/K(y))^2 \wedge (|X_1|/K(y))\} = 1.$$

Define $a_n = K(n/\log \log n) \log \log n$ for $n \geq 3$ and $L = \limsup S_n/a_n$. The relation $P(X_n > a_n \text{ i.o.}) = 0$ holds if and only if L is finite. When L is finite, then $1 \leq L \leq 1.5$. Both bounds are sharp (Klass [225, 226]).

7.5.5 Let $V(x) = P(X_1 < x)$, $G(x) = P(|X_1| > x)$, and

$$K(x) = x^{-2} \int_{|y| \leq x} y^2 \, dV(y).$$

Let $\{\gamma_n\}$ be a sequence of numbers such that $P(S_n \geq \gamma_n) \geq \varepsilon$ and $P(S_n \leq \gamma_n) \geq \varepsilon$ for some $\varepsilon > 0$. There exists a non-decreasing sequence of numbers $\{a_n\}$ satisfying the condition $0 < \limsup(S_n - \gamma_n)/a_n < \infty$ a.s. if and only if $\liminf_{x \to \infty} P(X_1 > x)/(G(x) + K(x)) = 0$ (Pruitt [371]).

7.5.6 If $E(X_1^+)^2 < \infty$, then there exists a sequence of numbers $\{a_n\}$ such that $\limsup(S_n - mS_n)/a_n = 1$ a.s. (Pruitt [371]).

7.5.7 Let $\{a_n\}$ be a non-decreasing sequence of numbers such that the series $\sum P(|X_1| \geq a_n)$ converges and $\liminf B_n/a_n > 0$, where $B_n^2 = 2nV(a_n) \log \log n$ and $V(x) = P(X_1 < x)$. Then $\limsup(S_n - \alpha_n)/B_n = 1$ a.s., where $\alpha_n = n \int_{|x| < a_n} x \, dV(x)$ (Maller [271]).

7.5.8 There exists no non-decreasing sequence of positive numbers $\{a_n\}$ such that $\sum_{k=n}^{\infty} a_k^{-2} \leq C n a_n^{-2}$ for all n and $\limsup |S_n - mS_n|/a_n = 1$ a.s. (Rogozin [390]).

7.5.9 Define $\sigma^2(x)$ and $j_n(x, y)$ in accordance with Subsection 6.8.23 (Chapter 6). Let $\delta > 0$, $\alpha, \beta \in \mathbb{R}$, $\gamma = \min(\alpha, \beta)$, $a_n \uparrow \infty$, $\sum_{k=n}^{\infty} a_k^{-3} = O(na_n^{-3})$. In order that

$$-\beta \leq \liminf(S_n - mS_n)/a_n \leq \limsup(S_n - mS_n)/a_n \leq \alpha \text{ a.s.},$$

it is necessary and sufficient that the following conditions be satisfied:

(A) $\sum_{n=1}^{\infty} P(|X_1| \geq a_n) < \infty$,
(B) the series in the condition (6.61) converges for every $\varepsilon > \gamma$ and $0 < x < 1 < y$. In order that

$$-\beta = \liminf(S_n - mS_n)/a_n \leq \limsup(S_n - mS_n)/a_n = \alpha \text{ a.s.,}$$

it is necessary and sufficient that the following conditions be satisfied: (A), (B), and
(C) for every $\varepsilon < \gamma$ there exist constants x and y such that $0 < x < 1 < y$ and the series in the condition (B) diverges (Martikainen [286]).

7.5.10 There exists a sequence of positive constants $\{a_n\}$ such that

$$-\infty < \liminf(S_n - mS_n)/a_n < \limsup(S_n - mS_n)/a_n < \infty \text{ a.s.}$$

if and only if the d.f. $V(x)$ of the random variable X_1 belongs to the domain of partial attraction of a normal distribution. (The necessity was proved by Rogozin [390] and Heyde [184], and the sufficiency was proved by Kesten [218].) There exists a sequence of positive constants $\{a_n\}$ such that

$$-\infty < \liminf S_n/a_n < \limsup S_n/a_n < \infty \text{ a.s.}$$

if and only if $V(x)$ belongs to the domain of partial attraction of either normal or degenerate at a non-zero point distribution (Martikainen [284]).

7.5.11 Let $\mathscr{L}_1$ be the set of sequences of numbers $\{a_n\}$ such that $n^{-1/2}a_n \uparrow \infty$ and

$$-\infty < \liminf(S_n - mS_n)/a_n < \limsup(S_n - mS_n)/a_n < \infty \text{ a.s.}$$

for some sequence of i.i.d. random variables $\{X_n\}$. Let $\mathscr{L}_2$ be the set of sequences of numbers $\{a_n\}$ such that $n^{-1/2}a_n \uparrow \infty$ and the condition $\sum_{n=1}^{\infty} P(|X_1| \geq a_n) < \infty$ is equivalent to the following condition: the sequence of i.i.d. random variables $\{X_n\}$ obeys the strong law of large numbers with the norming sequence $\{a_n\}$. In order that $\{a_n\} \in \mathscr{L}_1$, it is necessary and sufficient that the condition $\liminf a_n^2(n \log \log n)^{-1} > 0$ hold and the condition

(D) $$\sum_{k=n}^{\infty} a_k^{-2} = O(na_n^{-2})$$

do not hold. The condition $\{a_n\} \in \mathscr{L}_2$ is equivalent to the condition (D) (Martikainen [286]). It follows that the condition (6.24) of Theorem 6.9 (Chapter 6) cannot be weakened.

7.5.12 Let $\{X_n\}$ be a sequence of non-negative i.i.d. random variables. If the function $P(X_1 \geq x)$ is slowly varying at infinity, then for every positive sequence of constants $a_n \uparrow \infty$ either $\lim S_n/a_n = 0$ a.s. or $\limsup S_n/a_n = \infty$ a.s., according as $\sum_{n=1}^{\infty} P(X_1 \geq a_n)$ converges or diverges (Teicher [447]). Adler [1] obtained a generalization of this result in the case of independent non-identically distributed random variables.

7.5.13 Let $\{X_n\}$ be a sequence of non-negative i.i.d. random variables, and let $P(X_1 \geq x)$ vary slowly at infinity. Then the relation $\liminf c_n < \infty$, where $c_n = nP(X_1 \geq a_n)$ and $a_n \to \infty$, implies the relation $\liminf S_n/a_n = 0$ a.s. If $a_n \uparrow \infty$, then the relation $\limsup S_n/a_n = 0$ a.s. is equivalent to the condition $\sum P(X_1 \geq a_n) < \infty$. If $c_n \to \infty$ and $\sum n^{-1} c_{n+1} \exp\{-(1-\varepsilon)c_n\} < \infty$ for some $\varepsilon > 0$, then $S_n/a_n \to \infty$ a.s. (Mikosch [303, 304]).

7.5.14 Let the d.f. $V(x) = P(X_1 < x)$ belong to the domain of attraction of a stable law with the exponent α such that $0 < \alpha < 2$, $\alpha \neq 1$. If $1 < \alpha < 2$, we assume that $EX_1 = 0$. Suppose $P(X_1 \geq x) = o(P(X < -x))$ as $x \to \infty$. Then there exists a sequence of numbers $a_n \uparrow \infty$ such that $\liminf S_n/a_n = -\infty$ a.s., $\limsup S_n/a_n = 1$ a.s. for $\alpha \in (1, 2)$, and $\limsup S_n/a_n \in (-\infty, 0]$ a.s. for $\alpha \in (0, 1)$ (Mikosch [303]).

7.5.15 Suppose $Ee^{itX_1} = e^{-|t|^\gamma}$, where $0 < \gamma < 2$. Then

$$\limsup n^{-1/\gamma} |S_n|^{1/\log \log n} = e^{1/\gamma} \text{ a.s.}$$

(Chover [68]).

7.5.16 Suppose that $E|X_1|^\delta < \infty$ for some $\delta > 0$ and $EX_1 = 0$ if $\delta \geq 1$. We put $\delta_0 = \sup\{\delta : E|X_1|^\delta < \infty\}$ and $\varepsilon = \min(\delta_0, 2)$. Then

$$\limsup n^{-1/\varepsilon} |S_n|^{1/\log n} = 1 \text{ a.s.}$$

If $X_1 \geq 0$ a.s., then it is possible to replace $\limsup$ by $\lim$ (Mikosch [302]).

7.5.17 Let the d.f. $V(x) = P(X_1 < x)$ belong to the domain of normal attraction of a stable law with the exponent $\alpha < 2$. If $\alpha > 1$, we assume that $EX_1 = 0$. Then for every sequence of numbers $c_n \to \infty$ and every integer $k \geq 0$ we have

$$\limsup (\log_0 n \log_1 n \cdots \log_k n)^{-1/\alpha} |S_n|^{1/(c_n \log_k + 2 n)} = 1 \text{ a.s.},$$

where $\log_0 n = n$ and $\log_{k+1} n = \log(\log_k n)$ (Mikosch [302]).

7.5.18 If $EX_1 = 0$ and $\text{Var } X_1 = 1$, then

$$\liminf (n/\log \log n)^{-1/2} \max_{1 \leq k \leq n} |S_k| = 8^{-1/2} \pi \text{ a.s.}$$

(Jain and Pruitt [207]).

7.5.19 If $EX_1^2 = \infty$, then

$$\lim (n/\log \log n)^{-1/2} \max_{1 \leq k \leq n} |S_k| = \infty \text{ a.s.}$$

(Csáki [86]).

7.5.20 Suppose that $EX_1 = 0$ and $\operatorname{Var} X_1 = 1$. Let $f(x)$ be a Riemann-integrable on $(0, 1)$ real function. Then

$$\limsup(2n^3 \log\log n)^{-1/2} \sum_{k=1}^{n} f(k/n)S_k = \left(\int_0^1 F^2(x)\,dx\right)^{1/2} \quad \text{a.s.,}$$

where $F(x) = \int_x^1 f(t)\,dt$ for $x \in (0, 1)$. In particular, if $f(x) = x^\alpha$, where $\alpha > -1$, then

$$\limsup(2n^{2\alpha+3} \log\log n)^{-1/2} \sum_{k=1}^{n} k^\alpha S_k = \{(\alpha + 3/2)(\alpha + 2)\}^{-1/2} \quad \text{a.s.}$$

(Strassen [442]).

7.5.21 Suppose $EX_1 = 0$ and $\operatorname{Var} X_1 = 1$. Then

$$\limsup n^{-3/2}(2n \log\log n)^{-1/2} \sum_{k=1}^{n} |S_k| = 3^{-1/2} \quad \text{a.s.}$$

and

$$\limsup n^{-2}(2n \log\log n)^{-1} \sum_{k=1}^{n} S_k^2 = 4\pi^{-2} \quad \text{a.s.}$$

(Strassen [442]).

7.5.22 Let $EX_1 = 0$, $\operatorname{Var} X_1 = 1$, $0 \leq b \leq 1$, and

$$c_k = I(S_k > b(2k \log\log k)^{1/2}).$$

Then

$$\limsup n^{-1} \sum_{k=3}^{n} c_k = 1 - \exp\{-4(b^{-2} - 1)\} \quad \text{a.s.}$$

(Strassen [442]).

7.5.23 Suppose that $EX_1 = 0$, $\operatorname{Var} X_1 = 1$, and $Ee^{hX_1} < \infty$ for $|h| < h_0$ and some $h_0 > 0$. Let $\varphi(x)$ be a positive function defined on the positive half-line with a continuous derivative and such that $x^{-1/2}\varphi(x)$ does not decrease, $\varphi'(x)/\varphi'(y) \to 1$ as $y \to \infty$ and $x/y \to 1$, and moreover $\varphi(x) < x^\delta$ for all x and some $\delta < 3/5$. Then

$$P(S_m \geq \varphi(m) \text{ for some } m \geq n) \sim \int_n^\infty \varphi'(x) \exp\{-\varphi^2(x)/(2x)\}(2\pi x)^{-1/2}\,dx$$

as $n \to \infty$ (Strassen [444]).

7.5.24 Let $\{n_n; k = 1, 2, \ldots\}$ be a strictly increasing sequence of positive integers such that

$$\liminf_{k \to \infty} n_k/n_{k+1} > 0. \tag{7.46}$$

If $EX_1 = 0$ and $\text{Var } X_1 = \sigma^2 < \infty$, then
$$\limsup S_{n_k}/(2n_k \log \log n_k)^{1/2} = \sigma \text{ a.s.}$$
Conversely, if
$$P(\limsup |S_{n_k}|/(n_k \log \log n_k)^{1/2} < \infty) > 0,$$
then $\text{Var } X_1 < \infty$ and $EX_1 = 0$.

If we replace (7.46) by $\limsup n_k/n_{k+1} < 1$, then the conditions $EX_1 = 0$ and $\text{Var } X_1 = \sigma^2 < \infty$ imply the relation
$$\limsup S_{n_k}/(2n_k \log n_k)^{1/2} = \sigma \text{ a.s.};$$
conversely, if $P(\limsup |S_{n_k}|/(n_k \log n_k)^{1/2} < \infty) > 0$, then $\text{Var } X_1 < \infty$ and $EX_1 = 0$ (Gut [153]).

7.5.25 Let $\{X_n\}$ be a sequence of i.i.d. random variables on a probability space $(\Omega, \mathcal{A}, P)$. Suppose $EX_1 = 0$ and $\text{Var } X_1 = 1$. Then for every positive $y < 1$ and δ and almost all $\omega \in \Omega$ there exists $N_0 = N_0(y, \delta, \omega) < \infty$ such that every interval $N \leq \log n \leq N + N^{y+\delta}$ ($N \geq N_0$) contains at least one n satisfying the condition $S_n(\omega) \geq (2yn \log \log n)^{1/2}$ (Schatte [416] under the additional assumption $E|X_1|^3 < \infty$, and Nikitenko [314]).

7.5.26 Let $\{X_n\}$ be a sequence of random variables such that $\liminf EX_n^2 < \infty$. Then there exist a sub-sequence $\{X_{n_k}\}$ and random variables X and Y satisfying the conditions $Y \geq 0$, $EX^2 < \infty$, $EY < \infty$, and
$$\limsup_{N \to \infty} (2N \log \log N)^{-1/2} \sum_{k=1}^N (X_{n_k} - X) = Y^{1/2} \text{ a.s.}$$
This result is optimal in the following sense. If $\{a_n\}$ is an arbitrary sequence of positive numbers satisfying the condition $a_n \to \infty$, then there exists a sequence of random variables $\{X_n\}$ such that $EX_n^2 \leq a_n$ for all n and
$$\limsup(2n \log \log n)^{-1/2} \sum_{i=1}^n (X_i - X) = \infty \text{ a.s.}$$
for every random variable X, and the same holds for every sub-sequence of $\{X_n\}$ (Berkes [24]).

7.5.27 Let $\{Y_n\}$ be a sequence of random variables, and let $\{a_n\}$ be a sequence of positive numbers such that $a_n \to \infty$, $a_{n+1}/a_n \to 1$ and for every $\varepsilon > 0$ the inequality $a_{n+s} \geq a_n(1 - \varepsilon)$ holds for all $s \geq 1$ and all sufficiently large n (the latter condition is satisfied for any non-decreasing sequence $\{a_n\}$). Furthermore, we suppose that for every x in some interval $(1, 1 + \beta)$ there exists a positive constant C such that
$$P\left(\max_{1 \leq k \leq n} Y_k \geq xa_n\right) \leq C/\psi(a_n)$$

for some function $\psi \in \Psi_c$ and all sufficiently large n, where Ψ_c is defined in Section 6.6. Then

$$\limsup Y_n/a_n \leq 1 \text{ a.s.}$$

(Petrov [348]). Some applications of this proposition can be found in Petrov [348, 349].

In Subsections 7.5.28–7.5.42 we consider a sequence of independent not necessarily identically distributed random variables $\{X_n\}$. We write $S_n = \sum_{k=1}^n X_k$. In Subsections 7.5.28–7.5.37 we assume that $EX_n = 0$ and $\text{Var } X_n < \infty$ for all $n \in \mathbb{N}$, and we put $B_n = \sum_{k=1}^n \text{Var } X_k$.

7.5.28 There exist sequences of independent random variables satisfying the central limit theorem but not obeying the law of the iterated logarithm in the form (7.3). An example of such a sequence was constructed by Marcinkiewicz and Zygmund [276] for other aims.

7.5.29 There exists a sequence of independent random variables $\{X_n\}$ with zero means and finite variances such that $\text{Var } X_n \asymp 1$,

$$\sup_x |P(S_n < xB_n^{1/2}) - \Phi(x)| = O((\log B_n)^{-1}),$$

and the law of the iterated logarithm does not hold (Egorov [107]).

7.5.30 There exists a sequence of independent random variables $\{X_n\}$ satisfying the classical form of the law of the iterated logarithm and the condition $B_n/n \to C$, where C is a positive constant, but not satisfying the central limit theorem (Egorov [107]).

7.5.31 We put $t_n = (2 \log \log B_n)^{1/2}$. If $B_n \to \infty$, $B_n^{-1/2} t_n X_n \to 0$ a.s., and $B_n^{-1/2} t_n |X_n| \leq Y$ a.s. for all n and some random variable Y with $EY^2 < \infty$, then the relation (7.3) holds, even though it may fail if only one of the conditions does not hold. In particular, the conditions $B_n \to \infty$, $B_n^{-1/2} t_n X_n \to 0$ a.s., and $\sup_{n \geq 1} B_n^{-1/2} t_n |X_n| \leq C$ a.s. for some constant C imply (7.3). Moreover, when $B_n^{-1/2} t_n X_n \to 0$ a.s. and $EX_n^2/B_n \to 0$, then the central limit theorem implies the law of the iterated logarithm, but the converse is not always true (Tomkins [453]).

7.5.32 Let $B_n \to \infty$, and let there exist a sequence of positive constants $\{M_n\}$ such that $|X_n| \leq M_n$ a.s. and $M_n = O(B_n^{1/2}(\log \log B_n)^{-1/2})$. Then there exists a constant $L \geq 0$ such that

$$\limsup |S_n|/(2B_n \log \log B_n)^{1/2} = L \text{ a.s.}$$

(Egorov [107]).

7.5.33 Suppose that $B_n \to \infty$ and $B_n^{-1/2} t_n X_n \to 0$ a.s., where $t_n = (2 \log \log B_n)^{1/2}$. We put

$$H_n(x) = B_n^{-1} \sum_{k=1}^{n} E(X_k^2 I(|X_k| \leqslant x B_k^{1/2} t_k^{-1})),$$

$$H_- = \liminf H_n(x), \quad H_+ = \limsup H_n(x).$$

Then

$$H_- \leqslant \limsup S_n/(t_n B_n^{1/2}) \leqslant H_+ \text{ a.s.}$$

and H_- and H_+ do not depend on x (Tomkins [452]).

7.5.34 Suppose that $|X_n| \leqslant c_n B_n^{1/2}$ a.s. for $n \in \mathbb{N}$, where c_n are constants. We put $v = \limsup c_n (2 \log \log B_n)^{1/2}$. If $v < \infty$, then

$$0 < \limsup S_n/(2 B_n \log \log B_n)^{1/2} \leqslant 1 + \sum_{k=3}^{\infty} v^{k-2}/k! \text{ a.s.}$$

(Tomkins [450]).

7.5.35 Suppose that there exists a random variable X with finite variance such that

$$n^{-1} \sum_{k=1}^{n} P(|X_k| \geqslant x) \leqslant P(|X| \geqslant x)$$

for all sufficiently large n and x. If, moreover, $\liminf B_n/n > 0$, then the sequence $\{X_n\}$ obeys the law of the iterated logarithm, i.e. the relation (7.3) holds (Egorov [108]).

7.5.36 Suppose that $B_n \to \infty$, $\limsup B_{n+1}/B_n < \infty$, and

$$\sum_n (2B_n \log \log B_n)^{-p/2} E|X_n|^p < \infty$$

for some p in the interval $2 < p \leqslant 3$. Then the relation (7.3) holds (Wittmann [467]).

7.5.37 We put $V_n(x) = P(X_n < x)$,

$$L_n(x) = \int_{|y| \geqslant x} y^2 \, dV_n(y), \quad v_n = (B_n/\log \log B_n)^{1/2}, \quad a_n = (2B_n \log \log B_n)^{1/2}.$$

Suppose that $B_n \to \infty$. If

$$\sum_{k=1}^{n} L_k(\varepsilon v_n) = o(B_n) \text{ for every } \varepsilon > 0, \tag{7.47}$$

then $\limsup S_n/a_n \geqslant 1$ a.s. If $\operatorname{ess\,sup} X_n = o(v_n)$ and the condition (7.47) is satisfied, then $\limsup S_n/a_n = 1$ a.s. (Martikainen [287]).

7.5.38 Let $\{a_n\}$, $\{b_n\}$, and $\{c_n\}$ be sequences of positive numbers such that $a_n < c_n$, $c_n \uparrow \infty$, and there exists the limit $\lim b_n/c_n = \gamma$. We put $Y_n = X_n I(|X_n| < a_n)$. If

$$\limsup b_n^{-1} \sum_{k=1}^n (Y_k - EY_k) = 1 \text{ a.s.}$$

and

$$\sum_{n=1}^\infty \int_{|x| \geq a_n} x^2(x^2 + c_n^2)^{-1} \, dV_n(x) < \infty,$$

then

$$\limsup c_n^{-1} \sum_{k=1}^n \left(X_k - \int_{|x| < c_k} x \, dV_k(x) \right) = \gamma \text{ a.s.}$$

(Petrov [333]).

7.5.39 Suppose that $a_n \uparrow \infty$, $S_n/a_n \xrightarrow{P} 0$, and $X_n \leq \varepsilon_n a_n$ a.s. for $n \in \mathbb{N}$, where $\varepsilon_n \to 0$. Then the relation $\limsup S_n/a_n \leq 1$ a.s. is equivalent to the condition

(A) for every $\varepsilon > 0$, every $c > 1$, and every integer $r \geq 1$, there exists a sequence of positive integers $\{h_n\}$ such that

$$\sum_{n=r}^\infty e^{-(1+\varepsilon)c^n h_n} \prod_{k \in I_{n,r}} Ee^{h_k X_k} < \infty,$$

where $I_{n,r}$ is the set of all $k \in \mathbb{N}$ satisfying the inequalities $i_{n-r} < k \leq i_n$ and i_n is the greatest integer for which $a_{i_n} \leq c^n$.

The relation $\limsup S_n a_n = 1$ a.s. is equivalent to the set of the conditions (A) and

(B) for every $\varepsilon > 0$ there exist $c > 1$ and an integer $r \geq 1$ such that

$$\sum_{n=r}^\infty \inf_{h > 0} e^{-(1-\varepsilon)c^n h} \prod_{k \in I_{n,r}} Ee^{hX_k} = \infty$$

(Martikainen [281]).

7.5.40 Suppose that the condition of Kolmogorov's theorem (Theorem 7.1) are satisfied. Then

$$\liminf (B_n/\log \log B_n)^{-1/2} \max_{1 \leq k \leq n} |S_k| = 8^{-1/2}\pi \text{ a.s.}$$

(Martikainen [288]).

7.5.41 Suppose $a_n \uparrow \infty$ and condition (B) of Section 7.3 is satisfied. Moreover, we suppose that the following conditions are satisfied:

(A) for every $c > 1$ there exists the finite limit $\lim a_{c^n}/a_{c^{n-1}}$, where a_{c^n} stands for $a_{[c^n]}$;

(C) there exists a positive constant β such that

$$\sum_{n=1}^{\infty} P(S_{c^n} > (1+\varepsilon)a_{c^n}) < \infty$$

for every c in the interval $1 < c < 1 + \beta$ and every $\varepsilon > 0$. Then

$$\limsup S_n/a_n \leq 1 \text{ a.s.} \qquad (7.48)$$

If $\{X_n\}$ is a sequence of i.i.d. random variables satisfying the conditions (A) and (B), then (7.48) and (C) are equivalent. If $\{X_n\}$ is a sequence of i.i.d. random variables satisfying the conditions (A), (B), (C), and

(D) there exists a positive constant β such that

$$\sum_{n=1}^{\infty} P(S_{c^n} > (1-\varepsilon)a_{c^n}) = \infty$$

for every c in the interval $1 < c < 1 + \beta$ and every $\varepsilon > 0$, then

$$\limsup S_n/a_n = 1 \text{ a.s.} \qquad (7.49)$$

(Petrov [354]). Actually in the i.i.d. case when the conditions (A), (B), and (C) are satisfied, (D) is equivalent to (7.49).

7.5.42 Suppose that $EX_n = 0$, $\operatorname{Var} X_n = 1$, and $\sup_{n \geq 1} E|X_n|^3 < \infty$. Let $\{a_{nm}\}$ be a bounded triangular array of numbers. Let

$$T_n = \sum_{m=1}^{n} a_{nm} X_m, \quad t_n = \sum_{m=1}^{n} a_{nm}^2.$$

If $\liminf t_n/n > 0$, then $\limsup T_n/(2t_n \log \log t_n)^{1/2} \geq 1$ a.s. (Tomkins [449]). Upper and lower bounds in the classical law of the iterated logarithm for double sequences of independent random variables (in particular, for weighted sums of i.i.d. random variables) were obtained by Tomkins [449] and Stadtmüller [434].

Bibliography

1. Adler, A. (1988). On the law of the iterated logarithm for nonidentically distributed random variables. *Stoch. Anal. Appl.*, **6**, 117–27.
2. Alda, V. (1952). A note on Poisson's distribution. *Časopis mat. (Czechoslov. Math. J.)*, **2**, 243–6.
3. Aldous, D. J. (1977). Limit theorems for subsequences of arbitrarily-dependent sequences of random variables. *Z. Wahrsch. Verw. Geb.*, **40**, 59–82.
4. Amosova, N. N. (1976). Rate of convergence in the one-sided law of large numbers. *Lithuanian Math. J.*, **16**, 313–19.
5. Amosova, N. N. (1978). On probabilities of one-sided deviations of sums of independent random variables. *Mat. Zametki*, **24**, 123–31 (in Russian).
6. Ananjevsky, S. M. (1978). On concentration functions. *Vestnik Leningrad. Univ.*, No. 13, 7–13 (in Russian).
7. Ananjevsky, S. M. (1982). Generalized concentration functions. *Vestnik Leningrad. Univ.*, No. 19, 5–11 (in Russian).
8. Anděl, J. and Dupač, V. (1989). An extension of the Borel lemma. *Comment. Math. Univ. Carol.*, **30**, 403–4.
9. Arak, T. V. (1981). On the convergence rate in Kolmogorov's uniform limit theorem. *Th. Probab. Appl.*, **26**, 219–39, 437–51.
10. Arak, T. V. (1982). An improvement of the lower bound for the rate of convergence in Kolmogorov's limit theorem. *Th. Probab. Appl.*, **27**, 826–32.
11. Arak, T. V. and Zaitsev, A. Yu. (1988). Uniform limit theorems for sums of independent random variables, *Proc. Steklov Math. Inst.* **174**. American Mathematical Society, Providence, RI, 1988.
12. Araujo, A. and Giné, E. (1980). *The central limit theorem for real and Banach valued random variables.* Wiley, New York.
13. Arnold, B. C. (1978). Some elementary variations of the Lyapunov inequality. *SIAM J. Appl. Math.*, **35**, 117–18.
14. Barbour, A. D. (1986). Asymptotic expansions based on smooth functions in the central limit theorem. *Probab. Th. Rel. Fields*, **72**, 289–303.
15. Barbour, A. D. and Hall, P. (1984). Stein's method and the Berry–Esseen theorem. *Austral. J. Statist.*, **26**, 8–15.
16. Baum, L. E. and Katz, M. (1965). Convergence rates in the law of large numbers. *Trans. Amer. Math. Soc.*, **120**, 108–23.
17. Baum, L. E., Katz, M., and Stratton, H. H. (1971). Strong laws for ruled sums. *Ann. Math. Statist.*, **42**, 625–9.
18. Baxter, G. and Shapiro, J. M. (1960). On bounded infinitely divisible random variables. *Sankhya*, **A22**, 253–60.
19. Beesack, P. R. (1984). Inequalities for absolute moments of a distribution: from Laplace to von Mises. *J. Math. Anal. Appl.*, **98**, 435–57.
20. Bennett, G. (1962). Probability inequalities for the sum of independent random variables. *J. Amer. Statist. Ass.*, **57**, (297), 33–45.

21. Bentkus, V. and Kirsha, K. (1989). Estimates of closeness of a distribution function to the normal law. *Lithuanian Math. J.*, **29**, 321–32.
22. Bergström, H. (1949). On the central limit theorem in the case of not equally distributed random variables. *Skand. Aktuarietidskrift*, **32**, 37–62.
23. Berkes, I. (1972). A remark to the law of the iterated logarithm. *Studia Sci. Math. Hungar.*, **7**, 189–97.
24. Berkes, I. (1974). The law of the iterated logarithm for subsequences of random variables. *Z. Wahrsch. Verw. Geb.*, **30**, 209–15.
25. Berkes, I. (1990). An extension of the Komlós subsequence theorem. *Acta Math. Hungar.*, **55**, 103–10.
26. Berkes, I. and Dehling, H. (1989). Almost sure and weak invariance principles for random variables attracted by a stable law. *Probab. Th. Rel. Fields*, **83**, 331–53.
27. Berkes, I. and Dehling, H. (1991). *Strong approximation on log dense sets*. University of Groningen, Department of Mathematics.
28. Berkes, I. and Péter, E. (1986) Exchangeable random variables and the subsequence principle. *Probab. Th. Rel. Fields*, **73**, 395–413.
29. Bernstein, S. N. (1939). Notes on Lyapunov's limit theorem. *Dokl. Akad. Nauk SSSR*, **24**, 3–7 (in Russian).
30. Bernstein, S. N. (1946). *Theory of probability*, (4th edn), Gostekhizdat, Moscow–Leningrad (in Russian).
31. Bernstein, S. N. (1964). *Collected works*, Vol. 4. Nauka, Moscow (in Russian).
32. Berry, A. C. (1941). The accuracy of the Gaussian approximation to the sum of independent variates. *Trans. Amer. Math. Soc.*, **49**, 122–36.
33. Bestsennaya, E. V. and Utev, S. A. (1991). The least upper bound for an even moment of sums of independent random variables. *Siberian Math. J.*, **32**, 171–3 (in Russian).
34. Bhattacharya, R. N. and Ranga Rao, R. (1976). *Normal approximation and asymptotic expansions*. Wiley, New York.
35. Bickel, P. J. (1970). A Hàjek–Rényi extension of Lévy's inequality and some applications. *Acta Math. Acad. Sci. Hungar.*, **21**, 199–206.
36. Bikelis, A. (1966). Estimates of the remainder in the central limit theorem. *Litovsk. Mat. Sb.*, **6**(3), 323–46 (in Russian).
37. Bikelis, A. (1972). Limit theorems for sums of independent random variables. *Litovsk. Mat. Sb.*, **12**(4), 5–14 (in Russian).
38. Billingsley, P. (1968). *Convergence of probability measures*. Wiley, New York.
39. Bingham, N. H. (1986). Variants of the law of the iterated logarithm. *Bull. London Math. Soc.*, **18**, 433–67.
40. Bingham, N. H. and Goldie, C. M. (1982). Probabilistic and deterministic averaging. *Trans. Amer. Math. Soc.*, **269**, 453–80.
41. Bingham, N. H., Goldie, C. M., and Teugels, J. L. (1987). *Regular variation*. Cambridge University Press.
42. Binmore, K. G. and Stratton, H. H. (1969). A note on characteristic functions. *Ann. Math. Statist.*, **40**, 303–7.
43. Blum, J. R. and Rosenblatt, M. (1959). On the structure of infinitely divisible distributions. *Pacific J. Math.*, **9**, 1–8.
44. Boas, R. P. (1967). Lipschitz behavior and integrability of characteristic functions. *Ann. Math. Statist.*, **38**, 32–6.
45. Bondarko, V. M. (1980). Estimates in the central limit theorem for quantiles. *Th. Probab. Math. Statist.*, **20**, 21–5.

46. Book, S. A. (1975). An extension of the Erdös–Rényi new law of large numbers. *Proc. Amer. Math. Soc.*, **48**, 438–46.
47. Book, S. A. (1975). A version of the Erdös–Rényi law of large numbers for independent random variables. *Bull. Inst. Math. Acad. Sinica*, **3**, 199–211.
48. Book, S. A. (1976). Large deviation probabilities and the Erdös–Rényi law of large numbers. *Canad. J. Statist.*, **4**, 185–209.
49. Book, S. A. (1983). Large deviations and applications. In: *Encyclopedia of statistical sciences*, Vol. 4, pp. 476–80. Wiley, New York.
50. Borel, E. (1909). Les probabilités dénombrables et leurs applications arithmétiques. *Rend. Circ. Mat. Palermo*, **27**, 247–71.
51. Borovkov, A. A. (1972). Notes on inequalities for sums of independent variables. *Th. Probab. Appl.*, **17**, 556–7.
52. Brosamler, G. A. (1988). An almost everywhere central limit theorem. *Math. Proc. Cambridge Philos. Soc.*, **104**, 561–74.
53. Brown, B. M. (1970). Characteristic functions, moments, and the central limit theorem. *Ann. Math. Statist.*, **41**, 658–64.
54. Brown, B. M. (1972). Formulae for absolute moments. *J. Austral. Math. Soc.*, **13**, 104–6.
55. Buldygin, V. V. (1980). The strong law of large numbers and convergence of Gaussian sequences to zero. *Th. Probab. Math. Statist.*, **19**, 35–43.
56. Bulinsky, A. V. (1977). On normalization in the law of the iterated logarithm. *Th. Probab. Appl.*, **22**, 398–9.
57. Butzer, P. L. and Hahn, L. (1978). General theorems on rates of convergence in distribution of random variables, I, II. *J. Multivar. Anal.*, **8**, 181–201, 202–21.
58. Butzer, P. L. and Hahn, L. (1979). On the connection between the rates of norm and weak convergence in the central limit theorem. *Math. Nachr.*, **91**, 245–51.
59. Cacoullos, T. (1982). On upper and lower bounds for the variance of a function of a random variable. *Ann. Probab.*, **10**, 799–809.
60. Cantelli, F. P. (1917). Sulla probabilità come limite dela frequenza. *Rend. Accad. Naz. Lincei Roma*, **26**, 39–45.
61. Chatterjee, S. D. (1970). A general strong law. *Inventiones Math.*, **9**, 235–45.
62. Chatterjee, S. D. and Pakshirajan, R. P. (1957). On the unboundedness of infinitely divisible laws. *Sankhya*, **17**, 349–50.
63. Chebyshev, P. L. (1948). *The complete collection of works*, Vol. 3. Izdat. Akad. Nauk SSSR, Moscow–Leningrad (in Russian).
64. Chen, L. H. Y. (1982). An inequality for the multivariate normal distribution. *J. Multivar. Anal.*, **12**, 306–15.
65. Chen, R. (1976). A remark on the strong law of large numbers. *Proc. Amer. Math. Soc.*, **61**, 112–16.
66. Chernoff, H. (1981). A note on an inequality involving the normal distribution. *Ann. Probab.*, **9**, 533–5.
67. Chistyakov, G. P. (1990). On a problem of A. N. Kolmogorov. *Zap. Nauch. Sem. Leningrad. Otd. Mat. Inst.*, **184**, 289–319 (in Russian).
68. Chover, J. (1966). A law of the iterated logarithm for stable summands. *Proc. Amer. Math. Soc.*, **17**, 441–3.
69. Chow, Y. S. and Lai, T. L. (1973). Limiting behavior of weighted sums of independent random variables. *Ann. Probab.*, **1**, 810–24.
70. Chow, Y. S. and Lai, T. L. (1975). Some one-sided theorems on the tail distribution of sample sums with applications to the last time and largest excess of boundary crossings. *Trans. Amer. Math. Soc.*, **208**, 51–72.

71. Chow, Y. S. and Lai, T. L. (1978). Paley-type inequalities and convergence rates related to the law of large numbers and extended renewal theory. *Z. Wahrsch. Verw. Geb.*, **45**, 1–19.
72. Chow, Y. S. and Robbins, H. (1961). On sums of independent random variables with infinite moments and 'fair' games. *Proc. Nat. Acad. Sci. U.S.A.*, **47**, 330–5.
73. Chow, Y. S. and Zhang, C. H. (1986). A note on Feller's strong law of large numbers. *Ann. Probab.*, **14**, 1088–94.
74. Christoph, G. (1980). Über notwendige und hinreichende Bedingungen für Konvergenzgeschwindigkeitsaussagen im Falle einer stabilen Grenzverteilungen. *Z. Wahrsch. Verw. Geb.*, **54**, 29–40.
75. Christoph, G. (1987). Strong approximations for partial sums of random variables attracted to a stable law. *Math. Nachr.*, **134**, 289–94.
76. Christoph, G. (1991). On some differences in limit theorems with a normal or a nonnormal stable limit law. *Math. Nachr.*, **153**, 247–56.
77. Chung, K. L. (1947). Note on some strong law of large numbers. *Amer. J. Math.*, **69**, 189–92.
78. Chung, K. L. (1974). *A Course in probability theory*, (2nd edn). Academic Press, New York.
79. Chung, K. L. and Erdös, P. (1947). On the lower limit of sums of independent random variables. *Ann. Math.*, **48**, 1003–13.
80. Chung, K. L. and Erdös P. (1951). Probability limit theorems assuming only the first moment. *Mem. Amer. Math. Soc.*, No. 6, 1–19.
81. Chung, K. L. and Erdös, P. (1952). On the application of the Borel–Cantelli lemma. *Trans. Amer. Math. Soc.*, **72**, 179–86.
82. Cox, D. C. and Kemperman, J. H. B. (1983). Sharp bounds on the absolute moments of a sum of two i.i.d. random variables. *Ann. Probab.*, **11**, 765–71.
83. Cramér, H. (1938). Sur un nouveau théorème-limite de la théorie des probabilites. *Actual. Sci. Indust. (Paris)*, **736**, 5–23.
84. Cramér, H. (1946). *Mathematical methods of statistics*. Princeton University Press.
85. Cramér, H. (1962). *Random variables and probability distributions*, (2nd edn). Cambridge University Press.
86. Csáki, E. (1978). On the lower limits of maxima and minima of Wiener process and partial sums. *Z. Wahrsch. Verw. Geb.*, **43**, 205–21.
87. Csörgő, M. and Révész, P. (1981). *Strong approximations in probability and statistics*. Academic Press, New York.
88. Csörgő, S. (1979). Erdös–Rényi laws. *Ann. Statist.*, **7**, 772–87.
89. Csörgő, S. and Hall, P. (1984). The Komlós–Major–Tusnády approximations and their applications. *Austral. J. Statist.*, **26**, 189–218.
90. Csörgő, S., Tandori, K., and Totik, V. (1983). On the strong law of large numbers for pairwise independent random variables. *Acta Math. Hungar.*, **42**, 319–30.
91. Daley, D. J. (1977). Tighter bounds for the absolute third moments. *Scand. J. Statist.*, **4**, 183–4.
92. Darling, D. A. (1952). The influence of the maximum term in addition of independent random variables. *Trans. Amer. Math. Soc.*, **73**, 95–107.
93. Daugavet, A. I. (1980). On asymptotic expansions of some numerical characteristics of sums of independent random variables. *Vestnik Leningrad. Univ.*, No. 1, 16–22 (in Russian).
94. Daugavet, A. I. (1987). Bounds for the convergence rate of numerical characteristics in limit theorems with stable limit law. *Th. Probab. Appl.*, **32**, 531–5.

95. Daugavet, A. I. and Petrov, V. V. (1987). A generalization of the Esseen inequality for the concentration function. *J. Soviet Math.*, **36**, 473–6.
96. Daugavet, A. I. and Petrov, V. V. (1986). On estimates of the concentration function. In: *Rings and modulus. Limit theorems of probability theory* (ed. Z. I. Borevich and V. V. Petrov), vol. 1, pp. 164–9. Leningrad University Press (in Russian).
97. De Acosta, A. (1983). A new proof of the Hartman–Wintner law of the iterated logarithm. *Ann. Probab.*, **11**, 270–6.
98. Deheuvels, P. and Devroye, L. (1987). Limit laws of Erdös–Rényi type. *Ann. Probab.*, **15**, 1363–86.
99. Deheuvels, P., Devroye, L., and Lynch, J. (1986). Exact convergence rate in the limit theorems of Erdös–Rényi and Shepp. *Ann. Probab.*, **14**, 209–23.
100. Deheuvels, P. and Steinebach, J. (1987). Exact convergence rates in strong approximation laws for large increments of partial sums. *Probab. Th. Rel. Fields*, **76**, 369–93.
101. Dharmadhikari, S. W. and Jogdeo, K. (1969). Bounds on moments of certain variables. *Ann. Math. Statist.*, **40**, 1506–8.
102. Doeblin, W. (1939). Sur les sommes d'un grand nombres des variables aléatoires indépendantes. *Bull. Soc. Math. France*, **53**, 23–32, 35–64.
103. Doeblin, W. and Lévy, P. (1936). Sur les sommes de variables aléatoires indépendantes à dispersion bornées inférieurement. *C.R. Acad. Sci. Paris*, **202**, 2027–9.
104. Doob, J. L. (1953). *Stochastic processes*. Wiley, New York.
105. Dyson, F. J. (1953). Fourier transforms of distribution functions. *Canad. J. Math.*, **5**, 554–8.
106. Ebralidze, Sh. S. (1971). Inequalities for probabilities of large deviations in terms of pseudomoments. *Th. Probab. Appl.*, **16**, 737–41.
107. Egorov, V. A. (1969). On the law of the iterated logarithm. *Th. Probab. Appl.*, **14**, 693–9.
108. Egorov, V. A. (1971). A generalization of the Hartman–Wintner theorem on the law of the iterated logarithm. *Vestnik Leningrad. Univ.*, No. 7, 22–8 (in Russian).
109. Egorov, V. A. (1972). Some theorems on the strong law of large numbers and law of the iterated logarithm. *Th. Probab. Appl.*, **17**, 86–100.
110. Egorov, V. A. (1973). On the rate of convergence to the normal law which is equivalent to the existence of a second moment. *Th. Probab. Appl.*, **18**, 175–80.
111. Egorov, V. A. (1980). On the rate of convergence to a stable law. *Th. Probab. Appl.*, **25**, 180–7.
112. Eisenberg, B. and Gan Shixin. (1983). Uniform convergence of distribution functions. *Proc. Amer. Math. Soc.*, **88**, 145–6.
113. Erdös, P. (1949, 1950). On a theorem of Hsu and Robbins. *Ann. Math. Statist.*, **20**, 286–91; **21**, 138.
114. Erdös, P. and Rényi, A. (1959). On Cantor's series with convergent $\sum 1/q_n$. *Ann. Univ. Sci. Budapest, Sect. Math.*, **2**, 93–109.
115. Erdös, P. and Rényi, A. (1970). On a new law of large numbers. *J. Analyse Math.*, **23**, 103–11.
116. Erickson, K. B. (1973). The strong law of large numbers when the mean is undefined. *Trans. Amer. Math. Soc.*, **185**, 371–81.
117. Erickson, K. B. and Kesten, H. (1974). Strong and weak limit points of a normalized random walk. *Ann. Probab.*, **2**, 553–79.

118. Esseen, C.-G. (1942). On the Liapounoff limit of error in the theory of probability. *Arkiv Mat., Astr. och Fysik*, **28A**(2), 1–19.
119. Esseen, C.-G. (1945). Fourier analysis of distribution functions. A mathematical study of the Laplace–Gaussian law. *Acta Math.*, **77**, 1–125.
120. Esseen, C.-G. (1956). A moment inequality with an application to the central limit theorem. *Skand. Aktuarietidskrift*, **39**(3–4), 160–70.
121. Esseen, C.-G. (1965). On infinitely divisible one-sided distributions. *Math. Scand.*, **17**, 65–76.
122. Esseen, C.-G. (1968a). On the concentration function of a sum of independent random variables. *Z. Wahrsch. Verw. Geb.*, **9**, 290–308.
123. Esseen, C.-G. (1968b). On the remainder term in the central limit theorem. *Arkiv Mat.*, **8**, 7–15.
124. Esseen, C.-G. (1975). Bounds for the absolute third moments. *Scand. J. Statist.*, **2**, 149–52.
125. Etemadi, N. (1981). An elementary proof of the strong law of large numbers. *Z. Wahrsch. Verw. Geb.*, **55**, 119–22.
126. Etemadi, N. (1983a). On the laws of large numbers for nonnegative random variables. *J. Multivar. Anal.*, **13**, 187–93.
127. Etemadi, N. (1983b). Stability of sums of weighted nonnegative random variables. *J. Multivar. Anal.*, **13**, 361–5.
128. Fainleib, A. S. (1968). A generalization of Esseen's inequality and an application of it to probabilistic number theory. *Izv. Akad. Nauk SSSR, Ser. Mat.*, **32**, 859–79 (in Russian).
129. Feller, W. (1943). The general form of the so-called law of the iterated logarithm. *Trans. Amer. Math. Soc.*, **54**, 373–402.
130. Feller, W. (1946). A limit theorem for random variables with infinite moments. *Amer. J. Math.*, **68**, 257–62.
131. Feller, W. (1968). On the Berry–Esseen theorem. *Z. Wahrsch. Verw. Geb.*, **10**, 261–8.
132. Feller, W. (1967). On regular variation and local limit theorems. In: Proceedings 5th Berkeley Symposium on Mathematical Statistics and Probability, Vol. 2, Part 1, pp. 373–88. University of California Press, Berkeley and Los Angeles.
133. Feller, W. (1971). *An introduction to probability theory and its applications*, Vol. 2 (2nd edn). Wiley, New York.
134. Fisher, A. (n.d.) A pathwise central limit theorem for random walks (preprint).
135. Fishler, R. M. (1967). Borel–Cantelli type problems for mixing sets. *Acta Math. Acad. Sci. Hungar.*, **18**, 67–9.
136. Fisz, M. (1962). Infinitely divisible distributions: recent results and applications. *Ann. Math. Statist.*, **33**, 68–84.
137. Fisz, M. and Varadarajan, V. S. (1963). A condition for absolute continuity of infinitely divisible distribution function. *Z. Wahrsch. Verw. Geb.*, **1**, 335–9.
138. Frechét, M. and Shohat, J. (1931). A proof of the generalized second limit theorem in the theory of probability. *Trans. Amer. Math. Soc.*, **33**, 533–43.
139. Friedman, N., Katz, M., and Koopmans, L. H. (1966). Convergence rates for the central limit theorem. *Proc. Nat. Acad. Sci. U.S.A.*, **56**, 1062–5.
140. Frolov, A. N. (1990). The Erdös–Rényi law of large numbers when Cramér's condition is not fulfilled. *Vestnik Leningrad Univ., Math.*, **23**, No. 4, 30–5.
141. Frolov, A. N. (1991). On the Erdös–Rényi law of large numbers for non-identically distributed random variables under violation of Cramér's condition. *Vestnik Leningrad Univ., Math.* **24**, No. 2, 66–70.

142. Fuk, D. Kh. and Nagaev, S. V. (1971, 1976). Probability inequalities for sums of independent random variables. *Th. Probab. Appl.*, **16**, 643–60; **21**, 875.
143. Gaposhkin, V. F. (1972). Convergence and limit theorems for sequences of random variables. *Th. Probab. Appl.*, **17**, 379–400.
144. Gnedenko, B. V. (1939). On the theory of the domains of attraction of stable laws. *Uchen. Zap. Mosk. Univ.*, **30**, 61–81 (in Russian).
145. Gnedenko, B. V. (1949). On some properties of limit distributions for the normed sums. *Ukrain. Math. J.*, **1**, 3–8 (in Russian).
146. Gnedenko, B. V. (1962). *The theory of probability*. Chelsea, New York.
147. Gnedenko, B. V. and Kolmogorov, A. N. (1968). *Limit distributions for sums of independent random variables*, (2nd edn). Addison-Wesley, Reading, MA.
148. Godwin, H. J. (1964). *Inequalities on distribution functions*. Griffin, London.
149. Götze, F. and Hipp, C. (1978). Asymptotic expansions in the central limit theorem under moment conditions. *Z. Wahrsch. Verw. Geb.*, **42**, 67–87.
150. Goursat, E. (1910). *Cours d'analyse mathématique*, T. 1. Gautier-Villars, Paris.
151. Gradshteyn, I. S. and Ryzhik, I. M. (1980). *Tables of integrals, sums, series and products*. Academic Press, New York.
152. Grübel, R. (1983). Über unbegrenzt teilbare Verteilungen. *Arch. Math.*, **41**, 80–8.
153. Gut, A. (1986). Law of the iterated logarithm for subsequences. *Probab. Math. Statist.*, **7**, 27–58.
154. Hàjek, J. and Rényi, A. (1955). Generalization of an inequality of Kolmogorov. *Acta Math. Acad. Sci. Hungar.*, **6**, 281–3.
155. Hall, P. (1979). On the rate of convergence in the central limit theorem for distributions with regularly varying tails. *Z. Wahrsch. Verw. Geb.*, **49**, 1–11.
156. Hall, P. (1980a). On the limiting behaviour of the mode and median of a sum of independent random variables. *Ann. Probab.*, **8**, 419–30.
157. Hall, P. (1980b). Characterizing the rate of convergence in the central limit theorem. *Ann. Probab.*, **8**, 1037–48.
158. Hall, P. (1981a). On the rate of convergence to a stable law. *J. London Math. Soc.*, **23**, 179–92.
159. Hall, P. (1981b). Order of magnitude of moments of sums of random variables. *J. London Math. Soc.*, **24**, 562–8.
160. Hall, P. (1981c). Two-sided bounds on the rate of convergence to a stable law. *Z. Wahrsch. Verw. Geb.*, **57**, 349–64.
161. Hall, P. (1982a). On the rate of convergence in the weak law of large numbers. *Ann. Probab.*, **10**, 374–81.
162. Hall, P. (1982b). Bounds on the rate of convergence of moments in the central limit theorem. *Ann. Probab.*, **10**, 1004–18.
163. Hall, P. (1982c). *Rates of convergence in the central limit theorem*. Pitman, London etc.
164. Hall, P. (1983a). Fast rates of convergence in the central limit theorem. *Z. Wahrsch. Verw. Geb.*, **62**, 491–507.
165. Hall, P. (1983b). Two-sided bounds for nonuniform rates of convergence in the central limit theorem. *Z. Wahrsch. Verw. Geb.*, **65**, 61–72.
166. Hall, P. (1983c). Sets which determine the rate of convergence in the central limit theorem. *Ann. Probab.*, **11**, 355–61.
167. Hall, P. (1983d). Chi squared approximations to the distribution of a sum of independent random variables. *Ann. Probab.*, **11**, 1028–36.
168. Hall, P. (1983e). On the rate of convergence of moments in the central limit theorem for lattice distributions. *Trans. Amer. Math. Soc.*, **278**, 169–81.

169. Hall, P. (1984a). A leading term approach to asymptotic expansions in the central limit theorem. *Proc. London Math. Soc.*, **49**, 423–44.
170. Hall, P. (1984b). On unimodality and rates of convergence for stable laws. *J. London Math. Soc.*, **30**, 371–84.
171. Hall, P. and Barbour, A. D. (1984). Reversing the Berry–Esseen inequality. *Proc. Amer. Math. Soc.*, **90**, 107–10.
172. Hall, P. and Heyde, C. C. (1980). *Martingale limit theory and its applications.* Academic Press, New York.
173. Hall, P. and Nakata, T. (1986). On non-uniform and global descriptions of the rate of convergence of asymptotic expansions in the central limit theorem. *J. Austral. Math. Soc.*, **A41**, 326–35.
174. Hall, P. and Wightwick, J. C. H. (1986). Convergence determining sets in the central limit theorem. *Probab. Th. Rel. Fields*, **71**, 1–17.
175. Hardy, G. H., Littlewood, J. E., and Pólya, G. (1952). *Inequalities,* (2nd edn). Cambridge University Press.
176. Hartman, P. and Wintner, A. (1941). On the law of the iterated logarithm. *Amer. J. Math.*, **63**, 169–76.
177. Hartman, P. and Wintner, A. (1942). On the infinitesimal generators of integral convolutions. *Amer. J. Math.*, **64**, 273–98.
178. Hausdorff, F. (1914). *Grundzüge der Mengenlehre.* Teubner, Leipzig.
179. Heathcote, C. R. and Pitman, J. W. (1972). An inequality for characteristic functons. *Bull. Austral. Math. Soc.*, **6**, 1–9.
180. Hengartner, W. and Theodorescu, R. (1973). *Concentration functions.* Academic Press, New York.
181. Heyde, C. C. (1966). Some results on small-deviation probability convergence rates for sums of independent random variables. *Canad. J. Math.*, **18**, 656–65.
182. Heyde, C. C. (1967). On the influence of moments on the rate of convergence to the normal distribution. *Z. Wahrsch. Verw. Geb.*, **8**, 12–18.
183. Heyde, C. C. (1968). On almost sure convergence for sums of independent random variables. *Sankhya*, **A30**, 353–8.
184. Heyde, C. C. (1969a). A note concerning behaviour of iterated logarithm type. *Proc. Amer. Math. Soc.*, **23**, 85–90.
185. Heyde, C. C. (1969b). Some properties of metrics in a study on convergence to normality. *Z. Wahrsch. Verw. Geb.*, **11**, 181–92.
186. Heyde, C. C. (1973). On the uniform metric on the context of convergence to normality. *Z. Wahrsch. Verw. Geb.*, **25**, 83–95.
187. Heyde, C. C. (1975a). A nonuniform bound on convergence to normality. *Ann. Probab.*, **3**, 903–7.
188. Heyde, C. C. (1975b). A supplement to the strong law of large numbers. *J. Appl. Probab.*, **12**, 173–5.
189. Heyde, C. C. and Leslie, J. R. (1972). On the influence of moments on approximation by portion of a Chebyshev series in the central limit convergence. *Z. Wahrsch. Verw. Geb.*, **21**, 255–68.
190. Heyde, C. C. and Nakata, T. (1984). On the asymptotic equivalence of L_p metrics for convergence to normality. *Z. Wahrsch. Verw. Geb.*, **68**, 97–106.
191. Heyde, C. C. and Rohatgi, V. K. (1967). A pair of complementary theorems on convergence rates in the law of large numbers. *Proc. Cambridge Phil. Soc.*, **63**, 73–83.
192. Hipp, C. (1977). Edgeworth expansions for integrals of smooth functions. *Ann. Probab.*, **5**, 1004–11.

193. Ho Soo-Tong and Chen, L. H. Y. (1978). An L_p bound for the remainder in a combinatorial central limit theorem. *Ann. Probab.*, **6**, 231–49.
194. Hoeffding, W. (1961). On sequences of sums of independent random vectors. In: Proceedings, 4th Berkeley Symposium on Mathematical Statistics and Probability, Vol. 2, pp. 231–226. University of California Press, Berkeley and Los Angeles.
195. Hoeffding, W. (1963). Probability inequalities for sums of bounded random variables. *J. Amer. Statist. Assoc.*, **58**(301), 13–30.
196. Hsu, P. L. (1951a). Absolute moments and characteristic functions. *J. Chinese Math. Soc.*, **1**, 259–80.
197. Hsu, P. L. (1951b). A lemma on the coefficient of reduction of a sum of independent variates. *Sci. Record, Peking*, **4**, 197–200.
198. Hsu, P. L. (1958). The absolute continuity of the distribution functions in the class L. *Acta Sci. Natur. Univ. Pekinensis*, **4**, 145–50.
199. Hsu, P. L. (1983). *Collected papers*. Springer, New York.
200. Hsu, P. L. and Robbins, H. (1947). Complete convergence and the law of large numbers. *Proc. Nat. Acad. Sci. U.S.A.*, **33**, 25–31.
201. Hudson, W. N. and Tucker, H. G. (1975). On admissible translates of infinitely divisible distributions. *Z. Wahrsch. Verw. Geb.*, **32**, 65–72.
202. Ibragimov, I. A. (1966). On the accuracy of the approximation of distribution functions of sums of independent variables by the normal distribution. *Th. Probab. Appl.*, **11**, 559–79.
203. Ibragimov, I. A. (1967). On the Chebyshev–Cramér asymptotic expansions. *Th. Probab. Appl.*, **12**, 455–70.
204. Ibragimov, I. A. (1977). On determining an infinitely divisible distribution function by its values on a half-line. *Th. Probab. Appl.*, **22**, 384–90.
205. Ibragimov, I. A. and Linnik, Yu. V. (1971). *Independent and stationary sequences of random variables*. Wolters-Noordhoff, Groningen.
206. Ibragimov, I. A. and Presman, E. L. (1973). On the rate of convergence of the distributions of sums of independent random variables to accompanying distributions. *Th. Probab. Appl.*, **18**, 713–27.
207. Jain, N. C. and Pruitt, W. E. (1975). The other law of the iterated logarithm. *Ann. Probab.*, **3**, 1046–9.
208. Jesiak, B. and Rossberg, H.-J. (1981). New versions of the Lindeberg–Feller theorem. *Th. Probab. Appl.*, **26**, 845–9.
209. Johnson, W. B., Schechtman, G., and Zinn, J. (1985). Best constants in moment inequalities for linear combinations of independent and exchangeable random variables. *Ann. Probab.*, **13**, 234–53.
210. Karlin, S. and Studden, W. (1966). *Tchebysheff systems: with applications to analysis and statistics*. Wiley, New York.
211. Katz, M. (1963a). The probability in the tail of a distribution. *Ann. Math. Statist.*, **34**, 312–18.
212. Katz, M. (1963b). A note on the Berry–Esseen theorem. *Ann. Math. Statist.*, **34**, 1007–8.
213. Katz, M. (1968). A note on the weak law of large numbers. *Ann. Math. Statist.*, **39**, 1348–9.
214. Kawata, T. (1941). The function of mean concentration of a chance variable. *Duke Math. J.*, **8**, 666–77.
215. Kawata, T. (1972). *Fourier analysis in probability theory*. Academic Press, New York.

216. Kesten, H. (1969). A sharper form of the Doeblin–Lévy–Kolmogorov–Rogozin inequality for concentration functions. *Math. Scand.*, **25**, 133–44.
217. Kesten, H. (1970). The limit points of a normalized random walk. *Ann. Math. Statist.*, **41**, 1173–1205.
218. Kesten, H. (1972). Sums of independent random variables—without moment conditions. *Ann. Math. Statist.*, **43**, 701–32.
219. Khatuntseva, M. V. (1974). Estimates for quantiles in the central limit theorem. *Th. Probab. Appl.*, **19**, 598–604.
220. Khintchine, A. (1924). Über einen Satz der Wahrscheinlichkeitsrechnung. *Fund. Math.*, **6**, 9–20.
221. Khintchine, A. (1937). Zur Theorie der unbeschränktteilbaren Verteilungsgesetze. *Mat. Sb.*, **2**, 79–119.
222. Khintchine, A. (1938). *Limit laws for sums of independent random variables.* ONTI, Moscow–Leningrad.
223. Kingman, J. F. C. (1978). Uses of exchangeability. *Ann. Probab.*, **6**, 183–97.
224. Klaassen, C. A. J. (1985). On an inequality of Chernoff. *Ann. Probab.*, **13**, 966–74.
225. Klass, M. J. (1976, 1977). Toward a universal law of the iterated logarithm, I, II. *Z. Wahrsch. Verw. Geb.*, **36**, 165–78; **39**, 151–65.
226. Klass, M. J. (1984). The finite mean LIL bounds are sharp. *Ann. Probab.*, **12**, 907–11.
227. Klass, M. J. and Tomkins, R. J. (1984). On the limiting behavior of normed sums of independent random variables. *Z. Wahrsch. Verw. Geb.*, **68**, 107–20.
228. Kochen, S. B. and Stone, C. J. (1964). A note on the Borel–Cantelli lemma. *Illinois J. Math.*, **8**, 248–51.
229. Kolmogoroff, A. (1929). Über das Gesetz der iterierten Logarithmus. *Math. Ann.*, **101**, 126–35.
230. Kolmogoroff, A. (1930). Sur la loi forte des grands nombres. *C.R. Acad. Sci. Paris*, **191**, 910–12.
231. Kolmogoroff, A. (1933). *Grundbegriffe der Wahrscheinlichkeitsrechnung.* Springer, Berlin.
232. Kolmogorov, A. N. (1953). Recent works on limit theorems of probability theory. *Vestnik Mosk. Univ.*, (10), 20–38 (in Russian).
233. Kolmogorov, A. N. (1956). Two uniform limit theorems for sums of independent terms. *Th. Probab. Appl.*, **1**, 384–94.
234. Kolmogorov, A. N. (1958). Sur les propriétés des fonctions de concentration de M. P. Lévy. *Ann. Inst. Henri Poincaré*, **16**, 27–34.
235. Kolmogorov, A. N. (1963). On approximations of distributions of sums of independent terms by infinitely divisible distributions. *Trudy Mosk. Mat. Ob.*, **12**, 437–51 (in Russian).
236. Kolodyazhny, S. F. (1968). A generalization of a theorem of Esseen. *Vestnik Leningrad. Univ.*, No. 13, 28–33 (in Russian).
237. Komlós, J. (1967). A generalization of a problem of Steinhaus. *Acta Math. Acad. Sci. Hungar.*, **18**, 217–29.
238. Komlós, J., Major, P., and Tusnády, G. (1975). An approximation of partial sums of independent RV's, and the sample DF, I, II. *Z. Wahrsch. Verw. Geb.*, **32**, 111–31; **34**, 33–58.
239. Komlós, J. and Tusnády, G. (1975). On sequences of pure heads. *Ann. Probab.*, **3**, 608–17.
240. Kruglov, V. M. (1970). A note on infinitely divisible distributions. *Th. Probab. Appl.*, **15**, 319–24.

241. Kruglov, V. M. (1972). On the extension of the class of stable distributions. *Th. Probab. Appl.*, **17**, 685–94.
242. Kruglov, V. M. (1974). The behavior of sums of independent random variables. *Th. Probab. Appl.*, **19**, 374–9.
243. Kruglov, V. M. (1973). Convergence of numerical characteristics of sums of independent random variables and global theorems. *Lect. Notes Math.*, **330**, 255–86.
244. Kruglov, V. M. (1976). Global limit theorems. *Zap. Nauch. Sem. Leningrad. Otd. Mat. Inst.*, **61**, 84–101 (in Russian).
245. Kunisawa, K. (1949). On an analytical method in the theory of independent random variables. *Ann. Inst. Statist. Math.*, **1**, 1–77.
246. Lacey, M. T. and Philipp, W. (1990). A note on the almost everywhere central limit theorem. *Statist. Probab. Letters*, **9**, 201–5.
247. Laube, G. (1973). Weak convergence and convergence in the mean of distribution functions. *Metrika*, **20**, 103–5.
248. Laue, G. (1983). Characteristic functions of nonnegative random variables: the connection between $\operatorname{Re} f$ and $\operatorname{Im} f$. In Proceedings, 9th Prague Conference, *Information theory, statistical decision functions, random processes*, pp. 49–56. Akademia, Prague.
249. LeCam, L. (1965). On the distribution of sums of independent random variables. In *Bernoulli—Bayes—Laplace*, (anniversary volume, ed. J. Neyman and L. LeCam), pp. 179–202. Springer, Berlin.
250. Leslie, J. R. (1976). A refinement of the Osipov–Petrov bound for central limit convergence. *Z. Wahrsch. Verw. Geb.*, **35**, 231–5.
251. Lévy, P. (1925). *Calcul des probabilités*. Gautier-Villars, Paris.
252. Lévy, P. (1937). *Théorie de l'addition des variables aléatoires*. Gautier-Villars, Paris.
253. Lifshits, B. A. (1976). On the accuracy of approximation in the central limit theorem. *Th. Probab. Appl.*, **21**, 108–24.
254. Lin, Z. Y. (1990). The Erdös–Rényi laws of large numbers for non-identically distributed random variables. *Chin. Ann. Math.*, **11B**, 376–83.
255. Linnik, Yu. V. and Ostrovskii, I. V. (1977). *Decomposition of random variables and vectors*. American Mathematical Society, Providence, RI.
256. Loève, M. (1963). *Probability theory*, (3rd edn). Van Nostrand, Princeton.
257. Lukacs, E. (1970). *Characterisitc functions*, (2nd edn). Griffin, London.
258. Lukacs, E. (1975). *Stochastic convergence*. Academic Press, New York.
259. Lukacs, E. (1983). *Developments in characteristic function theory*. Griffin, London.
260. Lyapunov, A. M. (1901). Nouvelle forme du théorème sur la limite de probabilités. *Mém. Acad. Imp. Sci. St.-Petersbourg*, **12**(5), 1–24.
261. Lyapunov, A. M. (1954). *Collected works*, Vol. 1. Izdat. Akad. Nauk SSSR, Moscow, (in Russian).
262. Lynch, J. (1983). Some comments on the Erdös–Rényi law and a theorem of Shepp. *Ann. Probab.*, **11**, 801–2.
263. Maejima, M. (1978a). A non-uniform estimate in the central limit theorem for m-dependent random variables. *Keio Engineering Reports*, **31**(2), 15–20.
264. Maejima, M. (1978b). Some L_p versions for the central limit theorem. *Ann. Probab.*, **6**, 341–4.
265. Maejima, M. (1980). A note on the nonuniform rate of convergence to normality. *Yokohama Math. J.*, **28**, 97–106.

266. Major, P. (1976a). The approximation of partial sums of independent RV's. *Z. Wahrsch. Verw. Geb.*, **35**, 213–20.
267. Major, P. (1976b). Approximation of partial sums of i.i.d.r.v.s when the summands have only two moments. *Z. Wahrsch. Verw. Geb.*, **35**, 221–9.
268. Major, P. (1978). On the invariance principle for sums of independent identically distributed random variables. *J. Multivar. Anal.*, **8**, 487–517.
269. Major, P. (1979). An improvement of Strassen's invariance principle. *Ann. Probab.*, **7**, 55–61.
270. Makabe, H. (1960). A remark on the smoothness of the distribution function. *Yokohama Math. J.*, **8**, 59–68.
271. Maller, R. A. (1980). On the law of the iterated logarithm in the infinite variance case. *J. Austral. Math. Soc.*, **A30**, 5–14.
272. Maller, R. A. (1981). Some properties of stochastic compactness. *J. Austral. Math. Soc.*, **A30**, 264–77.
273. Manstavičius, E. (1982). Inequalities for the p-th moment, $0 < p < 2$, of a sum of independent random variables. *Lithuanian Math. J.*, **22**, 64–7.
274. Marcinkiewicz, J. (1964). *Collected papers.* Panstw. Wyd. Nauk., Warszawa.
275. Marcinkiewicz, J. and Zygmund, A. (1937a). Sur les fonctions indépendantes. *Fund. Math.*, **29**, 60–90.
276. Marcinkiewicz, J. and Zygmund, A. (1937b). Remarque sur la loi du logarithme itéré. *Fund. Math.*, **29**, 215–22.
277. Marcinkiewicz, J. and Zygmund, A. (1938). Quelques théorèmes sur les fonctions indépendantes. *Studia Math.*, **7**, 104–20.
278. Martikainen, A. I. (1977a). On the strong law of large numbers for ruled sums. *Vestnik Leningrad. Univ.*, No. 1, 56–65 (in Russian). (English transl. in *Vestnik Leningrad Univ. Math.*, 1982, **10**, 63–74.)
279. Martikainen, A. I. (1977b). Order of growth of ruled sums of random variables. *Vestnik Leningrad. Univ.*, No. 7, 57–62 (in Russian). (English transl. in *Vestnik Leningrad Univ. Math.*, 1982, **10**, 159–64.)
280. Martikainen, A. I. (1979a). Three theorems on limit superior of sums of independent random variables. *Vestnik Leningrad. Univ.*, No. 1, 45–51 (in Russian).
281. Martikainen, A. I. (1979b). An exponential criterium for the law of the iterated logarithm. *Zap. Nauch. Sem. Leningrad. Otd. Mat. Inst.*, **85**, 158–68 (in Russian). (English transl. in *J. Soviet Math.*, 1982, **20**, 2214–21.)
282. Martikainen, A. I. (1979c). On necessary and sufficient conditions for the strong law of large numbers. *Th. Probab. Appl.*, **24**, 813–20.
283. Martikainen, A. I. (1980a). A converse to the law of the iterated logarithm for a random walk. *Th. Probab. Appl.*, **25**, 361–2.
284. Martikainen, A. I. (1980b). A criterion for the strong relative stability of random walk. *Mat. Zametki*, **28**, 619–22 (in Russian).
285. Martikainen, A. I. (1983). One-sided versions of strong limit theorems. *Th. Probab. Appl.*, **28**, 48–64.
286. Martikainen, A. I. (1984). Criteria for strong convergence of normalized sums of independent random variables and their application. *Th. Probab. Appl.*, **29**, 502–16.
287. Martikainen, A. I. (1985). On one-sided law of the iterated logarithm. *Th. Probab. Appl.*, **30**, 736–49.
288. Martikainen, A. I. (1986). On Chung's law for nonidentically distributed random variables. *Th. Probab. Appl.*, **31**, 498–500.

289. Martikainen, A. I. and Petrov, V. V. (1977). On necessary and sufficient conditions for the law of the iterated logarithm. *Th. Probab. Appl.*, **22**, 16–23, 430.
290. Martikainen, A. I. and Petrov, V. V. (1980). On a theorem of Feller. *Th. Probab. Appl.*, **25**, 191–3.
291. Martikainen, A. I. and Petrov, V. V. (1990). On the Borel–Cantelli lemma. *Zap. Nauch. Sem. Leningrad. Otd. Mat. Inst.*, **184**, 200–7 (in Russian).
292. Martikainen, A. I. and Petrov, V. V. (1992). On strong limit theorems for sums of independent random variables. *Th. Probab. Appl.*, **37**, 81–6 (in Russian).
293. Matskyavichyus, V. K. (1983). A lower bound for the convergence rates in the central limit theorem. *Th. Probab. Appl.*, 1983, **28**, 596–601.
294. Mejlijson, I. (1973). The average of the values of a function at random points. *Israel J. Math.*, **15**, 193–203.
295. Mejzler, D. (1973). On a certain class of infinitely divisible distributions. *Israel J. Math.*, **16**, 1–19.
296. Michel, R. (1981). On the constant in the non-uniform version of the Berry–Esseen theorem. *Z. Wahrsch. Verw. Geb.*, **55**, 109–17.
297. Michel, R. (1983). On the influence of moments in nonuniform expansions of Chebyshev–Cramér type. *Statistics and Decisions*, **1**, 205–15.
298. Mijnheer, J. L. (1974). *Sample path properties of stable processes*. Math. Centrum, Amsterdam.
299. Mijnheer, J. L. (1980). A strong approximation of partial sums of i.i.d. random variables with infinite variance. *Z. Wahrsch. Verw. Geb.*, **32**, 1–7.
300. Mijnheer, J. L. (1983). Strong approximations of partial sums of i.i.d. random variables in the domain of attraction of a symmetric stable distribution. In Proceedings 9th Prague Conference, *Information theory, statistical decision functions, random processes*, pp. 83–89. Akademia, Prague.
301. Mijnheer, J. L. (1986). On the rate of convergence to a stable limit law. *Lithuanian Math. J.*, **26**, 255–9.
302. Mikosch, T. (1984). On the law of the iterated logarithm for independent random variables outside the domain of partial attraction of the normal law. *Vestnik Leningrad. Univ.*, No. 13, 35–9 (in Russian).
303. Mikosch, T. (1985). Two theorems on the law of the iterated logarithm without variance. *Vestnik Leningrad. Univ.*, No. 1, 110–12 (in Russian).
304. Mikosch, T. (1988). Iterated logarithm results for rapidly growing random walk. *Statistics*, **19**, 107–15.
305. Miroshnikov, A. L. and Rogozin, B. A. (1980). Inequalities for the concentration function. *Th. Probab. Appl.*, **25**, 176–80.
306. Miroshnikov, A. L. and Rogozin, B. A. (1982). Remarks on an inequality for the concentration function of sums of independent random variables. *Th. Probab. Appl.*, **27**, 848–50.
307. Mogyoródi, J. (1979). On an inequality of Marcinkiewicz and Zygmund. *Publ. Math. Debrecen*, **26**, 267–74.
308. Móri, T. F. and Székely, G. J. (1983). On the Erdös–Rényi generalization of the Borel–Cantelli lemma. *Studia Sci. Math. Hungar.*, **18**, 173–82.
309. Nagaev, S. V. (1965). Some limit theorems for large deviations. *Th. Probab. Appl.*, **10**, 214–35.
310. Nagaev, S. V. (1972). On necessary and sufficient conditions for the strong law of large numbers. *Th. Probab. Appl.*, **17**, 573–81.
311. Nagaev, S. V. (1979). Large deviations of sums of independent random variables. *Ann. Probab.*, **7**, 745–89.

312. Nagaev, S. V. and Pinelis, I. F. (1977). Some inequalities for the distribution of sums of independent random variables. *Th. Probab. Appl.*, **22**, 248–56.
313. Nagaev, S. V. and Rotar, V. I. (1973, 1976). On strengthening Lyapunov type estimates. *Th. Probab. Appl.*, **18**, 107–19; **21**, 220.
314. Nikitenko, T. V. (1990). On the distribution of values of sums of independent random variables. *Vestnik Leningrad. Univ. Ser. 1*, No. 2(8), 30–5 (in Russian).
315. Nikitenko, T. V. (1991). On the law of the iterated logarithm for independent identically distributed random variables. *Vestnik Leningrad. Univ.*, Ser. 1, No. 3 (15), 54–8 (in Russian).
316. Ohkubo, H. (1979). On the asymptotic tail behavior of infinitely divisible distributions. *Yokohama Math. J.*, **27**, 77–89.
317. Osipov, L. V. (1966). Refinement of Lindeberg's theorem. *Th. Probab. Appl.*, **11**, 299–302.
318. Osipov, L. V. (1967). Asymptotic expansions in the central limit theorem. *Vestnik Leningrad. Univ.*, No. 19, 45–62 (in Russian).
319. Osipov, L. V. (1968). On the closeness with which the distribution of the sum of independent random variables approximates the normal distributions. *Soviet Math. Doklady*, **9**, 233–6.
320. Osipov, L. V. (1971). On asymptotic expansions for the distribution function of a sum of random variables. *Th. Probab. Appl.*, **16**, 333–43.
321. Osipov, L. V. (1972). On asymptotic expansions of the distribution function of a sum of random variables with non-uniform estimates of the remainder. *Vestnik Leningrad. Univ.*, No. 1, 51–9 (in Russian).
322. Osipov, L. V. and Petrov, V. V. (1967). On an estimate of the remainder term in the central limit theorem. *Th. Probab. Appl.*, **12**, 281–6.
323. Paditz, L. (1989). On an analytical structure of the constant in the nonuniform version of the Esseen inequality. *Statistics*, **20**, 453–64.
324. Paulauskas, V. I. (1971). On a smoothing inequality. *Litovsk. Nat. Sb.*, **11**, 861–6. (English transl. in *Selected Transl. Math. Statist. Probab.*, 1977, **14**, 7–12. American Mathematical Society, Providence, RI.)
325. Paulauskas, V. I. (1972). Estimates of the rate of convergence in the central limit theorem for non-identically distributed summands. *Litovsk. Mat. Sb.*, **12**, 183–94 (in Russian).
326. Paulauskas, V. I. (1974). Estimates of the remainder in a limit theorem with a stable limit law. *Litovsk. Mat. Sb.*, **14**, 165–88 (in Russian).
327. Petrov, V. V. (1954). A generalization of Cramér's limit theorem. *Uspekhi Mat. Nauk*, **9**(4), 195–202. (English transl. in *Selected Transl. Math. Statist. Probab.*, 1966, **6**, 1–8. American Mathematical Society, Providence, RI.)
328. Petrov, V. V. (1960). On the central limit theorem for m-dependent random variables. In *Trudy vsesoyuzn. sov. po teorii veroyatn. i mat. statist.*, pp. 38–44. Izdat. Akad. Nauk ArmSSR, Erevan. (English transl. in *Selected Transl. Math. Statist. Probab.*, 1970, **9**, 83–8. American Mathematical Society, Providence, RI.)
329. Petrov, V. V. (1962). On some polynomials encountered in probability theory. *Vestnik Leningrad. Univ.*, No. 19, 150–3 (in Russian).
330. Petrov, V. V. (1965). An estimate of the deviation of the distribution function of a sum of independent random variables from the normal law. *Soviet Math. Doklady*, **6**, 242–4.
331. Petrov, V. V. (1966). On the connection between an estimate of the remainder term in the central limit theorem and the law of the iterated logarithm. *Th. Probab. Appl.*, **11**, 454–8.

332. Petrov, V. V. (1967). A generalization and sharpening of Bernstein's inequality. *Vestnik Leningrad. Univ.*, No. 19, 63–8 (in Russian).
333. Petrov, V. V. (1968). On the law of the iterated logarithm without assumptions about the existence of moments. *Proc. Nat. Acad. Sci. U.S.A.*, **59**, 1068–72.
334. Petrov, V. V. (1969). On the strong law of large numbers. *Th. Probab. Appl.*, **14**, 183–92.
335. Petrov, V. V. (1970). On an estimate of the concentration function of a sum of independent random variables. *Th. Probab. Appl.*, **15**, 701–3.
336. Petrov, V. V. (1973a). The order of growth of sums of dependent random variables. *Th. Probab. Appl.*, **18**, 348–50.
337. Petrov, V. V. (1973b). On the moments of distributions attracted to stable laws. *Th. Probab. Appl.*, **18**, 569–71.
338. Petrov, V. V. (1974a). Some theorems on the law of the iterated logarithm. *Zap. Nauch. Sem. Leningrad. Otd. Mat. Inst.*, **41**, 129–32. (English transl. in *J. Soviet Math.*, 1978, **9**, 99–102.)
339. Petrov, V. V. (1974b). The one-sided strong law of large numbers for ruled sums. *Vestnik Leningrad. Univ.*, No. 7, 55–9. (English transl. in *Vestnik Leningrad Univ. Math.*, 1979, **7**, 164–8.)
340. Petrov, V. V. (1975a). A generalization of an inequality of Lévy. *Th. Probab. Appl.*, **20**, 141–5.
341. Petrov, V. V. (1975b). An inequality for moments of a random variable. *Th. Probab. Appl.*, **20**, 391–2.
342. Petrov, V. V. (1975c). *Sums of independent random variables*. Springer, New York etc.
343. Petrov, V. V. (1975d). On the strong law of large numbers for a sequence of orthogonal random variables. *Vestnik Leningrad. Univ.*, No. 7, 52–7. (English transl. in *Vestnik Leningrad Univ. Math.*, 1980, **8**, 225–31.)
344. Petrov, V. V. (1976). Growth of sums of measurable functions. *Lithuanian Math. J.*, **16**, 115–17.
345. Petrov, V. V. (1978a). On the lower limit of the modulus of sums of independent random variables. *Vestnik Leningrad. Univ.*, No. 7, 49–54. (English transl. in *Vestnik Leningrad Univ. Math.*, 1983, **11**, 147–52.)
346. Petrov, V. V. (1978b). Remark on the lower limit for the modulus of sums of independent random variables. *Lithuanian Math. J.*, **18**, 528–31.
347. Petrov, V. V. (1979). A limit theorem for sums of independent non-identically distributed random variables. *Zap. Nauch. Sem. Leningrad. Otd. Mat. Inst.*, **85**, 188–92. (English transl. in *J. Soviet Math.*, 1982, **20**, 2232–5.)
348. Petrov, V. V. (1980). On the law of the iterated logarithm for sequences of dependent random variables. *Zap. Nauch. Sem. Leningrad. Otd, Mat. Inst.*, **97**, 186–94. (English transl. in *J. Soviet Math.*, 1984, **24**, 611–17.)
349. Petrov, V. V. (1982a). Sequences of m-orthogonal random variables. *Zap. Nauch. Sem. Leningrad. Otd. Mat. Inst.*, **119**, 198–202. (English transl. in *J. Soviet Math.*, 1984, **27**, 3136–40.)
350. Petrov, V. V. (1982b). Some inequalities for moments. *Izv. Akad. Nauk UzbekSSR, Ser. Fiz.-mat. Nauk*, No. 5, 31–4 (in Russian).
351. Petrov, V. V. (1987). *Limit theorems for sums of independent random variables*. Nauka, Moscow (in Russian).
352. Petrov, V. V. (1990). Some inequalities for moments of sums of independent random variables. In *Probability theory and mathematical statistics* (ed. B. Grigelionis *et al.*), Vol. 2, pp. 309–14. Mokslas–VSP, Vilnius–Utrecht.

353. Petrov, V. V. (1992a). Generalization of Rosenthal's inequalities. *Ann. Acad. Sci. Fennicae, Ser. A, Math.*, **17**, 117–21.
354. Petrov, V. V. (1992b). On the law of the iterated logarithm for sequences of independent random variables. *Zap. Nauch. Sem. Leningrad. Otd. Mat. Inst.*, **194**, 134–7 (in Russian).
355. Petrov, V. V. (1991). Limit theorems of classical type for sums of independent random variables. In *Itogi nauki i tekhniki. Current problems in mathematics. Fundamental directions*, Vol. 81, pp. 10–38. Akad. Nauk SSSR, Moscow (in Russian).
356. Petrov, V. V. (1993). On Feller's theorem on the strong law of large numbers. In *Rings and modulus. Limit theorems of probability theory* (ed. Z. I. Borevich and V. V. Petrov), Vol. 3, 217–21. St Petersburg Univ. Press (in Russian).
357. Petrov, V. V. and Shirokova, I. V. (1973). On exponential rate of convergence in the law of large numbers. *Vestnik Leningrad. Univ.*, No. 7, 155–7 (in Russian).
358. Pinelis, I. F. and Utev, S. A. (1984). Estimates of the moments of sums of independent random variables. *Th. Probab. Appl.*, **29**, 574–7.
359. Pitman, E. J. G. (1956). On the derivatives of a characteristic function at the origin. *Ann. Math. Statist.*, **27**, 1156–60.
360. Pólya, G. (1920). Über den zentralen Grenzwertsatz der Wahrscheinlichkeitsrechnung und das Momentproblem. *Math. Z.*, **8**, 171–80.
361. Pólya, G. (1949). Remark on characteristic functions. In *Proceedings, Berkeley Symposium on Mathematical Statistics and Probability*, pp. 115–123. Univ. California Press, Berkeley and Los Angeles.
362. Postnikova, L. P. and Yudin, A. A. (1978). A sharper form of an inequality for the concentration function. *Th. Probab. Appl.*, **23**, 359–62.
363. Prawitz, H. (1975). Weitere Ungleichungen für den absoluten Betrag einer charakteristischen Funktion. *Skand. Aktuarial J.*, 21–8.
364. Prohorov, Yu. V. (1950). On the strong law of large numbers. *Izv. Akad. Nauk SSSR, Ser. Mat.*, **14**, 523–36 (in Russian).
365. Prohorov, Yu. V. (1955). On sums of random variables having a common distribution. *Dokl. Akad. Nauk SSSR*, **105**, 645–7 (in Russian).
366. Prohorov, Yu. V. (1956). Convergence of random processes and limit theorems of probability theory. *Th. Probab. Appl.*, **1**, 157–214.
367. Prohorov, Yu. V. (1958). Strong stability of sums and infinitely divisible distributions. *Th. Probab. Appl.*, **3**, 141–53.
368. Prohorov, Yu. V. (1959a). An extremal problem in probability theory. *Th. Probab. Appl.*, **4**, 201–4.
369. Prohorov, Yu. V. (1959b). Remarks on the strong law of large numbers. *Th. Probab. Appl.*, **4**, 204–8.
370. Prohorov, Yu. V. (1960). On a uniform limit theorem of A. N. Kolmogorov. *Th. Probab. Appl.*, **5**, 98–106.
371. Pruitt, W. E. (1981). General one-sided laws of the iterated logarithm. *Ann. Probab.*, **9**, 1–48.
372. Pruitt, W. E. (1983). The class of limit laws for stochastically compact normed sums. *Ann. Probab.*, **11**, 962–9.
373. Pruitt, W. E. (1987). The contribution to the sum of the summand of maximum modulus. *Ann. Probab.*, **15**, 885–96.
374. Pruitt, W. E. (1990). The rate of escape of random walk. *Ann. Probab.*, **18**, 1417–61.

375. Rachev, S. T. (1991). *Probability metrics and the stability of stochastic models.* Wiley, Chichester.
376. Raikov, D. A. (1938). On the connection between the central limit law of probability theory and the law of large numbers. *Izv. Akad. Nauk SSSR, Ser. Mat.*, **2**, 323–38 (in Russian).
377. Ramachandran, B. (1967). *Advanced theory of characteristic functions.* Statist. Publ. Soc., Calcutta.
378. Ramachandran, B. (1969). On characteristic functions and moments. *Sankhya*, **A31**, 1–12.
379. Rao, K. S. and Kendall, D. G. (1950). On the generalized second limit-theorem in the calculus of probabilities. *Biometrika*, **37**, 224–30.
380. Rényi, A. (1970). *Probability theory.* Akadémiai Kiadó, Budapest.
381. Révész, P. (1965). On a problem of Steinhaus. *Acta Math. Acad. Sci. Hungar.*, **16**, 311–18.
382. Révész, P. (1967). *The laws of large numbers.* Academic Press, New York.
383. Révész, P. (1990). *Random walk in random and non-random environment.* World Scientific, Singapore etc.
384. Riedel, M. (1976). On the one-sided tails of infinitely divisible distributions. *Math. Nachr.*, **70**, 155–63.
385. Riedel, M. (1977). A new version of the central limit theorem. *Th. Probab. Appl.*, **22**, 183–4.
386. Robbins, H. (1953). On the equidistribution of sums of independent random variables. *Proc. Amer. Math. Soc.*, **4**, 786–99.
387. Rogozin, B. A. (1960). Remark on a paper by Esseen. *Th. Probab. Appl.*, **5**, 114–17.
388. Rogozin, B. A. (1961a). On an estimate of the concentration function. *Th. Probab. Appl.*, **6**, 94–7.
389. Rogozin, B. A. (1961b). On the increase of dispersion of sums of independent random variables. *Th. Probab. Appl.*, **6**, 97–9.
390. Rogozin, B. A. (1968). On the existence of exact upper sequences. *Th. Probab. Appl.*, **13**, 667–71.
391. Rosalsky, A. (1980). On the converse to the iterated logarithm law. *Sankhya*, **A42**, 103–8.
392. Rosén, B. (1961). On the asymptotic distribution of sums of independent identically distributed random variables. *Arkiv Math.*, **4**, 323–32.
393. Rosenthal, H. P. (1970). On the subspaces of L^p ($p > 2$) spanned by sequences of independent random variables. *Israel J. Math.*, **8**, 273–303.
394. Rosenthal, H. P. (1972). On the span in L^p of sequences of independent random variables. In Proceedings 6th Berkeley Symposium on Mathematical Statistics and Probability, Vol. 2, pp. 149–67. Univ. California Press, Berkeley and Los Angeles.
395. Rossberg, H.-J. (1974). On a problem of Kolmogorov concerning the normal distribution. *Th. Probab. Appl.*, **19**, 795–8.
396. Rossberg, H.-J., Jesiak, B., and Siegel, G. (1985). *Analytic methods in probability theory.* Akademie-Verlag, Berlin.
397. Rossberg, H.-J. and Siegel, G. (1975). Continuation of convergence in the central limit theorem. *Th. Probab. Appl.*, **20**, 866–8.
398. Rossberg, H.-J. and Siegel, G. (1981). One-sided characterization of the normal distribution in the set of infinitely divisible distributions. *Th. Probab. Appl.*, **26**, 392–9.

399. Rotar, V. I. (1975). On generalizations of the Lindeberg–Feller theorem. *Mat. Zametki*, **18**, 129–35 (in Russian).
400. Rotar, V. I. (1978). On non-classical estimates of approximation in the central limit theorem. *Mat. Zametki*, **23**, 143–53 (in Russian).
401. Rotar, V. I. (1982). On summation of independent terms in a non-classical situation. *Uspekhi Mat. Nauk*, **37**, (6), 137–56 (in Russian).
402. Rozovsky, L. V. (1974). On the convergence rate in the Lindeberg–Feller theorem. *Vestnik Leningrad. Univ.*, No. 11, 70–5. (English transl. in *Vestnik Leningrad Univ. Math.*, 1979, **7**, 83–9.)
403. Rozovsky, L. V. (1975). Asymptotic expansions in the central limit theorem. *Th. Probab. Appl.*, **20**, 794–804.
404. Rozovsky, L. V. (1977). On properties of asymptotic expansions. *Mat. Zametki*, **22**, 907–14 (in Russian).
405. Rozovsky, L. V. (1978a). On lower bound for the remainder in the central limit theorem. *Mat. Zametki*, **24**, 411–18 (in Russian).
406. Rozovsky, L. V. (1978b). On the precision of an estimate of the remainder term in the central limit theorem. *Th. Probab. Appl.*, **23**, 712–30.
407. Rozovsky, L. V. (1981a). On the convergence rate in the strong law of large numbers. *Th. Probab. Appl.*, **26**, 135–40.
408. Rozovsky, L. V. (1981b). Relation between rate of convergence in the weak and strong laws of large numbers. *Lithuanian Math. J.*, **21**, 75–84.
409. Sahanenko, A. I. (1985a). Convergence rate in the invariance principle for non-identically distributed variables with exponential moments. In *Advances in probability theory. Limit theorems for sums of random variables* (ed. A. A. Borovkov), pp. 2–73. Optimization Software Inc., Publication Division, New York.
410. Sahanenko, A. I. (1985b). Estimates in the invariance principle. *Trudy Inst. Mat. Sibirsk. Otd. Akad. Nauk SSSR*, **5**, 27–44.
411. Saulis, L. and Statulevicius, V. (1991). *Limit theorems for large deviations*. Kluwer, Dordrecht.
412. Savage, I. R. (1961). Probability inequalities of the Tchebysheff type. *J. Res. Nat. Bur. Standards*, **65B**, 211–22.
413. Sazonov, V. V. (1974a). On the estimation of moments of sums of independent random variables. *Th. Probab. Appl.*, **19**, 371–4.
414. Sazonov, V. V. (1974b). A new general estimate of the rate of convergence in the central limit theorem in R^k. *Proc. Nat. Acad. Sci. U.S.A.*, **71**, 118–21.
415. Sazonov, V. V. (1981). *Normal approximation—some recent advances*. Lecture Notes in Mathematics, Vol. 879.
416. Schatte, P. (1988a). On the value distribution of sums of random variables. *Th. Probab. Appl.*, **33**, 743–7.
417. Schatte, P. (1988b). On strong versions of the central limit theorem. *Math. Nachr.*, **137**, 249–56.
418. Scheffé, H. (1947). A useful convergence theorem for probability distributions. *Ann. Math. Statist.*, **18**, 434–8.
419. Seneta, E. (1976). *Regularly varying functions*. Lecture Notes in Mathematics, Vol. 508.
420. Sethuraman, J. (1970). Probabilities of deviations. In *Essays in probability and statistics*, (ed. R. C. Bose *et al.*), pp. 655–72. Univ. North Carolina Press, Chapel Hill.

421. Shao Qui-man. (1989). On a problem of Csörgő and Révész. *Ann. Probab.*, **17**, 809–12.
422. Shepp, L. A. (1964). A limit law concerning moving averages. *Ann. Math. Statist.*, **35**, 424–8.
423. Shiganov, I. S. (1982). On refinement of an upper bound for the constant in the remainder term of the central limit theorem. In *Problems of stability of stochastic models*, pp. 109–15. VINITI, Moscow (in Russian).
424. Shiryaev, A. N. (1984). *Probability*. Springer, New York.
425. Shuster, J. (1970). On the Borel–Cantelli problem. *Canad. Math. Bull.*, **13**, 273–5.
426. Simons, G. and Stout, W. F. (1978). A weak invariance principle with applications to domain of attraction. *Ann. Probab.*, **6**, 294–315.
427. Sirazdinov, S. H. and Gafurov, M. U. (1988). Asymptotics of the mean of a functional of a random walk. In *Lecture Notes in Mathematics*, Vol. 1299, pp. 474–81.
428. Skorohod, A. V. (1964). *Random processes with independent increments*. Nauka, Moscow (in Russian).
429. Skovoroda, B. F. and Mikosch, T. (1992). A strong law of large numbers for ruled sums. *Statist. Probab. Letters*, **13**, 129–37.
430. Smith, W. L. (1962). A note on characteristic functions which vanish identically in an interval. *Proc. Cambridge Phil. Soc.*, **58**, 430–2.
431. Smith, W. L. (1969). Some results using general moment functions. *J. Austral. Math. Soc.*, **10**, 429–41.
432. Spitzer, F. (1956). A combinatorial lemma and its application to probability theory. *Trans. Amer. Math. Soc.*, **82**, 323–39.
433. Spitzer, F. (1964). *Principles of random walk*. Van Nostrand, Princeton.
434. Stadtmüller, U. (1984). A note on the law of the iterated logarithm for weighted sums of random variables. *Ann. Probab.*, **12**, 35–44.
435. Steinebach, J. (1978). On a necessary condition for the Erdös–Rényi law of large numbers. *Proc. Amer. Math. Soc.*, **68**, 97–100.
436. Steinebach, J. (1984). Between invariance principles and Erdös–Rényi laws. In *Limit theorems in probability and statisitics* (ed. P. Révész), Vol. 2, pp. 981–1005. North Holland, Amsterdam.
437. Steutel, F. W. (1973). Some recent results in infinite divisibility. *Stoch. Processes Appl.*, **1**, 125–43.
438. Steutel, F. W. (1979). Infinite divisibility in theory and practice. *Scand. J. Statist.*, **6**(2), 57–62.
439. Stone, C. J. (1969). The growth of a random walk. *Ann. Math. Statist.*, **40**, 2203–6.
440. Stout, W. F. (1974). *Almost sure convergence*. Academic Press, New York.
441. Stout, W. F. (1979). Almost sure invariance principles when $EX_1^2 = \infty$. *Z. Wahrsch. Verw. Geb.*, **49**, 23–32.
442. Strassen, V. (1964). An invariance principle for the law of the iterated logarithm. *Z. Wahrsch. Verw. Geb.*, **3**, 211–26.
443. Strassen, V. (1966). A converse to the law of the iterated logarithm. *Z. Wahrsch. Verw. Geb.*, **4**, 265–8.
444. Strassen, V. (1967). Almost sure behavior of sums of independent random variables and martingales. In Proceedings, 5th Berkeley Symposium on Mathematical Statistics and Probability, Vol. 2, Part 1, pp. 315–343. Univ. California Press, Berkeley and Los Angeles.

445. Suchkov, A. P. and Ushakov, N. G. (1989). On fast decrease of the concentration functions of sums of independent random variables. *Th. Probab. Appl.*, **34**, 604–7.
446. Tanny, D. (1977). A new proof of Kesten's theorem on the growth of the sum of independent and identically distributed random variables. *Z. Wahrsch. Verw. Geb.*, **39**, 231–4.
447. Teicher, H. (1979). Rapidly growing random walk and an associated stopping time. *Ann. Probab.*, **7**, 1078–81.
448. Teicher, H. (1985). Almost certain convergence in double arrays. *Z. Wahrsch. Verw. Geb.*, **69**, 331–45.
449. Tomkins, R. J. (1974). On the law of the iterated logarithm for double sequences of random variables. *Z. Wahrsch. Verw. Geb.*, **30**, 303–14.
450. Tomkins, R. J. (1978). On the law of the iterated logarithm. *Ann. Probab.*, **6**, 162–8.
451. Tomkins, R. J. (1980). Limit theorems without moment hypotheses for sums of independent random variables. *Ann. Probab.*, **8**, 314–24.
452. Tomkins, R. J. (1983). Lindeberg functions and the law of the iterated logarithm. *Z. Wahrsch. Verw. Geb.*, **65**, 135–43.
453. Tomkins, R. J. (1992). Refinement of Kolmogorov's law of the iterated logarithm. *Statist. Probab. Letters*, **14**, 321–5.
454. Trautner, R. (1990). A new proof of the Komlós–Révész theorem. *Probab. Th. Rel. Fields*, **84**, 281–7.
455. Tsaregradskii, I. P. (1958). On uniform approximation of the binomial distribution by infinitely divisible distributions. *Th. Probab. Appl.*, **3**, 434–8.
456. Tucker, H. G. (1961). Best one-sided bounds for infinitely divisible random variables. *Sankhyā*, **A23**, 387–96.
457. Ul'yanov, V. V. (1976). A non-uniform estimate for the speed of convergence in the central limit theorem in R. *Th. Probab. Appl.*, **21**, 270–82.
458. Ul'yanov, V. V. (1978). On more precise convergence rate estimates in the central limit theorem. *Th. Probab. Appl.*, **23**, 660–3.
459. Utev, S. A. (1985). Extremal problems in moment inequalities. *Trudy Inst. Mat. Sibirsk. Otd. Akad. Nauk SSSR*, **5**, 56–75 (in Russian).
460. Van Beek, P. (1972). An approximation of Fourier methods to the problem of sharpening the Berry–Esseen inequality. *Z. Wahrsch. Verw. Geb.*, **23**, 187–96.
461. Van Kampen, E. R. (1940). Infinite product measures and infinite convolutions. *Amer. J. Math.*, **62**, 417–48.
462. Volodin, N. A. and Nagaev, S. V. (1977). A remark on the strong law of large numbers. *Th. Probab. Appl.*, **22**, 810–13.
463. Von Bahr, B. (1965). On the convergence of moments in the central limit theorem. *Ann. Math. Statist.*, **36**, 808–18.
464. Von Bahr, B. and Esseen, C.-G. (1965). Inequalities for the rth absolute moment of a sum of independent random variables, $1 \leqslant r \leqslant 2$. *Ann. Math. Statist.*, **36**, 299–303.
465. Von Mises, R. (1939). An inequality for the moments of a discontinuous distribution. *Skand. Aktuarietidskrift*, **22**, 32–6.
466. Wintner, A. (1947). *The Fourier transforms of probability distributions.* Baltimore.
467. Wittmann, R. (1985). A general law of iterated logarithm. *Z. Wahrsch. Verw. Geb.*, **68**, 521–43.
468. Wittmann, R. (1990). Summen unabhängiger Zufallsvariablen, die durch die Maximalterme dominiert werden. *Acta Math. Hungar.*, **56**, 225–8.

469. Wolfe, S. J. (1971a). On moments of infinitely divisible distribution functions. *Ann. Math. Statist.*, **42**, 2036–43.
470. Wolfe, S. J. (1971b). On the continuity properties of L functions. *Ann. Math. Statist.*, **42**, 2064–73.
471. Wolfe, S. J. (1975). On moments of probability distribution functions. In *Lecture Notes in Mathematics*, Vol. 457, pp. 306–16.
472. Wolfe, S. J. (1978). On the behavior of characteristic functions on the real line. *Ann. Probab.*, **6**, 554–62.
473. Woyczyński, W. A. (1974). Strong laws of large numbers in certain linear spaces. *Ann. Inst. Fourier*, **24**, 205–23.
474. Yamazato, M. (1978). Unimodality of infinitely divisible distribution function of class L. *Ann. Probab.*, **6**, 523–31.
475. Zaitsev, A. Yu. (1981). Some properties of N-fold convolutions of distributions. *Th. Probab. Appl.*, **26**, 148–52.
476. Zaitsev, A. Yu. (1983). On the accuracy of approximation of distributions of sums of independent random variables—which are nonzero with a small probability—by means of accompanying laws. *Th. Probab. Appl.*, **28**, 657–69.
477. Zaitsev, A. Yu. and Arak, T. V. (1983). On the rate of convergence in Kolmogorov's second uniform limit theorem. *Th. Probab. Appl.*, **28**, 351–74.
478. Zaremba, S. K. (1958). Note on the central limit theorem. *Math. Z.*, **69**, 295–8.
479. Zinger, A. A. (1965). On a class of limit distributions for normalized sums of independent random variables. *Th. Probab. Appl.*, **10**, 607–26.
480. Zinger, A. A. (1971). Limit laws for cumulative sums of independent random variables with a finite number of distribution types. *Th. Probab. Appl.*, **16**, 596–619.
481. Zolotarev, V. M. (1963). The analytic structure of infinitely divisible laws of class L. *Litovsk. Mat. Sb.*, **3**, 123–40. (English transl. in *Selected Transl. Math. Statist. Probab.*, 1981, **15**, 15–31. American Mathematical Society, Providence, RI.)
482. Zolotarev, V. M. (1967a). Some inequalities in the theory of probability and their applications to the sharpening of a theorem of A. M. Lyapunov. *Soviet Math. Doklady*, **8**, 1427–30.
483. Zolotarev, V. M. (1967b). A sharpening of the inequality of Berry–Esseen. *Z. Wahrsch. Verw. Geb.*, **8**, 332–42.
484. Zolotarev, V. M. (1973). Exactness of an approximation in the central limit theorem. In *Lecture Notes in Mathematics*, Vol. 330, pp. 531–43.
485. Zolotarev, V. M. (1986a). *One-dimensional stable distributions*. American Mathematical Society, Providence, RI.
486. Zolotarev, V. M. (1986b). *Contemporary theory of summation of independent random variables*. Nauka, Moscow (in Russian).
487. Zygmund, A. (1951). A remark on characteristic functions. In Proceedings, 2nd Berkeley Symposium on Mathematical Statistics and Probability, pp. 369–372. Univ. California Press, Berkeley and Los Angeles.

Author index

Adler, A. 257, 265
Alda, V. 135, 265
Aldous, D. J. 227, 265
Amosova, N. N. 138, 265
Ananjevsky, S. M. 46–7, 87, 265
Anděl, J. 238, 265
Arak, T. V. 107, 109–10, 265, 285
Araujo, A. 185, 265
Arnold, B. C. 37, 265

Barbour, A. D. 184, 194, 197, 265, 272
Baum, L. E. 137–8, 140, 229, 233, 265
Baxter, G. 47, 265
Beesack, P. R. 36, 265
Bennett, G. 78, 265
Bentkus, V. 42, 266
Bergström, H. 107, 184–5, 266
Berkes, I. 227, 238, 256, 260, 266
Bernstein, S. N. 58, 77, 120, 121, 135, 136, 266, 279
Berry, A. C. 150, 176–7, 181, 183–5, 191, 265–6, 270, 272–3, 277, 284–5
Bestsennaya, E. V. 77, 266
Bhattacharya, R. N. 42, 184–5, 197, 266
Bickel, P. J. 80, 86, 266
Bikelis, A. 184–5, 187, 196, 266
Billingsley, P. 44, 266
Bingham, N. H. 106, 233, 255, 266
Binmore, K. G. 40, 266
Blum, J. R. 49, 266
Boas, R. P. 39, 266
Bondarko, V. M. 198, 266
Book, S. A. 186, 235–6, 267
Borel, E. 1–5, 43, 199–201, 206, 211, 213, 215, 218, 224, 226, 238, 265, 267–8, 270, 274, 277
Borovkov, A. A. 79, 267, 282
Brosamler, G. A. 238, 267
Brown, B. M. 38, 267
Buldygin, V.V. 226, 267
Bulinsky, A. V. 267
Butzer, P. L. 197, 267

Cacoullos, T. 37, 267
Cantelli, F. P. 199–201, 206, 211, 213, 215, 218, 224, 226, 238, 267–8, 270, 274, 277
Chatterjee, S. D. 47, 227, 267

Chebyshev, P. L. 6, 7, 36, 53, 124, 132, 134, 185, 198, 200, 215, 267, 272–3, 277, 282
Chen, L. H. Y. 37, 184, 267, 273
Chen, R. 141, 267
Chernoff, H. 37, 234, 267, 274
Chistyakov, G. P. 192, 267
Chover, J. 258, 267
Chow, Y. S. 138, 231–2, 267
Christoph, G. 185, 238, 268
Chung, K. L. 1, 77, 226, 268, 276
Cox, D. C. 82, 268
Cramér, H. 8, 36, 40, 57–8, 178, 181, 185, 195, 198, 224–5, 235, 268, 270, 277–8
Csáki, E. 258, 268
Csörgö, M. 226, 235, 238, 255, 268, 283
Csörgö, S. 230, 235, 238, 268

Daley, D. J. 81, 268
Darling, D. A. 234, 268
Daugavet, A. I. 36, 197, 268–9
de Acosta, A. 255, 269
Deheuvels, P. 235, 269
Dehling, H. 238, 266
Devroye, L. 235, 269
Dharmadhikari, S. W. 83, 269
Doeblin, W. 49, 77, 107, 269, 274
Doob, J. L. 41, 80, 269
Dupač, V. 238, 265
Dyson, F. J. 44, 269

Ebralidze, Sh. S. 79, 269
Edgeworth, F. Y. 185, 198, 272
Egorov, V. A. 185, 193, 226, 261–2, 269
Eisenberg, B. 44, 269
Enger, J. 36
Erdös, P. 77, 137, 199, 226, 235, 267–9, 275, 277, 283
Erickson, K. B. 231, 269
Esseen, C.-G. 36, 47, 77, 81–2, 87, 142, 147, 149–50, 155, 176–7, 181, 183–5, 187, 191, 265, 269, 270, 272–4, 277, 281, 284–5
Etemadi, N. 226, 229, 270

Fainleib, A. S. 184, 270
Feller, W. 1, 41, 106, 110–11, 120–1, 123, 135, 184, 230, 270, 273, 277, 280, 282

Author index

Fisher, A. 238, 270
Fishler, R. M. 238, 270
Fisz, M. 36, 110, 270
Frechét, M. 43, 270
Friedman, N. 188, 193, 270
Frolov, A. N. 235, 270
Fuk, D. Kh. 77-8, 271

Gafurov, M. U. 141, 283
Gan Shixin 44, 269
Gaposhkin, V. F. 227, 271
Giné, E. 185, 265
Gnedenko, B. V. 1, 15, 25, 101, 105, 107, 110-11, 134, 185, 271
Godwin, H. J. 36, 271
Goldie, C. M. 106, 233, 266
Götze, F. 197, 271
Goursat, E. 170, 271
Gradsteyn, I. S. 62, 271
Grübel, R. 48, 271
Gut, A. 260, 271

Hahn, L. 197, 267
Hàjek, J. 54, 77, 268, 271
Hall, P. 40, 87, 139, 174, 184-5, 190, 194-8, 227, 238, 265, 268, 271-2
Hardy, G. H. 272
Hartman, P. 49, 248, 255, 269, 272
Hausdorff, F. 226, 272
Heathcote, C. R. 41, 272
Hengartner, W. 36, 272
Heyde, C. C. 77, 137, 139, 141, 184-5, 188-90, 193, 196, 226-7, 255, 257, 272
Hipp, C. 197, 271-2
Ho, Soo-Tong 184, 273
Hoeffding, W. 36, 78, 273
Hsu, P. L. 38, 110, 137, 269, 273
Hudson, W. N. 49, 273

Ibragimov, I. A. 49, 105, 107-8, 185-6, 188, 193, 195-6, 273

Jain, N. C. 258, 273
Jesiak, B. 108, 136, 273, 281
Jogdeo, K. 83, 269
Johnson, W. B. 83, 273

Karlin, S. 36, 273
Katz, M. 137-8, 140, 184, 229, 231, 265, 273
Kawata, T. 38-9, 44-5, 273
Kemperman, J. H. B. 82, 268
Kendall, D. G. 281
Kesten, H. 73, 77, 231, 234, 257, 269, 274, 284
Khatuntseva, M. V. 198, 274

Khintchine, A. 30, 34, 36, 49, 107, 111-12, 134, 255, 274
Kingman, J. F. C. 36, 274
Kirsha, K. 42, 266
Klaassen, C. A. J. 37, 274
Klass, M. J. 255-6, 274
Kochen, S. B. 226, 274
Kolmogorov, A. N. 25, 36, 52, 54, 68-9, 73, 77, 101, 105, 107, 134, 185, 205, 226, 239, 255, 263, 265, 267, 271, 274, 280-1, 284
Kolodyazhny, S. F. 185, 274
Komlós, J. 227, 235-7, 266, 268, 274, 284
Koopmans, L. H. 270
Kruglov, V. M. 48, 107, 111, 229, 274-5
Kunisawa, K. 45-7, 275

Lacey, M. T. 238, 275
Lai, T. L. 138, 232, 267-8
Laube, G. 36, 275
Laue, G. 41, 275
LeCam, L. 107, 275
Leslie, J. R. 184, 196, 272, 275
Lévy, P. 22, 30, 34-6, 40, 43-5, 47-9, 65, 77, 87, 99, 103, 105, 107, 111-12, 120, 126, 134, 176, 219, 266, 269, 274, 275, 279
Lifshits, B. A. 196, 275
Lin, Z. Y. 235, 275
Lindeberg, J. W. 123-4, 126, 135, 273, 278, 282
Linnik, Yu. V. 36, 105, 108, 185-6, 273, 275
Littlewood, J. E. 272
Loève, M. 1, 36, 107-8, 134, 228, 275
Lukacs, E. 1, 36, 41, 44, 275
Lyapunov, A. M. 7, 36, 62, 120, 126, 181, 184, 265-6, 270, 275, 278, 285
Lynch, J. 235, 275

Maejima, M. 42, 189, 275
Major, P. 236-8, 268, 274, 276
Makabe, H. 40, 276
Maller, R. A. 110, 256, 276
Manstavičius, E. 85, 276
Marcinkiewicz, J. 77, 82, 230, 255, 261, 276, 277
Markov, A. A. 134, 223
Martikainen, A. I. 226, 228-9, 232, 234, 238, 255-7, 262-3, 276-7
Matskyavichyus, V. K. 187, 277
Mejlijson, I. 233, 277
Mejzler, D. 108, 277
Michel, R. 185, 196, 277
Mijnheer, J. L. 108, 185, 238, 277
Mikosch, T. 234, 258, 277, 283
Mills, J. F. 182
Miroshnikov, A. L. 77, 277

Author index | 289

Mogyoródi, J. 82, 277
Mori, T. F. 226, 277

Nagaev, S. V. 36, 77–8, 185, 191, 226–8, 271, 277–8, 284
Nakata, T. 185, 190, 193, 198, 272
Nikitenko, T. V. 260, 278

O'Brien, G. L. 226
Ohkubo, H. 48–9, 278
Osipov, L. V. 158, 167, 173, 184–5, 194, 275, 278
Ostrovskii, I. V. 36, 275

Paditz, L. 185, 278
Pakshirajan, R. P. 47, 267
Paulauskas, V. I. 184–6, 191, 278
Péter, E. 227, 266
Petrov, V. V. 36–7, 77, 80, 81, 84, 101, 111, 137, 140–1, 173–4, 184–7, 224, 226, 229–30, 233–4, 238, 255, 261, 263–4, 269, 275, 277–80
Philipp, W. 238, 275
Pinelis, I. F. 77, 278, 280
Pitman, E. J. G. 39, 280
Pitman, J. W. 41, 272
Pólya, G. 36, 41, 272, 280
Postnikova, L. P. 77, 280
Prawitz, H. 41, 280
Presman, E. L. 107, 273
Prohorov, Yu. V. 44, 77, 80, 107, 109, 227–8, 280
Pruitt, W. E. 110, 234, 256, 273, 280

Rachev, S. T. 44, 281
Raikov, D. A. 135, 281
Ramachandran, B. 36, 48, 281
Ranga Rao, R. 42, 184–5, 197
Rao, K. S. 281
Rényi, A. 54, 77, 199, 226, 235, 267–9, 275, 277, 281
Révész, P. 226–7, 238, 255, 268, 281, 283–4
Riedel, M. 49, 136, 281
Robbins, H. 137, 231–2, 268–9, 273, 281
Rogozin, B. A. 68–9, 73, 77, 80, 192, 256–7, 274, 277, 281
Rohatgi, V. K. 139, 272
Rosalsky, A. 256, 281
Rosén, B. 36, 77, 281
Rosenblatt, M. 49, 266
Rosenthal, H. P. 59, 62–3, 77, 83, 280–1
Rossberg, H.-J. 49, 108, 136, 273, 281

Rotar, V. I. 107, 135, 191, 278, 282
Rozovsky, L. V. 140, 185, 187, 193–4, 196, 230, 282
Ryzhik, I. M. 62, 271

Sahanenko, A. I. 237, 282
Saulis, L. 186, 282
Savage, I. R. 36, 282
Sazonov, V. V. 85, 184–5, 191, 282
Schatte, P. 238, 260, 282
Schechtman, G. 83, 273
Scheffé, H. 44, 282
Seneta, E. 106, 282
Sethuraman, J. 186, 282
Shao, Qui-man 237, 283
Shapiro, J. M. 47, 265
Shepp, L. A. 235, 269, 275, 283
Shiganov, I. S. 185, 283
Shirokova, I. V. 137, 140, 280
Shiryaev, A. N. 1, 283
Shuster, J. 226, 283
Siegel, G. 49, 108, 136, 281
Simons, G. 238, 283
Sirazdinov, S. H. 141, 283
Skorohod, A. V. 86, 283
Skovoroda, B. F. 234, 283
Slutsky, E. E. 36
Smith, W. L. 41, 85, 283
Spitzer, F. 137, 226, 283
Stadtmüller, U. 264, 283
Statulevičius, V. 186, 282
Steinebach, J. 235, 238, 269, 283
Steutel, F. W. 36, 283
Stone, C. J. 226, 231, 274, 283
Stout, W. F. 226, 238, 255, 283
Strassen, V. 236–7, 248, 255, 259, 276, 283
Stratton, H. H. 40, 265–6
Studden, W. 36, 273
Suchkov, A. P. 87, 284
Székely, G. J. 226, 277

Tandori, K. 268
Tanny, D. 231, 284
Teicher, H. 232, 257, 284
Teugels, J. L. 106, 266
Theodorescu, R. 36, 272
Tomkins, R. J. 226, 255, 261–2, 264, 274, 284
Totik, V. 268
Trautner, R. 227, 284
Tsaregradskii, I. P. 107, 186, 284
Tucker, H. G. 47, 49, 273, 284
Tusnády, G. 235, 268, 274

Ul'yanov, V. V. 191, 284
Ushakov, N. G. 87, 284
Utev, S. A. 77, 266, 280, 284

van Beek, P. 184, 284
van Kampen, E. R. 226, 284
Varadarajan, V. S. 110, 270
Volodin, N. A. 226, 284
von Bahr, B. 82, 197, 284
von Mises, R. 37, 284

Wightwick, J. C. H. 194, 272
Wintner, A. 36, 49, 248, 255, 269, 272, 284
Wittmann, R. 234, 262, 284
Wolfe, S. J. 36, 39, 48, 108, 285
Woyczyński, W. A. 285

Yamazato, M. 108, 110, 285
Yudin, A. A. 77, 280

Zaitsev, A. Yu. 107, 109–10, 265, 285
Zaremba, S. K. 136, 285
Zhang, C. H. 231, 268
Zinger, A. A. 108, 285
Zinn, J. 83, 273
Zolotarev, V. M. 44, 107–8, 184–5, 191, 285
Zygmund, A. 43, 77, 82, 230, 255, 261, 276–7, 285

Subject index

almost surely 204
attraction 106–7
  partial 107

Bernstein's inequalities 58
Bernstein–Feller theorem 121
Berry–Esseen inequality 150, 176, 177, 181, 183
Borel field of events 1
Borel function 5
Borel set 1–4, 43
Borel–Cantelli lemma 199–201, 211, 215, 224, 226, 244–5, 250
Borel–Cantelli sequence of events 238

Cauchy–Buniakowski inequality 6
central limit theorem 118, 120, 142, 169
  global form 166, 168
central moment 6
characteristic function 9–13
  Lévy–Khintchine representation 30, 99
  Lévy representation 35
  Kolmogorov representation 36
Chebyshev's inequalities 6, 7
Chebyshev's theorem 134
Chebyshev–Hermite polynomials 171
Chernoff function 234
class L 101, 110
concentration functions 22, 45, 63, 225
constancy in the limit 98
convergence
  almost surely 204
  complete 16
  in probability 20, 127
  in variation 44
  weak 16–18
convolution 4
Cramér's condition 8, 57–8, 178, 181, 235
Cramér's condition (C) 12, 14, 195–6, 224–5
Cramér's series 178
cumulant 11, 178

density 2, 3
distribution 2, 3
  absolutely continuous 2
  Bernoulli 3, 150

binomial 3
continuous 2
degenerate 3, 127
discrete 2
infinitely divisible 28–36, 47–9, 91, 98–103, 107
lattice 2, 12, 13, 15
normal 3, 29, 112–13, 118
Poisson 3, 29, 31
singular 2
stable 103–5, 110–11
distribution function 2–4
domain of attraction 106–7

elementary events 1
  space of 1
Erdös–Rényi law 235
Esseen's inequality 149
Esseen's theorem 142
events 1
  mutually independent 4
expectation 5

Feller's theorem 120, 131

growth point 3, 12

Hàjek–Rényi inequality 54
Hartman–Wintner theorem 248, 255
Helly's theorem 16
Hölder's inequality 6, 72

independent events 4
  random variables 4–6
infinitely divisible distributions 28–36, 47–9, 91, 98–103, 107
infinite smallness 88
infinitely often 199

Jensen's inequality 6, 66

Kawata concentration function 44–5
Kesten's inequality 73, 77

Khintchine's theorem 91, 134
Kolmogorov's formula 36
  inequality 52, 54, 205
  metric 43
  spectral function 36
  theorem 131, 239
Kolmogorov–Rogozin inequality 68, 69, 73, 77
Komlós–Major–Tusnády approximation 236–7
Kronecker's lemma 209, 212–13, 249
Kunisawa concentration function 45–6

lattice distribution 2, 12–13, 15
law of the iterated logarithm 239, 248
law of large numbers
  strong 216, 228
  weak 127–8, 131
Lebesgue decomposition theorem 3
Lévy–Khintchine formula 30, 112
  spectral function 34
Lévy–Prohorov metric 44
Lévy's concentration function 22, 45, 87
  formula 35, 105
  inequality 51, 219
  metric 43, 111
  spectral function 35, 65, 103
  theorem 126
Lindeberg condition 124–6
Lindeberg ratio 124, 154, 247–8
Lindeberg–Feller theorem 123
lower class 239
Lyapunov's inequality 7, 62, 181, 202
Lyapunov's theorem 126, 155

Markov's theorem 134, 223
mean 5
median 9, 51
moment 5
  absolute 5
  absolute central 6
  central 6
  moment generating function 8
Mills ratio 182
Minkowski's inequality 6

Osipov's theorem 158, 167, 173

point of growth 3, 12
probabilities of large deviations 176, 184
probability function 2
  measure 1
  space 1–3
quantile 9, 50

random variables 1
  independent 4–6, 10
  orthogonal 230
  symmetric 10, 15
  symmetrized 10
random vectors 3
Rosenthal's inequality 59, 63, 83

σ-algebra of events 1
slowly varying function 106, 138
span of a lattice distribution 2
spectrum of a distribution 3
stable distribution 103–6
stable sequences of random variables 127
  strongly 216
strong limit theorems 112, 199
symmetrization inequalities 217

three-series theorem 205

uniform metric 43
upper class 239

variance 6

weak convergence 16–18
weak limit theorems 88, 112

UNIVERSITY LIBRARY